CASS LIBRARY OF SCIENCE CLASSICS

No. 4

General Editor: Dr. L. L. LAUDAN, University College London

THE ELECTRICAL RESEARCHES

OF THE HONOURABLE

HENRY CAVENDISH

Publisher's Note to the 1967 Edition

This is an exact reproduction of the 1879 edition of *The Electrical Researches of the Honourable Henry Cavendish*, edited by James Clerk Maxwell.

First published 1879 by Frank Cass and Company Limited

Published 2022 by Routledge
4 Park Square, Milton Park, Abingdon, Oxon OX14 4RN
605 Third Avenue, New York, NY 10017

Routledge is an imprint of the Taylor & Francis Group, an informa business

Publisher's Note

The publisher has gone to great lengths to ensure the quality of this reprint but points out that some imperfections in the original copies may be apparent.

ISBN 978-0-714-61057-3 (hbk)

THE
ELECTRICAL RESEARCHES
OF THE HONOURABLE
HENRY CAVENDISH

EDITED BY
J. CLERK MAXWELL

Routledge
Taylor & Francis Group
LONDON AND NEW YORK

THE ELECTRICAL RESEARCHES

OF THE HONOURABLE

HENRY CAVENDISH, F.R.S.

WRITTEN BETWEEN 1771 AND 1781,

EDITED FROM THE ORIGINAL MANUSCRIPTS

IN THE POSSESSION OF

THE DUKE OF DEVONSHIRE, K.G.,

BY

J. CLERK MAXWELL, F.R.S.

EDITED FOR THE SYNDICS OF THE UNIVERSITY PRESS.

Cambridge:
AT THE UNIVERSITY PRESS.

LONDON: CAMBRIDGE WAREHOUSE, 17, PATERNOSTER ROW.
CAMBRIDGE: DEIGHTON, BELL, AND CO.
LEIPZIG: F. A. BROCKHAUS.
1879

CONTENTS.

INTRODUCTION BY THE EDITOR.

FIRST PUBLISHED PAPER ON ELECTRICITY.

"An Attempt to explain some of the principal Phænomena of Electricity, by means of an Elastic Fluid." From the *Philosophical Transactions* for 1771 (p. 1—50).

Part I.

Part II.

Containing a comparison of the foregoing theory with Experiment.

PRELIMINARY PROPOSITIONS.

From the MS. in the possession of the Duke of Devonshire, No. 4.

APPENDIX.

From MS. No. 5.

THOUGHTS CONCERNING ELECTRICITY.

From MS. No. 18. *(Probably an early draft of the theory.)*

ACCOUNT OF THE EXPERIMENTS.

(1) Investigation of the Law of Force.

From the MS. No. 7 (apparently prepared for publication).

(2) Experiments on the Comparison of Charges.

From the MS. Nos. 9 and 10 (apparently prepared for publication).

(4) REPULSION AS SQUARE OF REDUNDANT FLUID.

From MS. No. 8.

SECOND PUBLISHED PAPER ON ELECTRICITY.

"AN ACCOUNT OF SOME ATTEMPTS TO IMITATE THE EFFECTS OF THE TORPEDO BY ELECTRICITY." From the *Phil. Trans.* for 1776 (pp. 196 to 225).

EXPERIMENTS IN 1771.

From MS. No. 12.

EXPERIMENTS IN 1772.

From MS. No. 13.

THE FOLLOWING INDEX IS BY CAVENDISH HIMSELF.

From MS. No. 14.

The pages refer to the MS., the numbers of the Articles to the present edition.

Index to Electrical Experiments, 1773.

[1] The two flat conductors between which the plate of air lies, or, in modern language, the electrodes.

HERE ENDS CAVENDISH'S INDEX.

M. [MEASURES.]

From MS. No. 19.

[1] Sea salt.

From MS. No. 13.

EXPERIMENTS WITH THE ARTIFICIAL TORPEDO.

From MS. No. 13.

RESISTANCE TO ELECTRICITY.

From MS. No. 20.

CALIBRATION OF TUBES.

Jan. 1781. *From MS. No.* 19.

RESISTANCE OF COPPER WIRE.

From MS. No. 19.

RESULT [OF COMPARISONS OF CHARGES].

From MS. No. 16.

RESULTS [ON RESISTANCE].

From MS. No. 20.

NOTES BY THE EDITOR.

FACSIMILES OF CAVENDISH'S FIGURES.

ERRATUM.

P. 137. Art. 283, line 2, for 1.53 read 1.153.

In the late Dr George Wilson's collection of Cavendish MSS. there is a drawing of which the opposite page is a reduced copy. The words "buried at Derby" are written in pencil on the margin.

Henry Cavendish was buried in the Devonshire Vault, All Saints' Church, Derby, but Mr J. Cooling, Jun., Churchwarden of All Saints, informs me that there is no slab or monument of any kind erected in memory of him there.

HENRY CAVENDISH ESQ^{R}

Eldest Son of the Right Honorable

LORD CHARLES CAVENDISH,

Third Son of William,

2ND ***DUKE OF DEVONSHIRE.***

Fellow of the Royal Society, and of the Society

OF ANTIQUARIES OF LONDON.

Trustee of the British Museum

and Foreign Associate in the First Class

of the INSTITUTE at Paris.

BORN October 10th 1731 } DIED February 24th 1810

THE

ELECTRICAL RESEARCHES

OF THE

HONOURABLE HENRY CAVENDISH, F.R.S.

INTRODUCTION*.

So little is known of the details of the life of Henry Cavendish, and so fully have the few known facts been given in the Life of Cavendish by Dr George Wilson†, that it is unnecessary here to repeat them except in so far as they bear on the history of his electrical researches.

He was born at Nice on the 10th October, 1731, he became a Fellow of the Royal Society in 1760, and was an active member of that body during the rest of his life. He died at Clapham on the 24th February, 1810.

His father was Lord Charles Cavendish, third son of William, second Duke of Devonshire, who married Lady Anne Grey, fourth daughter of the Duke of Kent. Henry was their eldest son. He had one brother, Frederick, who died 23rd February, 1812.

Of Lord Charles Cavendish we have the following notice by Dr Franklin‡. After describing an experiment of his on the passage of electricity through glass when heated to 400° F., he says,

> "It were to be wished that this noble philosopher would communi-
> "cate more of his experiments to the world, as he makes many, and
> "with great accuracy."

* By the Editor.

† Published in 1851 as the first volume of the Works of the Cavendish Society.

‡ *Franklin's Works*, edited by Jared Sparks, Boston, 1856, Vol. v, p. 383. See also Note 26 at the end of this book.

Lord Charles Cavendish has also recorded a very accurate series of observations* on the depression of mercury in glass tubes, and these have furnished the basis not only for the correction of the reading of barometers, &c., but for the verification of the theory of capillary action by Young, Laplace, Poisson and Ivory.

I think it right to notice the scientific work of Lord Charles Cavendish, because Henry seems to have been living with him during the whole period of his electrical researches. Some of the jottings of his electrical calculations are on torn backs of letters, one of which is addressed,

[The Ho]n^ble^ Mr Cavendish
at the R^t^ Honble
The Ld Charles
Cavendish's
Marlborough Street.

These calculations relate to the equivalent values of his trial plates when drawn out to different numbers of divisions. There is no date nor any part of the original letter.

The memoranda of some experiments similar to those in Art. 588, on the time of discharge of electricity through different bodies, are on the back of the usual Notice of the election of the Council and Officers of the Royal Society on the Thirtieth of November, 1774 (being St Andrew's Day) at Ten o'Clock in the Forenoon at the House of the Royal Society in Crane Court, Fleet Street. The address on the back of this letter is

To
The Hon Henry Cavendish
Gr^t^ Marlborough Street.

Dr Thomas Thomson, who was acquainted with Cavendish, says in his interesting sketch of him†,

"During his father's life-time he was kept in rather narrow circum-"stances, being allowed an annuity of £500 only, while his apartments

* *Phil. Trans.*, 1776, p. 382.

† *History of Chemistry*, Vol. I, p. 336, quoted in Wilson's *Life of Cavendish*, p. 159.

"were a set of stables, fitted up for his accommodation. It was during "this period that he acquired those habits of economy and those singular "oddities of character which he exhibited ever after in so striking a "manner."

The whole of the electric researches of which we are to give an account were made before the death of Lord Charles Cavendish, which took place in 1783. We must therefore suppose that they were made in Great Marlborough Street, and probably in the set of stables mentioned by Dr Thomson. He speaks of a "fore room and a back room" in Art. 469, and in Art. 335 he compares the size of the room in which he worked to that of a sphere 16 feet in diameter. The dimensions of his laboratory are of some importance in determining the electric capacity of bodies hung up in it, and by the foot-note to Art. 335 it would appear that the room was probably 14 feet high, which is somewhat lofty for "a set of stables," but I believe not much more than the height of some of the rooms in the dwelling-houses in Great Marlborough Street.

Let us then suppose that we have been admitted by Cavendish into his laboratory in Great Marlborough Street, as it was arranged for his electrical experiments in 1773, and let us make the best of an opportunity rarely, if ever, accorded to any scientific man of his own time, and examine the apparatus by which the electric fluid, instead of startling us with the brilliant phenomena, new instances of which were then every day being discovered, was made to submit itself, like everything else which entered that house, to be measured.

The largest piece of apparatus was the "machine for trying simple bodies" of which we have a description and sketch in Art. 241, and plans at Arts. 265 and 273. The framework of the machine is not represented in these figures.

We learn, however, from Dr Davy *, that

"Cavendish seemed to have in view, in construction, efficiency "merely, without attention to appearance. Hard woods were never "used, excepting when required. Fir-wood (common deal) was that "commonly employed."

* Wilson's *Life*, p. 178.

The bodies to be "tried" and the wires and vials for trying them were either supported on glass rods as shown in the sketch at Art. 239, or else hung by silk strings from a horizontal bar 7 feet 3½ inches from the floor as mentioned in Art. 466. The electrical connexions were made and broken at the proper times by means of silk strings passing over pullies attached to the horizontal bar.

One of the bodies, the charges of which Cavendish compared by means of this apparatus, was a globe 12·1 inches in diameter covered with tinfoil. This globe has historical interest as it was not only the standard of capacity with which Cavendish compared that of all other bodies, but it formed part of the apparatus by which he established that the electric repulsion varies inversely as the square of the distance.

There was also a set of circles of tin plate, one of 36 inches diameter, one of 18·5 and two of 9·3; and also square and oblong tin plates, and squared pieces of stone and slate, and a collection of cylinders and wires of different sizes.

There was another "machine," represented, with its framework, in Fig. 20, Art. 295, "for trying Leyden vials."

The "Leyden vials" were most of them flat plates of glass with circular coatings of tinfoil, one on each side. They were made in sets of three, any one of each set being nearly equal in capacity to the three of the former set taken together. Cavendish had thus a complete set of condensers of known capacity by means of which he measured the capacity of every piece of his apparatus, from the little wire which he used to connect his coated plates, and which he found to contain ·28 "inches of electricity," up to his battery of 49 jars, which contained 321000 "inches of electricity" *.

These "inches of electricity" can be directly compared with our modern measurements of electrostatic capacity. Indeed the only difference is that Cavendish's "inches of electricity" express the diameter of the sphere of equivalent capacity, while the modern measurements express the capacity by stating the radius of the same sphere in centimetres.

* About half a microfarad.

Of each of these plates of glass Cavendish has given a most minute description, so that each, if it were found, could be identified. Mr Cottrell, of the Royal Institution, has been kind enough to examine the catalogue of apparatus there, which contains Cavendish's Eudiometer and Registering Thermometer. No trace, however, of a set of glass plates could be found. It is possible, however, that if the plates were neatly packed up, their small bulk and their apparent uselessness may have enabled them to survive the periodical overhaulings of some less celebrated repository, and that they may yet gain an honourable place in the museum of historical instruments.

But we need not expect ever to discover a piece of apparatus of still greater historical interest—that by which Cavendish proved that the law of electric repulsion could not differ from that of the inverse square by more than $\frac{1}{50}$. It consisted of a pair of somewhat rickety wooden frames, to which two hemispheres of pasteboard were fastened by means of sticks of glass. By pulling a string these frames were made to open like a book, showing within the hemispheres the memorable globe of 12·1 inches diameter, supported on a glass stick as an axis. By pulling the string still more, the hemispheres were drawn quite away from the globe, and a pith ball electrometer was drawn up to the globe to test its "degree of electrification." A machine so bulky, so brittle, and so inelegant was not likely to last long, even in a lumber room. A facsimile of Cavendish's sketch of it is given at page 104. His own account of the experiment, in Arts. 217—234, is one of the most perfect examples of scientific exposition.

We might also notice the different electrometers, most of them consisting of a pair of cork or pith balls, mounted on straws or on linen threads, and some of them capable of having their weight altered by means of wires run into the straws; but though Cavendish had a wonderful power of making correct observations and getting accurate results with these somewhat clumsy instruments, we must confess that in these, the most vital organs of electric research, Cavendish showed less inventive genius than some of his contemporaries. When Lane and Henly brought out their respective electrometers, Cavendish compared their indications, and by stating in every case the distance at which Lane's electrometer discharged, he has enabled us to calculate in modern units every degree of electrification that he made use of. What was really needed for Cavendish's experiments was a sensitive electrometer. Cavendish did the best with the electrometers he found in existence, but he did not invent a better one.

It was not till 1785 that Coulomb began to publish the wonderful series of experiments, in which he got such good results with the torsion electrometer, an instrument constructed on the same principle as that with which Cavendish afterwards measured the attraction of gravitation; and it was not till 1787 that Bennett described in the *Philosophical Transactions* the gold leaf electrometer, by means of which Volta afterwards demonstrated the different electrification of the different metals.

The electrical machine, by Nairne, was one with a glass globe.

We should also notice the dividing engine, by Bird, for determining the thickness of the glass plates, and other small distances.

An attendant *, "Richard," appears occasionally, to help in turning the electrical machine, or in pulling the strings which made or broke the electrical connexions; and sometimes he is even asked his opinion as to the comparative strength of two electric shocks †. But there is no record of any other person having being admitted into the laboratory during the series of experiments to which we now refer.

The authority of Cavendish in electrical science was of course established by his paper of 1771, and accordingly we find him nominated by the Royal Society as one of a committee appointed in 1772 "to consider of a method for securing the powder magazine at Purfleet ‡."

A powder mill at Brescia having blown up in consequence of being struck by lightning, the Board of Ordnance applied to Mr Benjamin Wilson, F.R.S., who held the contract for the house-painting under the Board §, and who had some reputation as an

* Arts. 242, 560, 565. † Art. 511.

‡ See *Franklin's Works*, Vol. v, p. 430, note.

§ He also painted portraits of Franklin and of Gowin Knight, as well as of Garrick in various characters.

electrician, for a method to prevent a like accident to their magazines at Purfleet. Mr Wilson having advised a blunt conductor, and it being understood that Dr Franklin's opinion, formed upon the spot, was for a pointed one, the matter was referred, in 1772, to the Royal Society, and by them as usual to a committee, who after consultation presented a method conformable to Dr Franklin's theory *.

The Committee consisted of Cavendish, Dr, afterwards Sir William Watson; Dr Franklin, Mr J. Robertson (Clerk and Librarian to the Royal Society); Mr Wilson and Mr Delaval.

Dr Franklin read to the Committee a paper which is printed in his works, Vol. v, p. 435, but is not referred to in the report of the Committee, though the report is entirely in conformity with it†.

The Committee went down to Purfleet and examined all the buildings together, but I cannot trace any evidence that Cavendish did anything to modify the report, and Franklin does not mention him in any part of his writings, as one of the remarkable men with whom he was brought in contact.

The most noteworthy incident of the Committee was the dissent‡ of Mr Wilson, to which Mr Delaval adhered as regards that part of the report which recommended the conductors to be pointed. Mr Wilson followed up his dissent by a paper §, in which he gave his reasons for preferring blunt conductors; but the other four members of the Committee, Messrs Cavendish, Franklin, Watson, and Robertson, having heard and considered these objections, found no reason to change their opinion or vary from their Report ||.

But on the 15th May 1777, the Board House at Purfleet was struck by lightning, and some of the brickwork damaged. This being communicated by the Board of Ordnance to the Royal Society ¶, a Committee was appointed to examine the effects of the lightning and to report.

* *Phil. Trans.*, 1772, p. 42.

† The report is printed in *Franklin's Works*, Vol. v, p. 430, and is there stated to be "Drawn up by Benjamin Franklin, August 21, 1772." The paper on the utility of long, pointed rods is stated to have been read on August 27th, 1772.

‡ *Ib.*, p. 48. § *Ib.*, p. 49. || *Ib.*, p. 66. ¶ *Ib.*, 1778, p. 232.

The Committee consisted of Mr Henly, Mr Lane, Mr Nairne and Mr Planta, Secretary of the Royal Society. They reported in favour of making a channel all round the parapet and filling it with lead, and connecting this in four places with the main conductor on the roof of the building.

Mr Wilson, however, dissented from this Report, and communicated to the Royal Society an account of a most elaborate and indeed magnificent set of experiments conducted in the Pantheon, in which a cylinder 155 feet long, composed of 120 drums, and connected with a wire 4800 feet long, suspended on silk strings, was electrified, and the discharge made to strike a model of the Board House at Purfleet. The experiments were witnessed by King George III., and seem to have been very brilliant. The picture of the experiment, probably drawn by Mr Wilson, is, as a work of art, considerably above the average of the plates in the *Philosophical Transactions*.

The subject was referred to a larger Committee, consisting of Sir John Pringle, President of the Royal Society, Dr Watson, Henry Cavendish, W. Henly, Bishop Horsley, T. Lane, Lord Mahon, Edw. Nairne, and Dr Priestley.

They reported* in favour of having an additional number of conductors ten feet high, terminated with pieces of copper eighteen inches long, and as finely tapered and acutely pointed as possible.

"We give these directions," they conclude, "being persuaded, that "elevated rods are preferable to low conductors terminated in rounded "ends, knobs, or balls of metal; and conceiving, that the experiments "and reasons made and alledged to the contrary by Mr Wilson, are "inconclusive."

I have stated this incident at some length because it does not appear to have been noticed by Cavendish's biographers, and because it shows him cooperating with Franklin and others in an electrical investigation undertaken in the interest of the nation.

Cavendish's researches on the electric current have been hitherto very imperfectly known, as they are only alluded to in his celebrated paper on the Torpedo. The private investigations of Cavendish are contained in this volume, but the ex-

* *Phil. Trans.*, 1778, p. 313.

ternal events which were more or less connected with them, were as follows:

On July 1, 1773, Mr Walsh communicated to the Royal Society his paper "Of the Electric Property of the Torpedo. In a Letter from John Walsh, Esq., F.R.S., to Benjamin Franklin, Esq., LL.D., F.R.S., Ac. R. Par. Soc. Ext., &c."

The following extracts will indicate the chief points of electrical interest.

"The vigour of the fresh taken Torpedos at the Isle of Ré was not "able to force the torpedinal fluid across the minutest tract of air; "not from one link of a small chain, suspended freely, to another; not "through an almost invisible separation, made by the edge of a pen-"knife in a slip of tinfoil pasted on sealing-wax."

"The effect produced by the Torpedo when in air appeared, on "many repeated experiments to be about four times as strong as when "in water."

"The Torpedo, on this occasion, dispensed only the distinct instan-"taneous stroke, so well known by the name of the electric shock. "That protracted but lighter sensation, that Torpor or Numbness which "he at times induces, and from which he takes his name, was not then "experienced from the animal; but it was imitated with artificial elec-"tricity, and shewn to be producible by a quick succession of minute "shocks. This in the Torpedo may perhaps be effected by the suc-"cessive discharge of his numerous cylinders, in the nature of a running "fire of musketry; the strong single shock may be his general volley. "In the continued effect, as well as in the instantaneous, his eyes, usually "prominent, are withdrawn into their sockets."

Walsh shows that these phenomena "are in no ways repugnant to the laws of electricity," for "the same quantity of electric matter, according as it is used in a dense or rare state, will produce the different consequences."

"Let me here remark that the sagacity of Mr Cavendish in devising "and his address in executing electrical experiments, led him the first "to experience with artificial electricity, that a shock could be received "from a charge which was unable to force a passage through the least "space of air."

Walsh concludes his letter to Franklin in the following terms:—

"I rejoice in addressing these communications to You. He, who "predicted and shewed that electricity wings the formidable bolt of the "atmosphere, will hear with attention, that in the deep it speeds an "humbler bolt, silent and invisible: He, who analysed the electrified "Phial, will hear with pleasure that its laws prevail in animate Phials:

"He, who by Reason became an electrician, will hear with reverence "of an instinctive electrician, gifted in his birth with a wonderful "apparatus, and with the skill to use it*.

"However I may respect your talents as an electrician, it is certainly for knowledge of more general import that I am impressed "with that high esteem, with which I remain,

"Dear Sir,

"Your affectionate

"And obedient servant,

"JOHN WALSH."

This paper is followed in the *Philosophical Transactions* by "Anatomical Observations on the Torpedo," by John Hunter, F.R.S., in which the great anatomist describes the structure of the electric organs, in specimens of the fish furnished by Mr Walsh.

Considerable interest seems to have been excited by this account of the Torpedo, and several papers on the Torpedo and the Gymnotus are in the *Philosophical Transactions* for 1775, none of them, however, so valuable as the original one by Walsh.

The practical electricians, however, were by no means satis-

* That the electrical fishes still possess the power of exciting the imagination as well as the nerves of those who have felt their power may be seen from the following passage with which Prof. Du Bois Reymond begins his account of experiments on a living Malapterurus in the *Monatsberichte d. k. Acad. Berlin*, 28 Jan., 1858.

"Fast möchte man es, im Sinne Newton's, eine Anwandlung der Natur nennen, "dass es ihr gefallen hat, aus der Unzahl der Geschöpfe drei Fische, und zwar "der verschiedensten Art, nach Willkür herauszugreifen, um sie mit elektromotorischen Vorrichtungen von furchtbarer Gewalt als eine Waffe auszustatten, "neben welcher der Giftzahn der Klapperschlange, ja die nordamericanische Drehpistole, als eine plumpe und armselige Erfindung erscheint; eine Waffe die, ohne "ihren Träger der Gefahr blosszustellen, lautlos und mit Blitzesschnelle in die "Entfernung reicht, und minutenlang eine secundendicht gedrängte Reihe von "Geschossen schleudert, deren keines fehlen kann, weil alle auf allen Punkten des "Raumes gleichzeitig vorhanden sind."

In the *Journal of Anatomy and Physiology* for April, 1879, is a *Note on a Curious Habit of the Malapterurus Electricus*, by A. B. Stirling. The author attempted to feed Joe (the Malapterurus) with fresh worms, but he would not look at them. Another fish, however, called Dick (Clarias), swallowed them. When Joe considered that Dick had enjoyed his breakfast long enough, he swam up to him and gave him such a shock that the whole was disgorged, whereupon Joe swallowed it himself. When Dick at last succumbed to this treatment, Joe could no longer get his food prepared for him, and gave up eating altogether.

fied that the effects of these fishes were really produced by electricity.

"Mr Ronayne has made a curious remark upon the supposed elec-
"tricity of the torpedo: he says, 'if *that* could be proved, he does not
"'see why we might not have storms of thunder and lightning in the
"'depths of the ocean. Indeed, I must say, that when a Gentleman
"'can so far give up his reason as to believe the possibility of an
"'accumulation of electricity *among conductors* sufficient to produce
"'the effects ascribed to the Torpedo, he need not hesitate a moment
"'to embrace *as truths* the grossest contradictions that can be laid
"'before him*.'"

I am aware of only two occasions on which Cavendish, after he had settled his own opinion on any subject, thought it worth his while to set other people right who differed from him. One of these occasions was in 1778, when his experiments on the formation of nitric acid by the electric spark from phlogisticated and dephlogisticated air (nitrogen and oxygen) had been repeated without success by Van Marum with the great Teylerian electrical machine, and by Lavoisier and Monge, and when Cavendish "thought it right to take some measures to authenticate the truth of it." For this purpose he requested Mr Gilpin, clerk to the Royal Society, to repeat the experiment, and desired some of the gentlemen most conversant with these subjects to be present at putting the materials together, and at the examination of the produce†.

The other occasion, with which alone we are now concerned, is the only one in which the presence of visitors to Cavendish's laboratory is recorded. There can be no doubt that Cavendish had completely satisfied not only Mr Walsh, but what was more to the purpose, himself, that the electric phenomena of the torpedo are such as might arise from the discharge of a large quantity of electricity at a very feeble degree of electrification. It must therefore have been to satisfy other persons on this point that he took the trouble to construct an artificial torpedo of wood covered with leather, a rude model of the figure given

* Extract from MS. letter of W. Henly, dated 21 May, 1775, in the Canton Papers in the Royal Society's Library. Communicated to the editor by H. B. Wheatley, Esq.

† *Phil. Trans.* 1788.

by Walsh, with electric organs of pewter supplied with electricity from a battery of Leyden jars, by wires protected by glass tubes.

The accessories of this machine were equally unlike the kind of apparatus which Cavendish made when working for himself. The torpedo had a trough of salt water, the saltness of which was carefully adjusted, so as to be equal to that of the sea. It had also a basket to lie in, and a bed of sand to be buried in, and there were pieces of sole-leather, well soaked in salt water, which Cavendish placed between the torpedo and his hands, so that he might form some idea of what would happen if a traveller with wet shoes were to tread on a live torpedo half buried in wet sand.

It was on Saturday, 27th May, 1775, that Cavendish tried the effect of his Torpedo on a select company of men of science. We find in the Journal (Art. 601), the names of John Hunter, the great anatomist, Dr Joseph Priestley, chemist, electrician and expounder of human knowledge in general, Mr Thomas Ronayne, from Cork, the disbeliever in the electrical character of the torpedo, Mr Timothy Lane, apothecary and electrician, and Mr Edward Nairne, the eminent maker of philosophical instruments.

They got shocks from the torpedo to their complete satisfaction, and probably learnt a good deal about electricity, but it was neither to satisfy them nor to communicate to them his electrical discoveries, that Cavendish admitted them into his laboratory on this memorable occasion, but simply to obtain the testimony of these eminent men to the fact, that the shocks of the artificial torpedo agreed in a sufficient manner with Walsh's description of the effects of the live fish, to warrant the hypothesis that the shock of the real torpedo may also be an electrical phenomenon.

I have now related all that I have been able to ascertain of the external history of Cavendish, in so far as it bears on his electrical researches. We must in the next place consider the record of these researches—the two papers in the *Philosophical Transactions*, which are here reprinted, and the manuscripts now first published.

CAVENDISH'S WRITINGS ON ELECTRICITY.

In the *Philosophical Transactions* for 1771 there is a paper entitled "An attempt to explain some of the principal Phænomena of Electricity by Means of an Elastic Fluid: By the Honourable Henry Cavendish, F.R.S." [Read Dec. 19, 1771, and Jan. 9, 1772, pp. 584—677.] This paper and that on the Torpedo (*Phil. Trans.* 1776) are the only publications of Cavendish relating to electricity.

Dr George Wilson, however, in his Life of Cavendish * says,

"Besides his two published papers on electricity, Cavendish has "left behind him some twenty packets of manuscript essays, more or "less complete, on Mathematical and Experimental Electricity. These "papers are at present in the hands of Sir W. Snow Harris, who most "kindly sent me an abstract of them, with a commentary of great "value on their contents. It will I trust be made public.

"Sir W. states that Cavendish had really anticipated all those great "facts in common electricity which were subsequently made known to "the scientific world through the investigations and writings of the "celebrated Coulomb and other philosophers, and had also obtained the "more immediate results of experiments of a refined kind instituted in "our own day."

Sir William Thomson, to whom Sir William Snow Harris showed some of Cavendish's results, thus speaks of them in a note dated Plymouth, Monday, July 2, 1849.

"Sir William Snow Harris has been showing me Cavendish's un-"published MSS., put in his hands by Lord Burlington, and his work "upon them; a most valuable mine of results. I find already that "the capacity of a disc (circular) was determined experimentally by "Cavendish as $\frac{1}{1{\cdot}57}$ of that of a sphere of same radius. Now we "have capacity of disc $=\frac{2}{\pi}a=\frac{a}{1{\cdot}571}$!"

"It is much to be desired that those manuscripts of Cavendish "should be published complete; or, at all events, that their safe keeping "and accessibility should be secured to the world†."

* *Works of the Cavendish Society*, Vol. I. *Life of Cavendish*, by George Wilson, M.D., F.R.S.E., London, 1851, p. 469.

† Reprint of *Papers on Electrostatics and Magnetism*, § 235, foot-note.

The Cavendish Society, for whom Dr Wilson prepared his Life of Cavendish, with an account of his chemical researches, did not consider that it came within their design to publish his electrical researches.

Sir W. Harris, in whose hands the manuscripts were placed by the Earl of Burlington, died in 1867. He makes several references to them in his work on Frictional Electricity, edited after his death by Charles Tomlinson, F.R.S., and published in 1867*, but he did not live to edit the manuscripts themselves. Under these circumstances it was thought desirable by Sir W. Thomson, Mr Tomlinson, and other men of science, that something should be done to render the researches of Cavendish accessible.

They accordingly represented the state of the case to the Duke of Devonshire, to whom the manuscripts belong, and in 1874 he placed them in my hands.

I could find no trace of Sir W. Harris' commentary referred to by Dr Wilson, except that Dr Wilson mentions having returned it to Sir W. Harris.

On the inside of the lid of the box which contained the manuscripts was pasted a paper in the handwriting of Sir W. Harris, of which the following is a copy.

"The several parcels of manuscript papers by the late Mr Cavendish, "which the Earl of Burlington did me the honor to place in my hands "with a view to an examination and report on their contents may be "taken at 24 in number. Twenty of these contain sundry Philo-"sophical papers on Mathematical and Experimental Electricity, and "Four sundry other Papers relating to Meteorology.

"All these Papers are more or less confused as to systematic arrange-"ment, and require some considerable attention in decyphering. They "are in many instances rather notes of experiments and rough drafts "intended as a basis for more perfect productions than finished Philo-"sophical Papers.

"They are nevertheless extremely valuable and most interesting as "evidence of Mr Cavendish's great Philosophical † , and clearly "prove that he had anticipated nearly all those great facts in common "electricity which at a later period were made known to the scientific "world through the writings of Coulomb and the French philosophers.

* P. 23 (straw electrometer), p. 45 (globe and hemispheres), p. 58 (specific inductive capacity), p. 121 (measures of electricity), p. 208 (law of force), p. 223 (induction at a great distance).

† So in MS.

Papers on Electricity.

"Of the 20 parcels of papers on electricity 18 belong to the years "1771, 1772 & 1773, and have never yet appeared in print; the "two remaining parcels are dated 1775 and 1776, and are evidently "connected with the author's celebrated paper on the Torpedo pub-"lished in the *Royal Society's Transactions* for 1776. The papers "belonging to the years 1771, 1772 & 1773 consist of six papers on "Mathematical Electricity, nine experimental papers, one of Diagrams "and Figures, the remainder are of a miscellaneous character, and "contain some interesting Notes and Remarks and Thoughts on "Electricity."

On examining the 20 parcels of manuscripts I found their contents to be as follows:

No. 1. MS. p. 1—10.

Apparently an early form of the "Preliminary Propositions."

No. 2. MS. p. 1—31.

Draft of "Preliminary Propositions" as far as Prop. XXIII.

No. 3. MS. L. 3 to L. 23. Contains the same propositions in a less complete form and not numbered, also two drafts of the propositions on coated plates, each 12 pp., and 38 loose pages of drafts of propositions, and jottings of algebraical calculations.

No. 4. MS. p. 1—48. The fair copy of the "Preliminary Propositions." Props. XXIX. to XXXVI. Refers to figs. 1 to 10 of No. 15. See Arts. 140—174.

No. 5. MS. p. 1—20. "Appendix." Refers to fig. 11. See Arts 175—194.

No. 6. "Computations for explanation of experiments."

MS. p. 1—15. Drafts of the propositions.

16 pages of computations. "B. 17." Charge of a sphere within a concentric sphere. [This is placed here as a note at p. 166.]

"Attractions of elect. bodies more accurate," p. 1—4.

No. 7. MS. D. 1 to D. 13. Fair copy of First and Second Experiments. Refers to Figs. 12, 13. See Arts. 217—235.

Draft of do marked "DIA."

No. 8. MS. p. 1—7. Refers to Fig. 31. See Arts. 386—394.

No. 9. MS. p. 1—51. Continuation of Experiments. See Arts. 236—294.

No. 10. MS. p. 52—132. Part * of Experiments. See Arts. 295—385.

No. 11. MS. p. 1—8. IA. p. 10 A. 8, 9. p. 29 A. p. 32 A. 1 and 2. p. 57—64. p. 85, 86. p. 91—96. p. 103—108. p. 119—126. p. 133—138. p. 141, 142. p. 156—166.
All drafts of portions of the Account of Experiments.

No. 12. "Experiments 1771," MS. p. 1—24. See Arts. 438—465, also 14 loose sheets of calculations and measurements.

No. 13. "Experiments 1772," MS. p. 1—29. See Arts. 466—493.
M. 1 to M. 13. Measurements of glasses, &c. See Arts. 592—595.

No. 14. Experiment 1773, MS. p. 1—135. See Arts. 494—580.
Index to elect. exper. 1773 p. 1—8. See Contents.
Dimensions of trial plates, 4 pages.

No. 15. Figures and Diagrams.

1 to 10 refer to	Preliminary propositions	No. 4
11	Appendix	No. 3
12 „ 13 „ „	Exp. 1	No. 7
14 „ 19 „ „	Experiments, Part 1	No. 9
20 „ 27 „ „	„ Part 2	No. 10
30 „ „	Electrometer	No. 9
31 „ „	Repulsion	No. 8

No. 16. "Result." MS. p. 1—21. See Arts. 647—683.

No. 17. "Notes." 4 pp. notes to "Thoughts concerning Electricity." These are inserted in their proper places, Arts. 196—216.
MS. p. 1 to 15. Drafts of propositions for the paper of 1771, but founded on the theory stated in the "Thoughts." They are given in Note 18, p. 411.

No. 18. "Thoughts concerning Electricity," MS. p. 1—16. See Arts. 196—216.

No. 19. Resistance to Electricity, MS. p. 1—23. See Arts. 616—631. "Res." Results of ditto p. 1—4. See Arts. 684—696. Resistance of Copper wire, p. 1—38. See Arts. 636—646.

No. 20. Experiments with the artificial Torpedo, p. 1—26. See Arts. 576—615. M. 1 to M. 42. Measurement of Leyden

* So in MS.

jars and batteries and of thickness of plates. See Arts. 581—592. "Extract from Dr Williamson's exper. on elect. Eel made in july 1773" p. 1 to 14 + 4 pp. (See *Phil. Trans.*, 1775, p. 94.)

In Art. 349, p. 172 of this book, Cavendish uses the expression "when I wrote the second part* of this work." It appears from this that he meant it for a book, not a paper to be communicated to the Royal Society. Several portions of this book are contained in the manuscripts, but the order in which they were intended to be placed can be discovered only by help of the figures and diagrams, which are numbered from 1 to 31.

From these it appears that we must begin with No. 4 and No. 5, the Preliminary Propositions† and the Appendix‡. The Preliminary Propositions refer to the printed paper of 1771. The last proposition in that paper is numbered XXVII., and the first in the MS. is XXIX., so that one proposition appears to be missing, but as there are several drafts, in all of which the first proposition is numbered XXIX., it is probable either that Prop XXVIII. is not lost, but must be sought for among the enunciations in the second part of the printed paper, or else that Cavendish made a mistake in numbering his propositions.

The Lemmas, however, are numbered consecutively, the last in the printed paper being Lemma XI. and the first in the MS. Lemma XII.

The other mathematical manuscripts are either drafts of these propositions or jottings of calculations not intended for publication.

The paper entitled "Thoughts concerning electricity"§ (No. 18) is placed next. It forms a suitable introduction to the account of the experiments, as it indicates the leading ideas of Cavendish's researches. The paper has no date, but its contents show that it is an earlier form of the theory of electricity, which Cavendish had already abandoned before he wrote the paper cf 1771. The pro-

* This seems to refer to the second part of the paper in the *Phil. Trans.*, 1771, p. 670, or Art. 132 of this edition, and shows that this paper was intended to form the first part of the "Work."

† Arts. 140 to 174.

‡ Arts. 175 to 194.

§ Arts. 195 to 216.

positions in No. 17 belong to this form of the theory, and are given in Note 18.

We have next the account of the experiments, the order of which is

No. 7	Figs. 12 to 13	Exp. I. and II.	Arts. 217 to 235
No. 9	Figs. 14 „ 19	Exps. III. to VIII.	Arts. 236 „ 294
No. 10	Figs. 20 „ 30		Arts. 295 „ 385
No. 8	Fig. 31		Arts. 386 „ 394

The style in which these papers are written leaves no doubt that they were intended to form a book, and to be published. They are given here without any alteration except in the case of a few abbreviations the meaning of which is either obvious or is explained in some other part of the MS. I have also divided them into articles for the sake of more convenient reference. All additions to the MS. are enclosed in square brackets.

After this I have placed the paper on the Torpedo from the *Philosophical Transactions* for 1776. This, I think, is the whole of the "work" which is extant, but it is by no means a complete account of Cavendish's electrical researches. There are three forms in which Cavendish recorded the results of his experiments:

1st. A Journal containing notes of every observation as it was made, with the particulars of the experiments, and measurements of the apparatus.

2nd. "Results," containing a comparison of the different measures of quantities as recorded in the Journal, and a deduction of the most probable result. See Arts. 647—696.

3rd. An account of the experiments written for publication.

I have reproduced the journals for 1771* and 1772† entire, because they form a good example of Cavendish's method of work, and because they contain all the data of some of the most important electrical measurements.

The journal for 1773‡ is much larger than the others, and gives an account of many interesting and important researches.

Many pages of this journal, however, are filled with the details of the experiments for the comparison of the coated plates which

* Arts. 438 to 465. † Arts. 466 to 493. ‡ Arts. 494 to 580.

Cavendish used as standards of capacity. These experiments differ in no respect from those in the former journals, and all the conclusions which Cavendish deduced from them are stated by himself in the "Results." I have therefore thought it best to omit them from the journal, but to retain Cavendish's heading of each experiment and its date when known, and to make the numbers of the omitted articles run on continuously with those retained.

Many of the entries in the journals give the day of the week and of the month, but very few of them give the year. I have therefore ascertained in what years the stated days of the week and month coincided, and have inserted the most probable year within square brackets. It thus appears that the journal entitled "Experiments in 1773" begins with experiments made in October, 1772. Cavendish appears, however, to have got wrong in his reckoning for a good many days together during that month. See Art. 502.

It is somewhat difficult to account for the fact, that though Cavendish had prepared a complete description of his experiments on the charges of bodies, and had even taken the trouble to write out a fair copy, and though all this seems to have been done before 1774, and he continued to make experiments in electricity till 1781, and lived on till 1810, he kept his manuscript by him and never published it. It was not till 1784 that he communicated to the Poyal Society those "Experiments on Air," including the production of water and of nitric acid, the absorbing interest of which might perhaps account for some neglect of his electrical writings.

Cavendish cared more for investigation than for publication. He would undertake the most laborious researches in order to clear up a difficulty which no one but himself could appreciate, or was even aware of, and we cannot doubt that the result of his enquiries, when successful, gave him a certain degree of satisfaction But it did not excite in him that desire to communicate the discovery to others which, in the case of ordinary men of science, generally ensures the publication of their results. How completely these researches of Cavendish remained unknown to other men of science is shown by the external history of electricity.

Viscount Mahon, afterwards Lord Stanhope, a man of great ingenuity and fertility in invention, a pupil of Le Sage of Geneva, and the inventor of the printing press which bears his name, published in 1779 his *Principles of Electricity*. The theory developed in this book is that

"A positively electrified body surrounded by air will deposit "upon all the particles of that Air which shall come successively into "contact with it, a proportional part of its *superabundant* Electricity, "By which means, the *Air* surrounding that body will also become "*positively* electrified: that is to say, it will form round that positive "body, an electrical atmosphere, which will likewise be positive." (p. 7.)

"That the electrical *Density* of all such Atmospheres decreases, "when the distance from the charged Body is increased." (p. 14.)

He then proceeds to determine the law of the density of the electrical atmosphere, as it depends on the distance from the charged body. He assumes that if a cylinder with hemispherical ends is placed in the electrical atmosphere of a charged body, the density of the electricity at any part of the cylinder will depend on the density of the electrical atmosphere in contact with it.

He also shows by experiment that if the cylinder is insulated, and originally without charge, it does not become charged as a whole by being immersed in the electrical atmosphere of a charged body. Hence, when the electricity of the cylinder is disturbed, the whole positive charge on one portion of the surface is numerically equal to the whole negative charge on the other portion.

Now if the density (on the cylinder) were inversely as the distance from the charged body, a transverse section of the cylinder whose distance from the charged body is the geometric mean of the distances of the ends, would divide the charge into two equal parts (both of course of the same kind of electricity), but if the density were inversely as the square of the distance, the distance of the section which would bisect the charge would be the harmonic mean of the distance of the ends. In all this he tacitly confounds the point of bisection of the charge with the neutral point.

He then shows by experiment that the actual position of the neutral point agrees sufficiently well with the harmonic mean, but not with the geometric mean, and from this he concludes (p. 65),

"Consequently, it evidently appears, from what was said above, "that the Density of the Electricity, of the electrical Atmosphere (in "which the said Body *A*, *B* was immersed) was in the inverse Ratio of "the square of the Distance."

It is evident from this that Lord Mahon was entirely ignorant of everything which Cavendish had done.

About the close of the century Dr Thomas Young, whose acquaintance with all branches of science was as remarkable for its extent as for its profundity, says of this neutral point:

"It was from the situation of this point that Lord Stanhope first "inferred the true law of the electric attractions and repulsions, although "Mr Cavendish had before suggested the same law as the most probable "supposition." (Lecture LIII.)

The same writer, in his "Life of Cavendish," in the Supplement to the *Encyclopædia Britannica*, gives the following account of the first paper on electricity.

"3. *An Attempt to explain some of the principal Phenomena of "Electricity by means of an Elastic Fluid.* (*Phil. Trans.* 1771, p. 584.) "Our author's theory of electricity agrees with that which had been "published a few years before by Æpinus, but he has entered more "minutely into the details of calculation, showing the manner in which "the supposed fluid must be distributed in a variety of cases, and "explaining the phenomena of electrified and charged substances as "they are actually observed. There is some degree of unnecessary "complication from the great generality of the determinations: the "law of electric attraction and repulsion not having been at that time "fully ascertained, although Mr Cavendish inclines to the true sup-"position, of forces varying inversely as the square of the distance: "this deficiency he proposes to supply by future experiments, and leaves "it to more skilful mathematicians to render some other parts of the "theory still more complete. He probably found that the necessity "of the experiments, which he intended to pursue, was afterwards "superseded by those of Lord Stanhope and M. Coulomb; but he "had carried the mathematical investigation somewhat further at a "later period of his life, though he did not publish his papers; an "omission, however, which is the less to be regretted, as M. Poisson, "assisted by all the improvements of modern analysis, has lately treated "the same subject in a very masterly manner. The acknowledged im-"perfections, in some parts of Mr Cavendish's demonstrative reasoning, "have served to display the strength of a judgment and sagacity still "more admirable than the plodding labours of an automatical calcu-"lator. One of the corollaries* seems at first sight to lead to a mode "of distinguishing positive from negative electricity, which is not justi-"fied by experiment; but the fallacy appears to be referable to the "very comprehensive character of the author's hypothesis, which re-"quires some little modification to accommodate it to the actual cir-"cumstances of the electric fluid, as it must be supposed to exist in "nature."

No man was better able than Dr Young to appreciate the scientific merits of Cavendish, and it is evident that he spared no pains in obtaining the data from which he wrote this sketch of his life, yet this account of his electrical researches shows a complete ignorance of Cavendish's unpublished work, and this ignorance must have been shared by the whole scientific world.

Dr Young, as it appears from the above extract, was aware of the existence of unpublished papers by Cavendish relating to electricity, but he supposed that these papers were entirely mathematical, and that "he probably found that the necessity of the experiments which he intended to pursue was afterwards superseded by those of Lord Stanhope and M. Coulomb."

We now know that the unpublished mathematical papers were entirely subsidiary to the experimental ones, and it is plain from Art. 95 that Cavendish had actually made some of his experiments before the paper of 1771, and that all those on electrostatics were completed before the end of 1773.

The favourable reception which Lord Stanhope's very interesting and popular experiments met with may have influenced Cavendish not to publish his own, but his estimate of their value as a foundation for a theory of electricity may be gathered from the fact, that in his "Thoughts concerning Electricity," which appears to be his earliest writing on the subject, he devotes two pages (Arts. 195—198) to the refutation of the very theory of electric atmospheres which is the basis of Lord Stanhope's reasoning; whereas in the paper of 1771, which contains his more matured views, he does not even allude to that theory.

* Art. 49 and Note 1.

It was not till 1785 that the first of the seven electrical memoirs of M. Coulomb was published. The experiments recorded in these memoirs furnished the data on which the mathematical theory of electricity, as we now have it, was actually founded by Poisson, and it is impossible to overestimate the delicacy and ingenuity of his apparatus, the accuracy of his observations, and the sound scientific method of his researches; but it is remarkable, that not one of his experiments coincides with any of those made by Cavendish. The method by which Coulomb made direct measurements of the electric force at different distances, and that by which he compared the density of the surface-charge on different parts of conductors, are entirely his own, and were not anticipated by Cavendish. On the other hand, the very idea of the capacity of a conductor as a subject of investigation is entirely due to Cavendish, and nothing equivalent to it is to be found in the memoirs of Coulomb.

The leading idea which distinguishes the electrical researches of Cavendish from those of his predecessors and contemporaries, is the introduction of the phrase "degree of electrification" with a clear scientific definition, which shows that it is precisely equivalent to what we now call potential.

In his first published paper (1771), he begins at Art. 101 by giving a precise sense to the terms "positively and negatively electrified," which up to that time had been in common use, but were often confounded with the terms "over and under charged," and in Art. 102 he defines what is meant by the "degree of electrification."

We find the same idea, however, in the much earlier draft of his theory in the "Thoughts concerning Electricity," Art. 201, where the degree of electrification is boldly, if somewhat prematurely, explained in a physical sense, as the *compression*, or as we should now say, the *pressure*, of the electric fluid.

We can trace this leading idea through the whole course of the electrical researches.

He shows that when two charged conductors are connected by a wire they must be electrified in the same degree, and he devotes the greater part of his experimental work to the comparison of the charges of the two bodies when equally electrified.

He ascertained by a well-arranged series of experiments the ratios of the charges of a great number of bodies to that of a sphere 12·1 inches in diameter, and as he had already proved that the charges of similar bodies are in the ratio of their linear dimensions, he expressed the charge of any given body in terms of the diameter of the sphere, which, when equally electrified, would have an equal charge, so that when in his private journals he speaks of the charge of a body as being so many "globular inches," or more briefly, so many "inches of electricity," he means that the capacity of the body is equal to that of a sphere whose diameter is that number of inches.

In the present state of electrical science, the capacity of a body is defined as its charge when its potential is unity, and the capacity of a sphere as thus defined is numerically equal to its radius. Hence, when Cavendish says that a certain conductor contains n inches of electricity, we may express his result in modern language by saying that its electric capacity is $\frac{1}{2}n$ inches.

In his early experiments he seems to have endeavoured to obtain a number of conductors as different as possible in form, of which the capacities should be nearly equal. Thus we find him comparing a pasteboard circle of 19·4 inches in diameter with his globe of 12·1 inches in diameter, but finding the charge of the circle greater than that of the globe, he ever after uses a circle of tin plate, 18·5 inches in diameter, the capacity of which he found more nearly equal to that of the globe.

In like manner the first wire that he used was 96 inches long and 0·185 diameter, but afterwards he always used a wire of the same diameter, but 72 inches long, the capacity of which was more nearly equal to that of the globe.

He also provided himself with a set of glass plates coated with circles of tin-foil on both sides. These plates formed three sets of three of equal capacity, the capacities of the three sets being as 1, 3 and 9, with a tenth coated plate whose capacity was as 27.

Besides these he had "double" plates of very small capacity made of two plates of glass stuck together, and also other plates of wax and rosin, the inductive capacity of these substances being, as he had already found, less than that of glass; and jars of larger capacity, ranging up to his great battery of 49 jars, whose capacity was 321000 "inches of electricity." In estimating the capacity of his battery, he used the method of repeated touching with a body of small capacity. (Arts. 412, 441, 582.) This method is the same as that used by MM. Weber and Kohlrausch in their classical investigation of the ratio of the electric units*.

Thus the method of experimental research which Cavendish adhered to was the comparison of capacities, and the formation of a graduated series of condensers, such as is now recognised as the most important apparatus in electrostatic measurements.

We have next to consider the steps by which he established the accuracy of his theory, and the discoveries he made respecting the electrical properties of different substances.

Cavendish himself, in his description of his experiments, has shown us the order in which he wishes us to consider them. The first experiment† is that of the globe within two hemispheres, from which he proves that the electric force varies inversely as the square of the distance, or at least cannot differ from that ratio by more than a fiftieth part. The degree of accuracy of all the experiments was limited by the sensitiveness of the pith ball electrometer which he used. Bennett's gold leaf electrometer, which is much more sensitive, was not introduced till 1787, but in repeating the experiment we can now use Thomson's Quadrant electrometer, and thereby detect a deviation from the law of the inverse square not exceeding one in 72000. See Note 19.

The second experiment, Art. 235, is a repetition of the first with bodies of different shape.

The third experiment, Art. 265, shows that in comparing the charges of bodies, the place where the connecting wire touches the

* *Elektrodynamische Maasbestimmungen*, Abh. IV. p. 235.

† Arts. 217 to 235.

body, and the form of the connecting wire itself, are matters of indifference.

The fourth experiment, Art. 269, shows that the charges of bodies of the same shape and size, but of different substances, are equal.

The fifth, Art. 273, compares the charge of a large circle with that of two of half the diameter. According to the theory the charge of the large circle should be equal to that of the two small ones if they are at a great distance from each other, and equal to twice that of the small ones if they are close together. Cavendish tried them at three different distances and compared the results with his calculations.

The sixth experiment, Art. 279, compares one long wire with two of half the length and half the diameter, placed at different distances.

The seventh, Art. 281, compares the charges of a globe, a circle, a square, an oblong and three different cylinders, and the eighth, Art. 288, shows that the charge of the middle plate of three parallel plates is small compared with that of the two outer ones.

Cavendish next describes his experiments for comparison of the charges of coated plates of glass and other substances, but begins by examining the sources of error in measurements of this kind.

The first of these which he investigates is the spreading of electricity on the surface of the plates beyond the coatings of tin-foil. He distinguishes two kinds of this spreading, one a gradual creeping of the electricity over the surface of the glass, Art. 300, and the other instantaneous, Art. 307.

He attempted to check the first kind by varnishing the glass plates and by enclosing their edges in a thick frame of cement, but he found very little advantage in this method, and finally adopted the plan of performing all the operations of the experiment as quickly as possible, so as to allow very little time for the gradual spreading of the electricity.

He next investigated the instantaneous spreading of electricity on the glass near the edge of the coating. He noticed that at the instant of charging the plate in the dark, a faint light could be seen all round the edges. He also observed that after charging and discharging a coated plate of glass many times without cleaning it, a narrow fringed ring of dirt could be traced all round the coating, the space between this ring and the coating being clean, and in general about $\frac{1}{10}$ inch broad.

He also observed that the flash of light was stronger the first or second times of charging a plate than afterwards.

To determine how much the capacity of a coated plate was increased by this spreading of the electricity, he compared the capacity of a plate with a circular coating with that of the same plate with a new coating of nearly the same area, but cut into strips, so that its perimeter was very much greater than that of the circular coating.

In this way he found that if we suppose a strip of uniform breadth added to the coating all round its boundary, the capacity of this coating, supposing the electricity not to spread, will be equal to that of the actual coating as increased by the spreading of the electricity. The most probable breadth of this strip he found to be 0·07 inch for thick glass and 0·09 for thin.

When this correction was applied to the areas of the coatings of the different coated plates, the computed charges of plates made of the same kind of glass were found to be very nearly in the same ratio as their observed charges.

But the observed charges of coated plates were found to be always several times greater than the charges computed from their thickness and the area of their coatings, the ratio of the observed charge to the computed charge being for plate glass about 8·2, for crown glass about 8·5, for shellac about 4·47, and for bees' wax about 3·5. Thus Cavendish not only anticipated Faraday's discovery of the Specific Inductive Capacity of different substances, but measured its numerical value in several substances.

The values of the specific inductive capacity of various substances as determined by different modern observers are compared with those found by Cavendish in the table in Note 27.

To make it certain, however, that the difference between the observed and calculated capacities of coated plates really arose from the nature of the plate and not from some error in the theory,

Cavendish determined the capacity of a "plate of air," that is to say a condenser consisting of two circles of tinfoil on glass with air between them. The capacity of a plate of air was found to be much less than that of a plate of glass or of wax of the same dimensions, but it seemed to be about $\frac{1}{11}$ in excess of the calculated value. This discrepancy will be discussed in Note 17.

These may be considered the principal results of the investigations with coated plates, but the following list of collateral experimental researches will show how thoroughly Cavendish went to work.

A question of fundamental importance in the theory of dielectrics is whether the electric induction is strictly proportional to the electromotive force which produces it, or in other words, is the capacity of a condenser made of glass or any other dielectric the same for high and for low potentials?

The form in which Cavendish stated this question was as follows*:—"Whether the charge of a coated plate bears the same proportion to that of a simple conductor, whether the electrification is strong or weak."

Cavendish, who explained the fact that the capacity of a glass plate is greater than that of an air plate, by supposing that the electricity is free to move within certain portions of the glass, supposed that when the plate was more strongly electrified the electricity would be able to penetrate further into the glass, and that therefore its charge would be greater in proportion to that of a simple conductor or a plate of air the stronger the degree of electrification.

But according to the experiments he made to answer this question† a coated plate and a simple conductor whose charges were equal for the usual degree of electrification remained sensibly equal for higher and lower degrees, and if, as appeared probable from the experiments on the spreading of electricity at the edge of the coating, this spreading extended further for high degrees of electrification than for low, it would be necessary to admit that the charge of a glass plate became less in proportion to that of a simple conductor as the degree of electrification increased. Cavendish, however, concluded that the experiments were hardly accurate

* Art. 526. † Arts. 355—365.

enough to warrant the deduction from them of so improbable a conclusion.

He also found that the result of the comparison of a coated plate and a simple conductor was the same whether they were charged positively or negatively.

He tried whether the capacity of a plate of rosin altered with the temperature, but he could not find that it did*. In glass he found that the capacity increased as the temperature rose, but the most decided increase did not occur till the glass began to conduct somewhat freely. Cavendish therefore does not consider the experiment quite decisive†.

He found that the apparent capacity of a Florence flask‡ was greater when it continued charged a good while than when it was charged and discharged immediately, and he found that the same was the case with a coated globe of glass. This phenomenon, which Faraday called "electric absorption," has recently been carefully studied in different kinds of glass by Dr Hopkinson§. It is connected with the long-known phenomenon of the "residual charge," and the existence of such phenomena in many dielectrics renders it difficult to obtain consistent values of their inductive capacities; for the more rapidly the charging and discharging is effected the lower is the apparent value of the capacity. It is for this reason that condensers of glass cannot be used as standards of capacity when accurate measurements are desired.

Franklin had shown‖ that the charge of a glass condenser resides in the glass and not in the coatings, for when the coatings were removed they were found to be without charge, and when new coatings were put in their place the condenser thus reconstructed was found to be charged.

Cavendish tried whether this was the case with a charged plate of air, by lifting one of the electrodes and changing the air between them and then replacing the electrode. He found that the charge was not altered during these operations, and concluded that the charge resides, not in the air, but in the metal plates.

* Art. 523. † Art. 366. ‡ Art. 523.

§ *Phil. Trans.* 1877, p. 599.

‖ *Franklin's Works*, ed. Sparks, Vol. v., p. 201.

In Arts. 336 to 339 we find a most ingenious method of determining by experiment the effect of the floor, walls and ceiling of a room, and of other surrounding objects, in increasing the apparent capacity of a conductor placed in a given position in the room. The method consists in measuring the capacities of two conductors of the same shape but of different dimensions, the centre of each being at the given point in the room. If the experiment had been made with the conductors at an infinite distance from all other bodies their capacities would have been in the ratio of their corresponding dimensions, but the effect of surrounding objects is to make their capacities vary in a higher ratio than that of their dimensions, and from the measured ratio of the two capacities, the correction for the effect of surrounding objects on the capacity of any small body may be calculated.

Cavendish also verified by experiment what he had already proved theoretically, that the capacity of two condensers is not sensibly altered when they are placed near to each other or far apart.

But besides this series of experiments on electric capacity, another course of experiments on electric resistance was going on between 1773 and 1781, the knowledge of which seems never to have been communicated to the world.

In his paper on the Torpedo in the *Philosophical Transactions* for 1776 (Art. 398) he alludes to "some experiments of which I "propose shortly to lay an account before this Society," but he never followed up this proposal by divulging the method by which he obtained the results which he proceeds to state—"that iron "wire conducts about 400 million times better than rain or dis- "tilled water*," and that "sea water, or a solution of one part of "salt in 30 of water conducts 100 times, and a saturated solution "of sea-salt about 720 times better than rain water."

Such was the reputation of Cavendish for scientific accuracy, that these bare statements seem to have been accepted at once,

* This is equivalent to saying that iron wire conducts 555,555 times better than saturated solution of sea salt. A comparison of the experiments of Matthiessen on iron with those of Kohlrausch on solutions of sodium chloride at 18°C. would make the ratio 451,390. The resistance of iron increases and that of the solution diminishes as the temperature rises, and at a temperature of about 11°C. the ratio of the resistances would agree with that given by Cavendish.

and soon found their way into the general stock of scientific information, although no one, as far as I can make out, has ever conjectured by what method Cavendish actually obtained them, more than forty years before the invention of the galvanometer, the only instrument by which any one else has ever been able to compare electric resistances.

We learn from the manuscripts now first published, that Cavendish was his own galvanometer. In order to compare the intensity of currents he caused them to pass through his own body, and by comparing the intensity of the sensations he felt in his wrist and elbows, he estimated which of the two shocks was the more powerful.

As Cavendish does not appear to have prepared an account of these experiments in the manner in which he usually wrote out what he intended to publish, it may be well to describe them here, as we collect them from different parts of his Journals.

The conductors to be compared were for the most part solutions of common salt of known strength or of other substances. These solutions were placed in glass tubes, more than a yard long, bent near one end. The tubes had been previously calibrated by means of mercury.

Two wires were run into the tube, probably through holes in corks at each end, to serve as electrodes. The length of the effective column of the liquid could be altered by sliding the wire in the straight part of the tube.

In order to send electric discharges of equal quantity and equal electromotive force through two different tubes Cavendish chose six jars of nearly equal capacity from "Nairne's last battery." The two tubes to be compared were placed so that the wires run into their bent ends communicated with the outside of this battery of six jars. The wires run into the straight ends of the tubes were fastened to two separately insulated pieces of tinfoil. The six jars were then all charged at once by the same conductor till the gauge electrometer indicated the proper degree of electrification. The conductor was then removed, so that the six jars remained with their inside coatings insulated from each other and equally charged.

Cavendish then taking two pieces of metal, one in each hand, touched with one the tinfoil belonging to one of the tubes to be compared, and then with the other touched the knob of jar No. 1, so as to receive a shock, the charge passing through his body and the first tube.

He next laid one of the metals on the tinfoil of the second tube, and then touching with the other the knob of jar No. 2, he received a second shock, the discharge passing through his body and the second tube.

In this way he took six shocks, making them pass alternately through the first and the second tube, and proceeded to record his impression whether the intensity of the shock through the second tube was greater or less than that of the shock through the first, and concluded that the tube which gave the greater shock had the smaller resistance.

He then adjusted the wire in one of the tubes so as to make the resistance more nearly equal to that of the other, and repeated the experiment, always recording his impression of the result, till he found that one adjustment made the shock of the second tube sensibly greater than that of the first, and that another adjustment made it sensibly less.

From the result of the whole series of experiments he judged what adjustment would make the two shocks exactly equal.

Instead of using six jars only, he seems latterly to have used the whole battery, electrifying one row to a given degree and then communicating this charge to the whole battery, and taking the discharge of one row at a time through the tubes alternately. He seems to have found some advantage in thus using a discharge of greater quantity and smaller electromotive force.

The accuracy which Cavendish attained in the discrimination of the intensity of shocks is truly marvellous, whether we judge by the consistency of his results with each other, or whether we compare them with the latest results obtained with the aid of the galvanometer, and with all the precautions which experience has shown to be necessary in measuring the resistance of electrolytes.

One of the most important investigations which Cavendish

made in this way was to find, as he expressed it, "what power of the velocity the resistance is proportional to *."

Cavendish means by "resistance" the whole force which resists the current, and by "velocity" the strength of the current through unit of area of the section of the conductor.

(In modern language the word resistance is used in a different sense, and is measured by the force which resists a current of unit strength.)

By four different series of experiments on the same solution in wide and in narrow tubes, Cavendish found that the resistance (in his sense) varied as the

1·08, 1·03, 0·976, and 1·00

power of the velocity.

This is the same as saying that the resistance (in the modern sense) varies as the

0·08, 0·03, −0·024

power of the strength of the current in the first three sets of experiments, and in the fourth set that it does not vary at all.

This result, obtained by Cavendish in January, 1781, is an anticipation of the law of electric resistance discovered independently by Ohm and published by him in 1827. It was not till long after the latter date that the importance of Ohm's law was fully appreciated, and that the measurement of electric resistance became a recognised branch of research. The exactness of the proportionality between the electromotive force and the current in the same conductor seems, however, to have been admitted, rather because nothing else could account for the consistency of the measurements of resistance obtained by different methods, than on the evidence of any direct experiments.

Some doubts, however, having been suggested with respect to the mathematical accuracy of Ohm's law, the subject was taken up by the British Association in 1874, and the experiments of Professor Chrystal, by which the exactness of the law, as it relates to metallic conductors, was tested by currents of every degree of

* Arts. 574, 575, 620, 686.

intensity, are contained in the Report of the British Association for 1876.

The laws of the strength of currents in multiple and divided circuits are accurately stated by Cavendish in Arts. 417, 597, 598.

Cavendish applied the same method of experiment to compare the resistance of the same liquid at different temperatures*, and he found that "salt in 69 [of water] conducts 1·97 times better in heat of 105 than in that of 58½." He also found that "the proportion of the resistance of saturated solution and salt in 999 to each other seems not much altered by varying heat from 50 to 95."

Kohlrausch, who has made a most extensive series of experiments on the resistance of electrolytes, gives results from which it appears that the ratio of the resistances of salt in 69 at 105° F. and at 58½° F. would be 1·59. He also finds that the temperature coefficient for solutions of salt alters very little with the strength. See Note 33.

Cavendish also tested the resistance of solutions of salt of strengths varying from saturation to one in 20000 of distilled water, and arrived at the result, which Kohlrausch has shown to be nearly accurate, that for weak solutions the product of the resistance into the percentage of salt is nearly constant.

Of all substances, that for which different observers have given the most different measures of resistance is pure water.

It has been found indeed that the presence of the minutest trace of impurity in water diminishes its resistance enormously. Thus Kohlrausch found that it was necessary to use water quite freshly distilled in platinum vessels, for if placed in a glass vessel it rapidly diminished in resistance by dissolving a minute quantity of the glass, and a few minutes exposure to the air of the laboratory, by impregnating the water with a trace of tobacco smoke, was found sufficient to spoil it for a determination of resistance. Kohlrausch indeed estimates that the electric conductivity which he observed in the purest water he could obtain might be accounted for by the presence of no more than one ten millionth part of hydrochloric acid, a quantity which no chemical analysis

* Art. 691.

could detect. Hence the hypothesis that water is a non-conductor of electricity, if not true, cannot be disproved.

Some of these remarkable properties of water were detected by Cavendish. He found that the resistance of pump water was $4\frac{1}{6}$ times less than that of rain water, and that of rain water was 2·4 times less than that of distilled water*.

In January 1777, he found that salt in 2999 conducted about 70 or 90 times better than some water distilled in the preceding summer but only about 25 or 50 times better than the distilled water used in the year 1776†, and that the conductivity of distilled water increased by standing two or three hours in a glass tube‡.

He also found that in order to make the conducting powers of his weakest solutions of salt agree with the hypothesis that they are as the quantity of salt in them, it would be necessary only to suppose that his distilled water contained one part of salt in 120000§.

It was found that distilled water impregnated with fixed air from oil of vitriol and marble conducted $2\frac{1}{2}$ times better than the same water deprived of its air by boiling‖, and that the presence of absorbed air in a weak solution of salt seemed to increase its conductivity¶.

In order to find whether electricity is resisted in passing out of one medium into another in perfect contact with it, Cavendish prepared a tube containing 8 columns of saturated solution of sea salt enclosed between columns of mercury. He found that the shock was diminished in passing through a mixed column in which the length of salt water was 21·8 inches as much as in passing through a single column of the same size whose length was 22·94 inches**.

The difference would have been far greater if the comparison had been made with an ordinary galvanometer and continued currents which rapidly produce polarization, but with the small quantities of electricity which Cavendish used, the effect of polarization would hardly be sensible.

* Art. 525. † Art. 690. ‡ Art. 621.
§ Art. 630. ‖ Arts. 625, 693. ¶ Art. 692. ** Art. 578.

He also made a compound conductor consisting of 40 bits of tin soldered together. The shock through this appeared to be of the same strength as through a single piece of the same size. This experiment however is not of much value, as the resistance of the conductor was far too small compared with that of Cavendish's body to give good results*

We now come to a very remarkable set of experiments which Cavendish made on a series of salts and acids in order to determine their relative electric resistance. They are recorded in Arts. 626, 627 and 694, and are dated Jan. 13 and 15, 1777.

The strength of the different solutions was such, as Cavendish tells us, "that the quantity of acid in each should be equivalent to that in a solution of salt in 29 of water."

The total weight of each solution was 3 pounds 10 ounces and 12 pennyweights, or 1116 pennyweights Troy. The quantity of each substance when reduced to pennyweights is in every case very nearly the equivalent weight of that substance in the system adopted at present, in which the equivalent weight of hydrogen is taken as unity†.

Now these experiments were made in 1777, and it is difficult to see from what source, other than determinations of his own, he could have derived these numbers. Wenzel's *Lehre von den Verwandschaften* was published in 1777. I have not been able to consult the work itself, but from the account of it given in Kopp's *Geschichte der Chemie*, the equivalent numbers seem to have been larger than those used by Cavendish. Richter's *Anfangsgründe der Stöchyometrie* was not published till 1792.

It is difficult to account for the agreement not only of the ratios but of the absolute numbers given by Cavendish with those of the modern system, in which the equivalent weight of hydrogen is taken as unity. I can only conjecture from several parts of his

* Art. 579. The resistance of a man's body, from one hand to the other, varies from about 1000 ohms when the hands are well wetted with salt water, to about 12000 when the hands are dry. When the outer skin is removed by a blister, the resistance is very much diminished. The resistance of the compound conductor was probably a fraction of an ohm. See Note 31.

† See Note 34.

paper on Factitious Airs (*Phil. Trans.* 1766), that Cavendish was accustomed to compare the quantity of fixed air from different carbonates with that from 1000 grains of marble. Now the modern equivalent weight of marble is 100, so that if Cavendish took 100 pennyweights as the equivalent weight of marble, the equivalents of other substances would be as he has given them. This I think is more likely than that he should have selected inflammable air as his standard substance at a time when even his own experiments left it doubtful whether inflammable air was always of the same kind.

In his journal, Cavendish writes down these equivalent weights just as a modern chemist might do, without a hint that a list of these numbers was not at that time one of the things which every student of chemistry ought to know by heart. It is only by comparing the date of these researches with the dates of the principal discoveries in chemistry, that we become aware, that in the incidental mention of these numbers we have the sole record of one of those secret and solitary researches, the value of which to other men of science Cavendish does not seem to have taken into account, after he had satisfied his own mind as to the facts.

I take this opportunity of expressing my thanks to the many friends who have given me assistance in preparing this edition, and in particular to Mr C. Tomlinson, who gave me valuable information about the manuscripts; to Mrs Sime, who lent me a manuscript book of letters, &c., relating to Cavendish, collected by her brother, the late Dr George Wilson; to Mr W. Garnett, of St John's College, Cambridge, who copied out Arts. 236—294; and Mr W. N. Shaw, of Emmanuel College, who took the photographs from which the facsimile figures were executed; to Mr H. B. Wheatley, who furnished me with information connected with the history of the Royal Society; to Prof. Dewar, Mr P. T. Main, Mr G. F. Rodwell, and Dr E. J. Mills, who gave me information on chemical subjects; and Mr Dew Smith and Mr F. M. Balfour, of Trinity College, and Prof. Ernst von Fleischl, of Vienna, who gave me information about electrical fishes, and the physiological effect of electricity.

P. S. 14th June, 1879.

Just before sending this sheet to press I have received from Mr Robert H. Scott, F.R.S., a small packet marked "Cavendish Papers," which had been sent to the Meteorological Office by Sir Edward Sabine.

These papers relate entirely to magnetism, and do not fall within the scope of this volume, though they may supply important materials for the magnetic history of the earth, and are in all respects excellent specimens of Cavendish's scientific procedure.

I shall therefore only mention a few particulars in which these papers throw some additional light on Cavendish's life and work.

The descriptions of Cavendish by Cuvier, Young, Thomson and Wilson agree in representing him as living in London, and regularly attending the meetings of the Royal Society, but in other respects leading an isolated life, very much detached from the interests, whether social or scientific, of other men.

It has also been hinted that Lord Charles Cavendish, who, as we have already seen, was himself addicted to scientific pursuits, did not entirely approve of his son's devotion to science, or at least, for some reason or other, restricted him in the means of carrying on his work.

In these manuscripts, however, we have the details of a laborious series of observations undertaken to determine the errors of the variation compass and the dipping needle belonging to the Royal Society, and on Sept. 16, 1773, we find "Observations of needle in Garden by Father and Self," and a "Comparison of Society's compass in house and in soc[iety's] garden with Father's compass in room."

It appears, therefore, that Lord Charles Cavendish not only placed his instruments at his son's disposal, but made observations of the compass in concert with him, and that these observations were undertaken in order to make the instruments belonging to the Royal Society more available for accurate measurements. In the same Journal there are also "Measures taken for setting Dr Knight's magnets so that their poles shall be equidistant from var[iation] comp[ass] and dipp[ing] need[le] in 1775." The results of this enquiry are briefly stated by Cavendish in his paper on the Instruments belonging to the Royal Society in the "Philosophical

Transactions" for 1776. In the same volume there is an account of Dr Knight's great Magazines of magnets by Dr Fothergill.

A considerable portion of the MS. is taken up with "Directions for using the Dipping Needle," written out at greater or less length (probably according to the scientific capacity of the recipient) "for Captain Pickersgill," "for Captain Bayley," "for Dalrymple" [Hydrographer to the Honourable East India Company] &c.

There is also a treatise of 26 pages "On the different forms of construction of dipping needles."

Besides this, there is a series of observations of the magnetic variation and also of the dip, at various times, from 1773 to August 1809 (Cavendish died Feb. 24, 1810).

These observations were made for the most part only in the summer months, but during that time were carried on with the greatest regularity, and results for each year calculated from them.

We also find the record of "Trials of Nairne's needle in different parts of England in August, 1778."

It was tried "in Garden, Aug. 8. In Garden of Observatory at Oxford, Aug. 14. At Birmingham, in Bowling-green, Aug. 15. At Towcester, in Garden, Aug. 17. At St Ives, in Garden, Aug. 18. At Ely, in Garden, Aug. 18. At London, Aug. 19 and 22." From these trials he finds that "Lines of equal dip should seem to run about 44° to south of west, and dip should increase about 42′ by going 1° to N.W."

There is a long and valuable series of experiments on the magnetic properties of forged iron, blistered steel, and cast iron. "Some bars were got from Elwell 31¾ inch long, 2·1 broad, and about ·5 thick. On May 29, 1776, one of each was made magnetical, the marked end being the south pole. In trying the experiment the bars were placed perpendicularly against a wall 25 inches distant from the center of the needle, 91°¼ to west of usual magnetic north, either the top or bottom of the bar being always on a level with the needle. They were kept constantly with the marked end upwards till after the observations of June 30, after which they were kept with the marked end downwards."

Cavendish determines in every case the "fixed magnetism" and the "moveable magnetism" of the bar, and also its magnetism when "struck 100 times on an anvil, falling 1·6 inches by its weight, and tried immediately after."

There are also 23 pages of experiments on the effect of heat on magnets, and a mathematical investigation of the bending of the dipping-needle by its own weight as affecting the determination of the dip, together with measurements of the elasticity of steel and of glass.

AN ATTEMPT

TO EXPLAIN SOME OF THE PRINCIPAL

PHÆNOMENA OF ELECTRICITY,

BY MEANS OF AN ELASTIC FLUID *.

BY THE

HONOURABLE HENRY CAVENDISH, F.R.S.

* From the *Philosophical Transactions of the Royal Society* for 1771, Vol. LXI. pp. 584—677.

[*Read* Dec. 19, 1771 and Jan. 9, 1772.]

AN ATTEMPT TO EXPLAIN SOME OF THE PRINCIPAL PHÆNOMENA OF ELECTRICITY, BY MEANS OF AN ELASTIC FLUID.

1] SINCE I first wrote the following paper, I find that this way of accounting for the phænomena of electricity is not new. Æpinus, in his *Tentamen Theoriæ Electricitatis et Magnetismi**, has made use of the same, or nearly the same hypothesis that I have; and the conclusions he draws from it agree nearly with mine, as far as he goes. However, as I have carried the theory much farther than he has done, and have considered the subject in a different, and, I flatter myself, in a more accurate manner, I hope the Society will not think this paper unworthy of their acceptance.

2] The method I propose to follow is, first, to lay down the hypothesis; next, to examine by strict mathematical reasoning, or at least, as strict reasoning as the nature of the subject will admit of, what consequences will flow from thence; and lastly, to examine how far these consequences agree with such experiments as have yet been made on this subject. In a future paper, I intend to give the result of some experiments I am making, with intent to examine still further the truth of this hypothesis, and to find out the law of the electric attraction and repulsion.

HYPOTHESIS.

3] There is a substance, which I call the electric fluid, the particles of which repel each other and attract the particles of all other matter with a force inversely as some less power of the distance than the cube: the particles of all other matter also, repel each other, and attract those of the electric fluid, with a force

* [Petropoli, 1759.]

varying according to the same power of the distances. Or, to express it more concisely, if you look upon the electric fluid as matter of a contrary kind to other matter, the particles of all matter, both those of the electric fluid and of other matter, repel particles of the same kind, and attract those of a contrary kind, with a force inversely as some less power of the distance than the cube.

4] For the future, I would be understood never to comprehend the electric fluid under the word matter, but only some other sort of matter.

5] It is indifferent whether you suppose all sorts of matter to be indued in an equal degree with the foregoing attraction and repulsion, or whether you suppose some sorts to be indued with it in a greater degree than others; but it is likely that the electric fluid is indued with this property in a much greater degree than other matter; for in all probability the weight of the electric fluid in any body bears but a very small proportion to the weight of the matter; but yet the force with which the electric fluid therein attracts any particle of matter must be equal to the force with which the matter therein repels that particle; otherwise the body would appear electrical, as will be shewn hereafter.

To explain this hypothesis more fully, suppose that 1 grain of electric fluid attracts a particle of matter, at a given distance, with as much force as n grains of any matter, lead for instance, repel it: then will 1 grain of electric fluid repel a particle of electric fluid with as much force as n grains of lead attract it; and 1 grain of electric fluid will repel 1 grain of electric fluid with as much force as n grains of lead repel n grains of lead *.

6] All bodies in their natural state with regard to electricity, contain such a quantity of electric fluid interspersed between their particles, that the attraction of the electric fluid in any small part of the body on a given particle of matter shall be equal to the repulsion of the matter in the same small part on the same particle. A body in this state I call saturated with electric fluid: if the body contains more than this quantity of electric fluid, I call it overcharged: if less, I call it undercharged. This is the hypothesis; I now proceed to examine the consequences which will flow from it.

* [Note 1.]

Fig. 1.

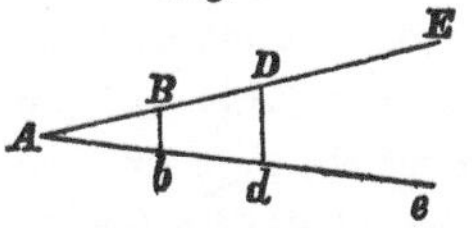

7] Lemma I. Let EAe (Fig. 1) represent a cone continued infinitely; let A be the vertex, and Bb and Dd planes parallel to the base; and let the cone be filled with uniform matter, whose particles repel each other with a force inversely as the n power of the distance. If n is greater than 3, the force with which a particle at A is repelled by $EBbe$ or all that part of the cone beyond Bb is as $\frac{1}{AB^{n-3}}$.

For supposing AB to flow, the fluxion of $EBbe$ is proportional to $-\dot{AB} \times AB^2$, and the fluxion of its repulsion on A is proportional to $\frac{-\dot{AB}}{AB^{n-2}}$; the fluent of which is $\frac{1}{(n-3)\,AB^{n-3}}$; which when AB is infinite is equal to nothing; consequently the repulsion of $EBbe$ is proportional to $\frac{1}{(n-3)\,AB^{n-3}}$ or to $\frac{1}{AB^{n-3}}$.

8] Cor. If AB is infinitely small, $\frac{1}{AB^{n-3}}$ is infinitely great; therefore the repulsion of that part of the cone between A and Bb, on A, is infinitely greater than the repulsion of all that beyond it.

9] Lemma II. By the same method of reasoning it appears, that if n is equal to 3, the repulsion of the matter between Bb and Dd on a particle at A, is proportional to the logarithm of $\frac{AD}{AB}$; consequently, the repulsion of that part is infinitely small in respect of that between A and Bb, and also infinitely small in respect of that beyond Dd.

10] Lemma III. In like manner, if n is less than 3, the repulsion of the part between A and Bb on A is proportional to AB^{3-n}: consequently the repulsion of the matter between A and Bb on A, is infinitely small in respect of that beyond it.

11] Cor. It is easy to see from these three lemmata, that, if the electric attraction and repulsion had been supposed to be inversely as some higher power of the distance than the cube; a particle could not have been sensibly affected by the repulsion of any fluid, except what was placed close to it. If the repulsion was inversely as the cube of the distance, a particle could not be sensibly affected by the repulsion of any finite quantity of fluid, except what was close to it. But as the repulsion is supposed to be inversely as some power of the distance less than the cube, a particle may be sensibly affected by the repulsion of a finite quantity of fluid, placed at any finite distance from it.

12] DEF. If the electric fluid in any body is by any means confined in such manner that it cannot move from one part of the body to the other, I call it immoveable: if it is able to move readily from one part to another, I call it moveable.

13] PROP. I. A body overcharged with electric fluid attracts or repels a particle of matter or fluid, and is attracted or repelled by it, with exactly the same force as it would, if the matter in it, together with so much of the fluid as is sufficient to saturate it, was taken away, or as if the body consisted only of the redundant fluid in it. In like manner an undercharged body attracts or repels with the same force, as if it consisted only of the redundant matter; the electric fluid, together with so much of the matter as is sufficient to saturate it, being taken away.

This is evident from the definition of saturation.

14] PROP. II. Two over or undercharged bodies attract or repel each other with just the same force that they would, if each body consisted only of the redundant fluid in it, if overcharged, or of the redundant matter in it, if undercharged.

For, let the two bodies be called A and B; by the last proposition the redundant substance in B impels each particle of fluid and matter in A, and consequently impels the whole body A, with the same force that the whole body B impels it: for the same reason the redundant substance in A impels the redundant substance in B, with the same force that the whole body A impels it. It is shewn therefore, that the whole body B impels the whole body A, with the same force that the redundant substance in B impels the whole body A, or with which the whole body A impels the redundant substance in B; and that the whole body A impels the redundant substance in B, with the same force that the redundant substance in A impels the redundant substance in B; therefore the whole body B impels the whole body A, with the same force with which the redundant substance in *A* impels the redundant substance in *B*, or with which the redundant substance in *B* impels the redundant substance in *A*.

15] COR. Let the matter in all the rest of space, except in two given bodies, be saturated with immoveable fluid; and let the fluid in those two bodies be also immoveable. Then, if one of the bodies is saturated, and the other either over or undercharged, they will not at all attract or repel each other.

If the bodies are both overcharged, they will repel each other.

If they are both undercharged, they will also repel each other.

If one is overcharged and the other undercharged, they will attract each other.

N.B. In this corollary, when I call a body overcharged, I would be understood to mean, that it is overcharged in all parts, or at least nowhere undercharged: in like manner, when I call it undercharged, I mean that it is undercharged in all parts, or at least nowhere overcharged.

16] PROP. III. If all the bodies in the universe are saturated with electric fluid, it is plain that no part of the fluid can have any tendency to move.

17] PROP. IV. If the quantity of electric fluid in the universe is exactly sufficient to saturate the matter therein, but unequally dispersed, so that some bodies are overcharged and others undercharged; then, if the electric fluid is not confined, it will immediately move till all the bodies in the universe are saturated.

For supposing that any body is overcharged, and the bodies near it are not, a particle at the surface of that body will be repelled from it by the redundant fluid within; consequently some fluid will run out of that body; but if the body is undercharged, a particle at its surface will be attracted towards the body by the redundant matter within, so that some fluid will run into the body.

N.B. In Prob. IV. Case 3, there will be shewn an exception to this proposition; there may perhaps be some other exceptions to it: but I think there can be no doubt, but what this proposition must hold good in general.

18] Lemma IV. Let *BDE*, *bde*, and $\beta\delta\epsilon$ (Fig. 2) be concentric spherical surfaces, whose center is *C*: if the space* *Bb* is filled with uniform matter, whose particles repel with a force inversely as the square of the distance, a particle placed anywhere within the space *Cb*, as at *P*, will be repelled with as much force in one direction as another, or it will not be impelled in any direction. This is demonstrated in Newton, *Princip.* Lib. I. Prop. 70. It follows also from his demonstration, that if the repulsion is inversely as some higher power of the distance than the square, the particle *P* will be impelled towards the center; and if the repulsion is inversely as some lower power than the square, it will be impelled from the center.

Fig. 2.

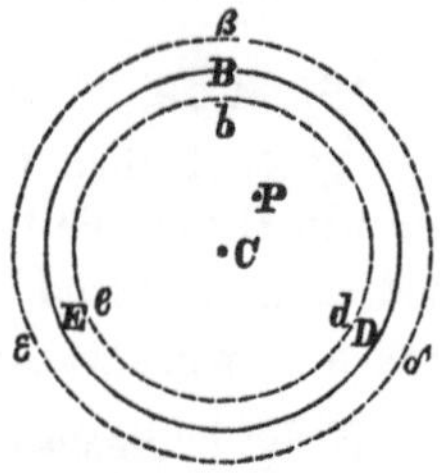

19] Lemma V. If the repulsion is inversely as the square of the distance, a particle placed anywhere without the sphere *BDE*, is repelled by that sphere, and also by the space *Bb*, with the same force that it would if all the matter therein was collected in the center of the sphere; provided the density of the matter therein is everywhere the same at the same distance from the center. This is easily deduced from Prop. 71, of the same book, and has been demonstrated by other authors.

20] Prop. V. Problem 1. Let the sphere *BDE* be filled with uniform solid matter, overcharged with electric fluid: let the fluid therein be moveable, but unable to escape from it: let the fluid in the rest of infinite space be moveable, and sufficient to saturate the matter therein; and let the matter in the whole of infinite space, or at least in the space $B\beta$, whose dimensions will be given below, be uniform and solid; and let the law of the electric attraction and repulsion be inversely as the square of the distance: it is required to determine in what manner the fluid will be disposed both within and without the globe.

Take the space *Bb* such, that the interstices between the particles of matter therein shall be just sufficient to hold a quantity of electric fluid, whose particles are pressed close together, so as to

* By the space *Bb* or $B\beta$, I mean the space comprehended between the spherical surfaces *BDE* and *bde*, or between *BDE* and $\beta\delta\epsilon$: by the space *Cb* or $C\beta$, I mean the spheres *bde* or $\beta\delta\epsilon$.

touch each other, equal to the whole redundant fluid in the globe, besides the quantity requisite to saturate the matter in *Bb*; and take the space *Bβ* such, that the matter therein shall be just able to saturate the redundant fluid in the globe: then, in all parts of the space *Bb*, the fluid will be pressed close together, so that its particles shall touch each other; the space *Bβ* will be intirely deprived of fluid; and in the space *Cb*, and all the rest of infinite space, the matter will be exactly saturated.

For, if the fluid is disposed in the above-mentioned manner, a particle of fluid placed anywhere within the space *Cb* will not be impelled in any direction by the fluid in *Bb*, or the matter in *Bβ*, and will therefore have no tendency to move: a particle placed anywhere without the sphere *βδε* will be attracted with just as much force by the matter in *Bβ*, as it is repelled by the redundant fluid in *Bb*, and will therefore have no tendency to move: a particle placed anywhere within the space *Bb*, will indeed be repelled towards the surface, by all the redundant fluid in that space which is placed nearer the center than itself; but as the fluid in that space is already pressed as close together as possible, it will not have any tendency to move; and in the space *Bβ* there is no fluid to move, so that no part of the fluid can have any tendency to move.

Moreover, it seems impossible for the fluid to be at rest, if it is disposed in any other form; for if the density of the fluid is not everywhere the same at the same distance from the center, but is greater near *b* than near *d*, a particle placed anywhere between those two points will move from *b* towards *d*; but if the density is everywhere the same at the same distance from the center, and the fluid in *Bb* is not pressed close together, the space *Cb* will be overcharged, and consequently a particle at *b* will be repelled from the center, and cannot be at rest: in like manner, if there is any fluid in *Bβ*, it cannot be at rest: and, by the same kind of reasoning, it might be shewn, that, if the fluid is not spread uniformly within the space *Cb*, and without the sphere *βδε*, it cannot be at rest.

21] Cor. I. If the globe *BDE* is undercharged, everything else being the same as before, there will be a space *Bb*, in which the matter will be intirely deprived of fluid, and a space *Bβ*, in which the fluid will be pressed close together; the matter in *Bb* being equal to the whole redundant matter in the globe, and the redundant fluid in $B\beta$, being just sufficient to saturate the matter in Bb: and in all the rest of space the matter will be exactly saturated. The demonstration is exactly similar to the foregoing.

22] COR. II. The fluid in the globe *BDE* will be disposed in exactly the same manner, whether the fluid without is immoveable, and disposed in such manner, that the matter shall be everywhere saturated, or whether it is disposed as above described; and the fluid without the globe will be disposed in just the same manner, whether the fluid within is disposed uniformly, or whether it is disposed as above described.

23] PROP. VI. PROBLEM 2. To determine in what manner the fluid will be disposed in the globe *BDE*, supposing everything as in the last problem, except that the fluid on the outside of the globe is immoveable, and disposed in such manner as everywhere to saturate the matter, and that the electric attraction and repulsion is inversely as some other power of the distance than the square.

I am not able to answer this problem accurately; but I think we may be certain of the following circumstances.

24] CASE 1. Let the repulsion be inversely as some power of the distance between the square and the cube, and let the globe be overcharged.

It is certain that the density of the fluid will be everywhere the same, at the same distance from the center. Therefore, first, There can be no space as Cb, within which the matter will be everywhere saturated; for a particle at b is impelled towards the center, by the redundant fluid in Bb, and will therefore move towards the center, unless Cb is sufficiently overcharged to prevent it Secondly, The fluid close to the surface of the sphere will be pressed close together; for otherwise a particle so near to it, that the quantity of fluid between it and the surface should be very small, would move towards it; as the repulsion of the small quantity of fluid between it and the surface would be unable to balance the repulsion of the fluid on the other side. Whence, I think, we may conclude, that the density of the fluid will increase gradually from the center to the surface, where the particles will be pressed close together: whether the matter exactly at the center will be overcharged, or only saturated, I cannot tell.

25] Cor. For the same reason, if the globe be undercharged, I think we may conclude, that the density of the fluid will diminish gradually from the center to the surface, where the matter will be intirely deprived of fluid.

26] Case 2. Let the repulsion be inversely as some power of the distance less than the square; and let the globe be overcharged.

There will be a space *Bb*, in which the particles of the fluid will be everywhere pressed close together; and the quantity of redundant fluid in that space will be greater than the quantity of redundant fluid in the whole globe *BDE*; so that the space *Cb*, taken all together, will be undercharged: but I cannot tell in what manner the fluid will be disposed in that space.

For it is certain, that the density of the fluid will be everywhere the same, at the same distance from the center. Therefore, let *b* be any point where the fluid is not pressed close together, then will a particle at *b* be impelled towards the surface, by the redundant fluid in the space *Bb*; therefore, unless the space *Cb* is undercharged, the particle will move towards the surface.

27] Cor. For the same reason, if the globe is undercharged, there will be a space *Bb*, in which the matter will be intirely deprived of fluid, the quantity of matter therein being more than the whole redundant matter in the globe; and, consequently, the space *Cb*, taken all together, will be overcharged*.

28] Lemma VI. Let the whole space comprehended between two parallel planes, infinitely extended each way, be filled with uniform matter, the repulsion of whose particles is inversely as the square of the distance; the plate of matter formed thereby will repel a particle of matter with exactly the same force, at whatever distance from it it be placed.

For suppose that there are two such plates, of equal thickness, placed parallel to each other, let *A* (Fig. 3) be any point not placed in or between the two plates: let *BCD* represent any part of the nearest plate: draw the lines *AB*, *AC*, and *AD*, cutting the furthest plate in

Fig. 3.

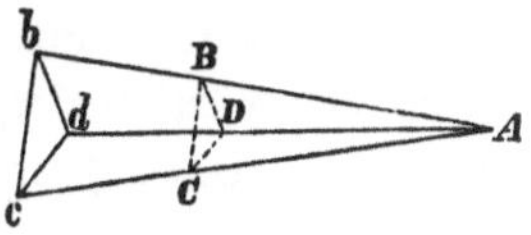

* [Note 2.]

b, *c*, and *d*; for it is plain that if they cut one plate, they must, if produced, cut the other: the triangle *BCD* is to the triangle *bcd*, as AB^2 to Ab^2; therefore a particle of matter at *A* will be repelled with the same force by the matter in the triangle *BCD*, as by that in *bcd*. Whence it appears, that a particle at *A* will be repelled with as much force by the nearest plate, as by the more distant; and consequently, will be impelled with the same force by either plate, at whatever distance from it it be placed.

29] COR. If the repulsion of the particles is inversely as some higher power of the distance than the square, the plate will repel a particle with more force, if its distance be small than if it be great; and if the repulsion is inversely as some lower power than the square, it will repel a particle with less force, if its distance be small than if it be great.

30] PROP. VII. PROB. 3. In Fig. 4, let the parallel lines *Aa*, *Bb*, &c. represent parallel planes infinitely extended each way: let the spaces* *AD* and *EH* be filled with uniform solid matter: let the electric fluid in each of those spaces be moveable and unable to escape: and let all the rest of the matter in the universe be saturated with immoveable fluid; and let the electric attraction and repulsion be inversely as the square of the distance. It is required to determine in what manner the fluid will be disposed in the spaces *AD* and *EH*, according as one or both of them are over or undercharged.

Fig. 4.

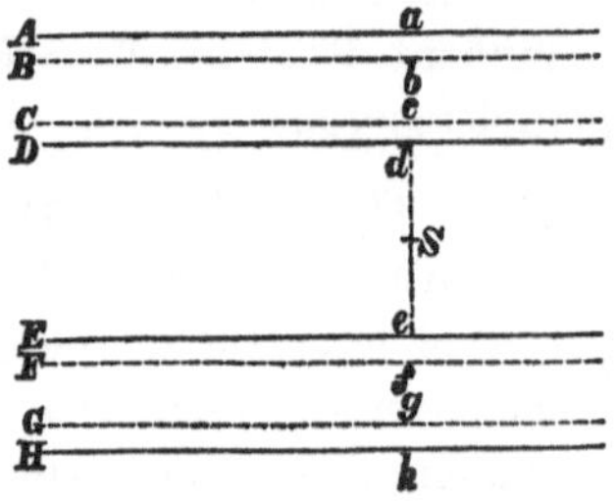

Let *AD* be that space which contains the greatest quantity of redundant fluid, if both spaces are overcharged, or which contains the least redundant matter, if both are undercharged; or, if one is overcharged, and the other undercharged, let *AD* be the overcharged one. Then, first, There will be two spaces, *AB* and *GH*, which will either be intirely deprived of fluid, or in which the particles will be pressed close together; namely, if the whole quantity of fluid in *AD* and *EH* together, is less than

* By the space *AD* or *AB*, &c. I mean the space comprehended between the planes *Aa* and *Dd*, or between *Aa* and *Bb*.

sufficient to saturate the matter therein, they will be intirely deprived of fluid; the quantity of redundant matter in each being half the whole redundant matter in *AD* and *EH* together: but if the fluid in *AD* and *EH* together is more than sufficient to saturate the matter, the fluid in *AB* and *GH* will be pressed close together; the quantity of redundant fluid in each being half the whole redundant fluid in both spaces. Secondly, In the space *CD* the fluid will be pressed close together; the quantity of fluid therein being such, as to leave just enough fluid in *BC* to saturate the matter therein. Thirdly, The space *EF* will be intirely deprived of fluid; the quantity of matter therein being such that the fluid in *FG* shall be just sufficient to saturate the matter therein: consequently, the redundant fluid in *CD* will be just sufficient to saturate the redundant matter in *EF*; for as *AB* and *GH* together contain the whole redundant fluid or matter in both spaces, the spaces *BD* and *EG* together contain their natural quantity of fluid; and therefore, as *BC* and *FG* each contain their natural quantity of fluid, the spaces *CD* and *EF* together contain their natural quantity of fluid. And fourthly, The spaces *BC* and *FG* will be saturated in all parts.

For, first, If the fluid is disposed in this manner, no particle of it can have any tendency to move: for a particle placed anywhere in the spaces *BC* and *FG*, is attracted with just as much force by *EF*, as it is repelled by *CD*; and it is repelled or attracted with just as much force by *AB*, as it is in a contrary direction by *GH*, and, consequently, has no tendency to move. A particle placed anywhere in the space *CD*, or in the spaces *AB* and *GH*, if they are overcharged, is indeed repelled with more force towards the planes *Dd*, *Aa* and *Hh*, than it is in the contrary direction; but as the fluid in those spaces is already as much compressed as possible, the particle will have no tendency to move.

Secondly, It seems impossible that the fluid should be at rest, if it is disposed in any other manner: but as this part of the demonstration is exactly similar to the latter part of that of Problem the first, I shall omit it.

31] Cor. I. If the two spaces *AD* and *EH* are both overcharged, the redundant fluid in *CD* is half the difference of the redundant fluid in those spaces: for half the difference of the redundant fluid in those spaces, added to the quantity in *AB*, which is half the sum, is equal to the whole quantity in *AD*. For a like reason, if *AD* and *EH* are both undercharged, the redundant matter in *EF* is half the difference of the redundant matter in those spaces; and if *AD* is overcharged, and *EH* undercharged, the redundant fluid in *CD* exceeds half the redundant fluid in *AD*, by a quantity sufficient to saturate half the redundant matter in *EH*.

32] Cor. II It was before said, that the fluid in the spaces *AB* and *GH* (when there is any fluid in them) is repelled against the planes *Aa* and *Hh*; and, consequently, would run out through those planes, if there was any opening for it to do so. The force with which the fluid presses against the planes *Aa* and *Hh*, is that with which the redundant fluid in *AB* is repelled by that in *GH*; that is, with which half the redundant fluid in both spaces is repelled by an equal quantity of fluid. Therefore, the pressure against *Aa* and *Hh* depends only on the quantity of redundant fluid in both spaces together, and not at all on the thickness or distance of those spaces, or on the proportion in which the fluid is divided between the two spaces. If there is no fluid in *AB* and *GH*, a particle placed on the outside of the spaces *AD* and *EH*, contiguous to the planes *Aa* or *Hh*, is attracted towards those planes by all the matter in *AB* and *GH*, *id est*, by all the redundant matter in both spaces; and, consequently, endeavours to insinuate itself into the space *AD* or *EH*; and the force with which it does so depends only on the quantity of redundant matter in both spaces together. The fluid in *CD* also presses against the plane *Dd*, and the force with which it does so is that with which the redundant fluid in *CD* is attracted by the matter in *EF*.

33] Cor. III. If *AD* is overcharged, and *EH* undercharged: and the redundant fluid in *AD* is exactly sufficient to saturate the redundant matter in *EH*, all the redundant fluid in *AD* will be collected in the space *CD*, where it will be pressed close together: the space *EF* will be intirely deprived of fluid, the quantity of matter therein being just sufficient to saturate the redundant fluid in *CD*, and the spaces *AC* and *FH* will be everywhere saturated. Moreover, if an opening is made in the planes *Aa* or *Hh*, the fluid within the spaces *AD* or *EH* will have no tendency to run out thereat, nor will the fluid on the outside have any tendency to run in at it: a particle of fluid too placed anywhere on the outside of both spaces, as at *P*, will not be at all attracted or repelled by those spaces, any more than if they were both saturated; but a particle placed anywhere between those spaces, as at *S*, will be repelled from *d* towards *e*; and if a communication was made between the two spaces, by the canal *de*, the fluid would run out of *AD* into *EH*, till they were both saturated.

34] PROP. VIII. PROB. 4. To determine in what manner the fluid will be disposed in the space *AD*, supposing that all the rest of the universe is saturated with immoveable fluid, and that the electric attraction and repulsion is inversely as some other power of the distance than the square.

I am not able to answer this Problem accurately, except when the repulsion is inversely as the simple or some lower power of the distance; but I think we may be certain of the following circumstances.

35] CASE 1. Let the repulsion be inversely as some power of the distance between the square and the cube, and let *AD* be overcharged.

First, It is certain that the density of the fluid must be everywhere the same, at the same distance from the planes *Aa* and *Dd*. Secondly, There can be no space as *BC*, of any sensible breadth, in which the matter will not be overcharged. And thirdly, The fluid close to the planes *Aa* and *Dd* will be pressed close together. Whence, I think, we may conclude, that the density of the fluid will increase gradually from the middle of the space to the outside, where it will be pressed close together. Whether the matter exactly in the middle will be overcharged, or only saturated, I cannot tell.

36] CASE 2. Let the repulsion be inversely as some power of the distance between the square and the simple power, and let *AD* be overcharged.

There will be two spaces *AB* and *DC*, in which the fluid will be pressed close together, and the quantity of redundant fluid in each of those spaces will be more than half the redundant fluid in *AD*; so that the space *BC*, taken all together, will be undercharged; but I cannot tell in what manner the fluid will be disposed in that space. The demonstrations of these two cases are exactly similar to those of the two cases of Prob. 2.

37] CASE 3. If the repulsion is inversely as the simple or some lower power of the distance, and *AD* is overcharged, all the fluid will be collected in the spaces *AB* and *CD*, and *BC* will be intirely deprived of fluid. If *AD* contains just fluid enough to saturate it, and the repulsion is inversely as the distance, the fluid will remain in equilibrio, in whatever manner it is disposed; provided its density is everywhere the same at the same distance from the planes *Aa* and *Dd*: but if the repulsion is inversely as some less power than the simple one, the fluid will be in equilibrio, whether it is either spread uniformly, or whether it is all collected in that plane which is in the middle between *Aa* and *Dd*, or whether it is all collected in the spaces *AB* and *CD*; but not, I believe, if it is disposed in any other manner.

The demonstration depends upon this circumstance; namely, that if the repulsion is inversely as the distance, two spaces *AB* and *CD*, repel a particle placed either between them, or on the outside of them, with the same force as if all the matter of those spaces was collected in the middle plane between them.

It is needless mentioning the three cases in which *AD* is undercharged, as the reader will easily supply the place.

38] Though the four foregoing problems do not immediately tend to explain the phænomena of electricity, I chose to insert them; partly because they seem worth engaging our attention in themselves; and partly because they serve, in some measure, to confirm the truth of some of the following propositions, in which I am obliged to make use of a less accurate kind of reasoning.

39] In the following propositions, I shall always suppose the bodies I speak of to consist of solid matter, confined to the same spot, so as not to be able to alter its shape or situation by the attraction or repulsion of other bodies on it: I shall also suppose the electric fluid in these bodies to be moveable, but unable to escape, unless when otherwise expressed. As for the matter in all the rest of the universe, I shall suppose it to be saturated with immoveable fluid. I shall also suppose the electric attraction and repulsion to be inversely as any power of the distance less than the cube, except when otherwise expressed.

40] By a canal, I mean a slender thread of matter, of such kind that the electric fluid shall be able to move readily along it, but shall not be able to escape from it, except at the ends, where it communicates with other bodies. Thus, when I say that two bodies communicate with each other by a canal, I mean that the fluid shall be able to pass readily from one body to the other by that canal*.

41] PROP. IX. If any body at a distance from any over or undercharged body be overcharged, the fluid within it will be lodged in greater quantity near the surface of the body than near the center. For, if you suppose it to be spread uniformly all over the body, a particle of fluid in it, near the surface, will be repelled towards the surface by a greater quantity of fluid than that by which it is repelled from it; consequently, the fluid will flow towards the surface, and make it denser there: moreover, the particles of fluid close to the surface will be pressed close together; for otherwise, a particle placed so near it, that the quantity of redundant fluid between it and the surface should be very small, would move towards it; as the small quantity of redundant fluid between it and the surface would be unable to balance the repulsion of that on the other side.

From the four foregoing problems it seems likely, that if the electric attraction or repulsion is inversely as the square of the distance, almost all the redundant fluid in the body will be lodged close to the surface, and there pressed close together, and the rest of the body will be saturated. If the repulsion is inversely as some power of the distance between the square and the cube, it is likely that all parts of the body will be overcharged: and if it is inversely as some less power than the square, it is likely that all parts of the body, except those near the surface, will be undercharged.

42] COR. For the same reason, if the body is undercharged, the deficiency of fluid will be greater near the surface than near

* [Note 3.]

the center, and the matter near the surface will be entirely deprived of fluid. It is likely too, if the repulsion is inversely as some higher power of the distance than the square, that all parts of the body will be undercharged: if it is inversely as the square, that all parts, except near the surface, will be saturated: and if it is inversely as some less power than the square, that all parts, except near the surface, will be overcharged.

43] PROP. X. Let the bodies *A* and *D* (Fig. 5) communicate

Fig. 5.

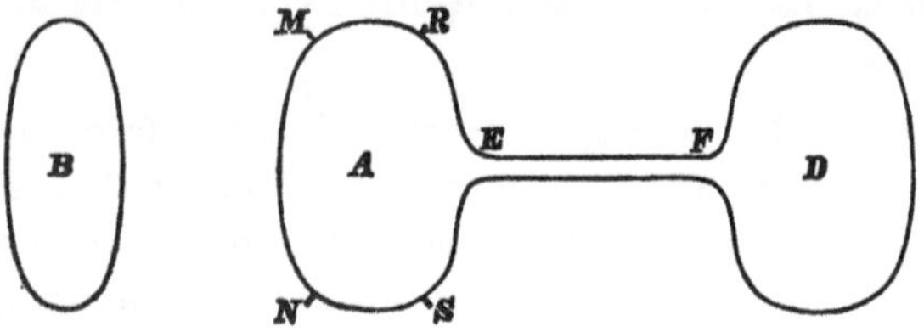

with each other by the canal *EF*; and let one of them, as *D*, be overcharged; the other body *A* will be so also.

For as the fluid in the canal is repelled by the redundant fluid in *D*, it is plain, that unless *A* was overcharged, so as to balance that repulsion, the fluid would run out of *D* into *A*.

In like manner, if one is undercharged, the other must be so too.

44] PROP. XI. Let the body *A* (Fig. 6) be either saturated or over or undercharged; and let the fluid within it be in equilibrio. Let now the body *B*, placed near it, be rendered overcharged, the fluid within it being supposed immoveable, and disposed in such manner, that no part of it shall be undercharged; the fluid in *A* will no longer be in equilibrio, but will be repelled from *B*: therefore, the fluid will flow from those parts of *A* which are nearest to *B*, to those which are more distant from it; and, consequently, the part adjacent to *MN* (that part of the surface of *A* which is turned towards *B*) will be made to contain less electric fluid than it did before, and that adjacent to the opposite surface *RS* will contain more than before.

Fig. 6.

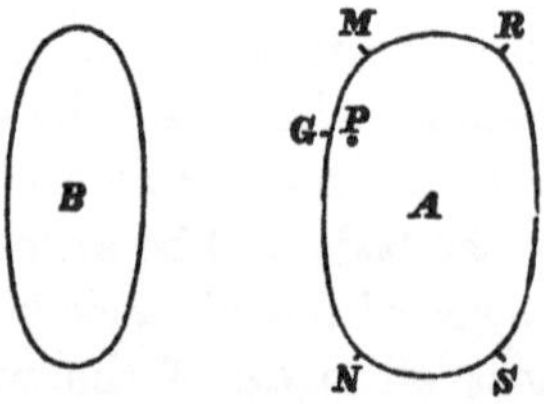

It must be observed, that when a sufficient quantity of fluid has flowed from *MN* towards *RS*, the repulsion which the fluid in the part adjacent to *MN* exerts on the rest of the fluid in *A*, will be so much weakened, and the repulsion of that in the part near *RS* will be so much increased, as to compensate the repulsion of *B*, which will prevent any more fluid flowing from *MN* to *RS*.

The reason why I suppose the fluid in *B* to be immoveable is, that otherwise a question might arise, whether the attraction or repulsion of the body *A* might not cause such an alteration in the disposition of the fluid in *B*, as to cause some parts of it to be undercharged; which might make it doubtful, whether *B* did on the whole repel the fluid in *A*. It is evident, however, that this proposition would hold good, though some parts of *B* were undercharged, provided it did on the whole repel the fluid in *A*.

45] Cor. If *B* had been made undercharged, instead of overcharged, it is plain that some fluid would have flowed from the further part *RS* to the nearer part *MN*, instead of from *MN* to *RS*.

46] Prop. XII. Let us now suppose that the body *A* communicates by the canal *EF*, with another body *D*, placed on the contrary side of it from *B*, as in Fig. 5; and let these two bodies be either saturated, or over or undercharged; and let the fluid within them be in equilibrio. Let now the body *B* be overcharged: it is plain that some fluid will be driven from the nearer part *MN* to the further part *RS*, as in the former proposition; and also some fluid will be driven from *RS*, through the canal, to the body *D*; so that the quantity of fluid in *D* will be increased thereby, and the quantity in *A*, taking the whole body together, will be diminished; the quantity in the part near *MN* will also be diminished; but whether the quantity in the part near *RS* will be diminished or not, does not appear for certain; but I should imagine it would be not much altered.

47] Cor. In like manner, if *B* is made undercharged, some fluid will flow from *D* to *A*, and also from that part of *A* near *RS*, to the part near *MN*.

48] Prop. XIII. Suppose now that the bodies *A* and *D* communicate by the bent canal *MPNnpm* (Fig. 7) instead of the straight one *EF*: let the bodies be either saturated or over or undercharged as before; and let the fluid be at rest; then if the body B

Fig. 7.

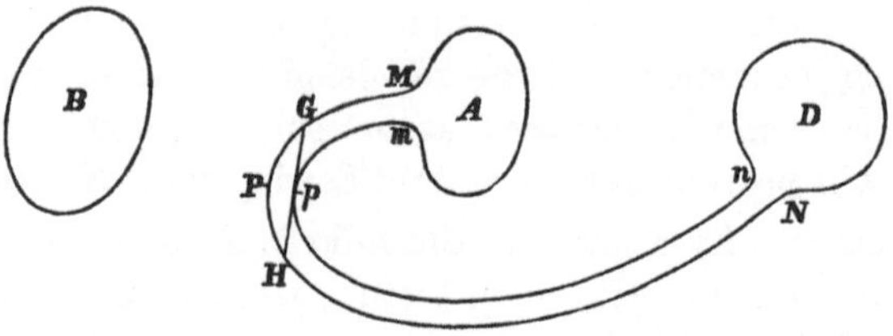

is made overcharged, some fluid will still run out of A into D; provided the repulsion of B on the fluid in the canal is not too great.

The repulsion of B on the fluid in the canal will at first drive some fluid out of the leg $MPpm$ into A, and out of $NPpn$ into D, till the quantity of fluid in that part of the canal which is nearest to B is so much diminished, and its repulsion on the rest of the fluid in the canal is so much diminished also as to compensate the repulsion of B: but as the leg $NPpn$ is longer than the other, the repulsion of B on the fluid in it will be greater; consequently some fluid will run out of A into D, on the same principle that water is drawn out of a vessel through a siphon.

49] But if the repulsion of B on the fluid in the canal is so great, as to drive all the fluid out of the space $GPHpG$, so that the fluid in the leg $MGpm$ does not join to that in $NHpn$; then it is plain that no fluid can run out of A into D; any more than water will run out of a vessel through a siphon, if the height of the bend of the siphon above the water in the vessel, is greater than that to which water will rise in vacuo.

50] Cor. If B is made undercharged, some fluid will run out of D into A; and that though the attraction of B on the fluid in the canal is ever so great.

51] Prop. XIV. Let ABC (Fig. 8) be a body overcharged with immoveable fluid, uniformly spread; let the bodies near ABC on the outside be saturated with immoveable fluid; and let D be a body inclosed within ABC, and communicating by the canal DG with other distant bodies saturated with fluid; and let the fluid in D and the canal and

Fig. 8.

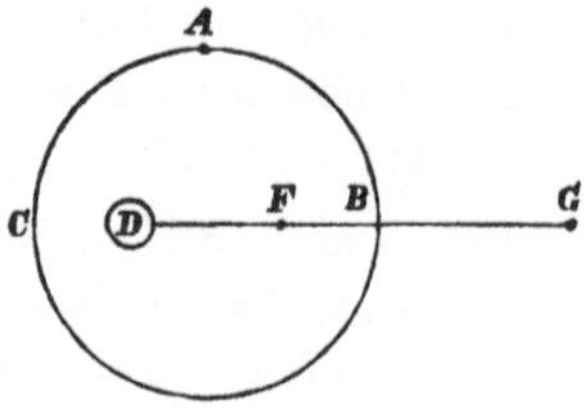

those bodies be moveable; then will the body *D* be rendered undercharged.

For let us first suppose that *D* and the canal are saturated, and that *D* is nearer to *B* than to the opposite part of the body, *C*; then will all the fluid in the canal be repelled from *C* by the redundant fluid in *ABC*; but if *D* is nearer to *C* than to *B*, take the point *F*, such that a particle placed there would be repelled from *C* with as much force as one at *D* is repelled towards *C*; the fluid in *DF*, taking the whole together, will be repelled with as much force one way as the other; and the fluid in *FG* is all of it repelled from *C*: therefore in both cases the fluid in the canal, taking the whole together, is repelled from *C*; consequently some fluid will run out of *D* and the canal, till the attraction of the unsaturated matter therein is sufficient to balance the repulsion of the redundant fluid in *ABC*.

52] PROP. XV. If we now suppose that the fluid on the outside of *ABC* is moveable; the matter adjacent to *ABC* on the outside will become undercharged. I see no reason however to think that that will prevent the body *D* from being undercharged; but I cannot say exactly what effect it will have, except when *ABC* is spherical and the repulsion is inversely as the square of the distance; in this case it appears by Prob. I. that the fluid in the part *DB* of the canal will be repelled from *C*, with just as much force as in the last proposition; but the fluid in the part *BG* will not be repelled at all: consequently *D* will be undercharged, but not so much as in the last proposition.

53] COR. If *ABC* is now supposed to be undercharged, it is certain that *D* will be overcharged, provided the matter near *ABC* on the outside is saturated with immoveable fluid; and there is great reason to think that it will be so, though the fluid in that matter is moveable.

54] PROP. XVI. Let *AEFB* (Fig. 9) be a long cylindric body, and *D* an undercharged body; and let the quantity of fluid in *AEFB* be such, that the part near *EF* shall be saturated. It appears from what has been said before, that the part near *AB* will be overcharged; and more-

Fig. 9.

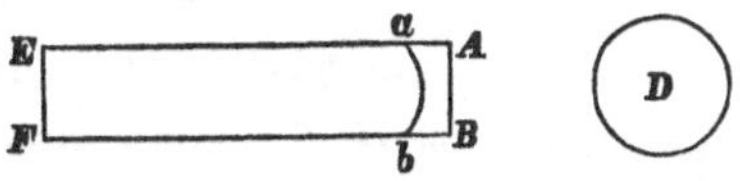

over there will be a certain space, as $AabB$, adjoining to the plane AB, in which the fluid will be pressed close together; and the fluid in that space will press against the plane AB, and will endeavour to escape from it; and by Prop. II. the two bodies will attract each other: now I say that the force with which the fluid presses against the plane AB, is very nearly the same with which the two bodies attract each other in the direction EA; provided that no part of $AEFB$ is undercharged.

Suppose so much of the fluid in each part of the cylinder as is sufficient to saturate the matter in that part, to become solid; the remainder, or the redundant fluid remaining fluid as before. In this case the pressure against the plane AB must be exactly equal to that with which the two bodies attract each other in the direction EA: for the force with which D attracts that part of the fluid which we supposed to become solid, is exactly equal to that with which it repels the matter in the cylinder; and the redundant fluid in $EabF$ is at liberty to move, if it had any tendency to do so, without moving the cylinder; so that the only thing which has any tendency to impel the cylinder in the direction EA is the pressure of the redundant fluid in $AabB$ against AB; and as the part near EF is saturated, there is no redundant fluid to press against the plane EF, and thereby to counteract the pressure against AB. Suppose now all the electric fluid in the cylinder to become fluid; the force with which the two bodies attract each other will remain exactly the same; and the only alteration in the pressure against AB, will be, that that part of the fluid in $AabB$, which we at first supposed solid and unable to press against the plane, will now be at liberty to press against it; but as the density of the fluid when its particles are pressed close together may be supposed many times greater than when it is no denser than sufficient to saturate the matter in the cylinder, and consequently the quantity of redundant fluid in $AabB$ many times greater than that which is required to saturate the matter therein, it follows that the pressure against AB will be very little more than on the first supposition.

N.B. If any part of the cylinder is undercharged, the pressure against AB is greater than the force with which the bodies attract. If the electric repulsion is inversely as the square or some higher power of the distance, it seems very unlikely that any part of the cylinder should be undercharged; but if the repulsion is inversely as some lower power than the square, it is not improbable but some part of the cylinder may be undercharged.

55] LEMMA VII. Let AB (Fig. 10) represent an infinitely thin flat circular plate, seen edgeways, so as to appear to the eye as a straight line; let C be the center of the circle; and let DC passing through C, be perpendicular to the plane of the plate; and let the plate be of uniform thickness, and consist of uniform matter, whose particles repel with a force inversely as the n power of the distance; n being greater than one, and less than three: the repulsion of the plate on a particle at D is proportional to $\frac{DC}{DC^{n-1}} - \frac{DC}{DA^{n-1}}$; provided the thickness of the plate and size of the particle D is given.

Fig. 10.

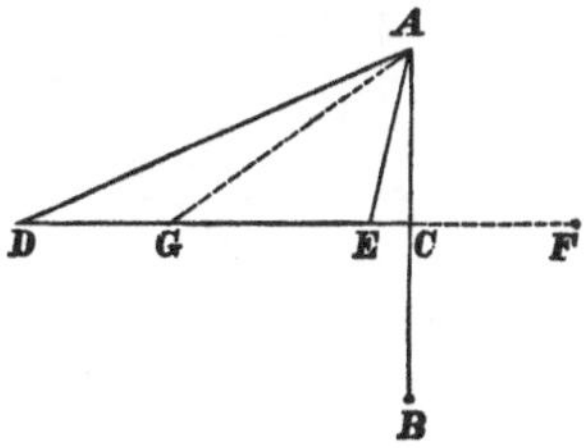

For if CA is supposed to flow, the corresponding fluxion of the quantity of matter in the plate is proportional to $CA \times C\dot{A}$; and the corresponding fluxion of the repulsion of the plate on the particle D, in the direction DC, is proportional to

$$\frac{CA \times C\dot{A}}{DA^n} \times \frac{DC}{DA}, \quad = \frac{D\dot{A} \times DC}{DA^n};$$

for $D\dot{A} : C\dot{A} :: CA : DA$; the variable part of the fluent of which is $\frac{-DC}{(n-1)\,DA^{n-1}}$: whence the repulsion of the plate on the particle D is proportional to $\frac{DC}{(n-1)\,DC^{n-1}} - \frac{DC}{(n-1)\,DA^{n-1}}$, or to $\frac{DC}{DC^{n-1}} - \frac{DC}{DA^{n-1}}$.

56] COR. If DC^{n-1} is very small in respect of CA^{n-1}, the particle D is repelled with very nearly the same force as if the diameter of the plate was infinite.

57] LEMMA VIII. Let L and l represent the two legs of a right-angled triangle, and h the hypothenuse; if the shorter leg l is so much less than the other, that l^{n-1} is very small in respect of L^{n-1}, $h^{3-n} - L^{3-n}$ will be very small in respect of l^{3-n}.

For $$h^{3-n} = (L^2 + l^2)^{\frac{3-n}{2}} = L^{3-n}\left(1 + \frac{l^2}{L^2}\right)^{\frac{3-n}{2}}$$

$$= L^{3-n}\left[1 + \frac{(3-n)\,l^2}{2L^2} - \frac{(3-n)\,(n-1)\,l^4}{8L^4}, \text{ \&c.}\right]$$

therefore

$$h^{3-n} - L^{3-n} = \frac{(3-n)\,l^2}{2L^{n-1}} - \frac{3-n \times n-1 \times l^4}{8L^{n+1}}, \text{ \&c.}$$

$$= \frac{l^{3-n} \times 3-n \times l^{n-1}}{2L^{n-1}} - \frac{l^{3-n} \times 3-n \times n-1 \times l^{n+1}}{8L^{n+1}}, \text{ \&c.}$$

which is very small in respect of l^{3-n}; as l^{n-1} is by the supposition very small in respect of L^{n-1}.

58] Lemma IX. Let DG now represent the axis of a cylindric or prismatic column of uniform matter; and let the diameter of the column be so small, that the repulsion of the plate AB on it shall not be sensibly different from what it would be, if all the matter in it was collected in the axis: the force with which the plate repels the column is proportional to

$$DC^{3-n} + AC^{3-n} - DA^{3-n};$$

supposing the thickness of the plate and base of the column to be given.

For, if DC is supposed to flow, the corresponding fluxion of the repulsion is proportional to

$$\frac{D\dot{C}}{DC^{n-2}} - \frac{DC \times D\dot{C}}{DA^{n-1}} = \frac{D\dot{C}}{DC^{n-2}} - \frac{D\dot{A}}{DA^{n-2}};$$

the fluent of which, $\dfrac{AC^{3-n} + DC^{3-n} - DA^{3-n}}{3-n}$, vanishes when DC vanishes.

59] Cor. I. If the length of the column is so great that AC^{n-1} is very small in respect of DC^{n-1}, the repulsion of the plate on it is very nearly the same as if the column was infinitely continued.

For by Lemma VIII. $AC^{3-n} + DC^{3-n} - DA^{3-n}$ differs very little in this case from AC^{3-n}; and if DC is infinite, it is exactly equal to it.

60] Cor. II. If AC^{n-1} is very small in respect of DC^{n-1}, and the point E be taken in DC such that EC^{n-1} shall be very small in respect of AC^{n-1}, the repulsion of the plate on the small part of the column EC, is to its repulsion on the whole column DC, very nearly as EC^{3-n} to AC^{3-n}.

61] Lemma X. If we now suppose all the matter of the plate to be collected in the circumference of the circle, so as to form an infinitely slender uniform ring, its repulsion on the column DC will be less than when the matter is spread uniformly all over the plate, in the ratio of

$$\frac{(3-n)\,AC^2}{2}\times\left(\frac{1}{AC^{n-1}}-\frac{1}{DA^{n-1}}\right) \text{ to } DC^{3-n}+AC^{3-n}-DA^{3-n}.$$

For it was before said, that if the matter of the plate be spread uniformly, its repulsion on the column will be proportional to $DC^{3-n}+AC^{3-n}-DA^{3-n}$, or may be expressed thereby; let now AC, the semidiameter of the plate, be increased by the infinitely small quantity $A\dot{C}$; the quantity of matter in the plate will be increased by a quantity, which is to the whole, as $2A\dot{C}$ to AC; and the repulsion of the plate on the column will be increased by

$$(3-n)\,A\dot{C}\times AC^{2-n}-A\dot{C}\times\frac{AC}{DA}\times(3-n)\times DA^{2-n},$$

$$=(3-n)\times A\dot{C}\times AC\times\left(\frac{1}{AC^{n-1}}-\frac{1}{DA^{n-1}}\right):$$

therefore if a quantity of matter, which is to the whole quantity in the plate as $2A\dot{C}$ to AC be collected in the circumference, its repulsion on the column DC will be to that of the whole plate as

$$3-n\times A\dot{C}\times AC\times\left(\frac{1}{AC^{n-1}}-\frac{1}{DA^{n-1}}\right),\text{ to } DC^{3-n}+AC^{3-n}-DA^{3-n};$$

and consequently the repulsion of the plate when all the matter is collected in its circumference, is to its repulsion when the matter is spread uniformly, as

$$\frac{3-n\times AC^2}{2}\times\left(\frac{1}{AC^{n-1}}-\frac{1}{DA^{n-1}}\right),\text{ to } DC^{3-n}+AC^{3-n}-DA^{3-n}.$$

62] Cor. I. If the length of the column is so great, that AC^{n-1} is very small in respect of DC^{n-1}, the repulsion of the plate, when all the matter is collected in the circumference, is to its repulsion when the matter is spread uniformly, very nearly as $\frac{3-n \times AC^{3-n}}{2}$ to AC^{3-n}, or as $3-n$ to 2.

63] Cor. II. If EC^{n-1} is very small in respect of AC^{n-1}, the repulsion of the plate on the short column EC, when all the matter in the plate is collected in its circumference, is to its repulsion when the matter is spread uniformly, very nearly as

$$\frac{3-n \times n-1 \times EC^2}{4AC^{n-1}} \text{ to } EC^{3-n},$$

or as $3-n \times n-1 \times EC^{n-1}$ to $4AC^{n-1}$; and is therefore very small in comparison of what it is when the matter is spread uniformly.

For by the same kind of process as was used in Lemma VIII., it appears, that if EC^2 is very small in respect of AC^2,

$$AC^2 \times \left(\frac{1}{AC^{n-1}} - \frac{1}{EA^{n-1}}\right)$$

differs very little from $\frac{n-1 \times EC^2}{2EA^{n-1}}$, or from $\frac{n-1 \times EC^2}{2AC^{n-1}}$; and if EC^{n-1} is very small in respect of AC^{n-1}, EC^2 is *à fortiori* very small in respect of AC^2.

64] Cor. III. Suppose now that the matter of the plate is denser near the circumference than near the middle, and that the density at and near the middle is to the mean density, or the density which it would everywhere be of if the matter was spread uniformly, as δ to 1; the repulsion of the plate on EC will be less than if the matter was spread uniformly, in a ratio approaching much nearer to that of δ to 1, than to that of equality.

65] Cor. IV. Let everything be as in the last corollary, and let π be taken to one, as the force with which the plate actually repels the column DC, (DC^{n-1} being very great in respect of AC^{n-1}), is to the force with which it would repel it, if the matter was spread uniformly; the repulsion of the plate on EC will be to its repulsion on DC, in a ratio between that of $EC^{3-n} \times \delta$ to $AC^{3-n} \times \pi$, and that of EC^{3-n} to $AC^{3-n} \times \pi$, but will approach much nearer to the former ratio than to the latter.

66] Lemma XI. In the line *DC* produced, take *CF* equal to *CA*: if all the matter of the plate *AB* is collected in the circumference, its repulsion on the column *CD*, infinitely continued, is equal to the repulsion of the same quantity of matter collected in the point *F*, on the same column.

For the repulsion of the plate on the column in the direction *CD*, is the same, whether the matter of it be collected in the whole circumference, or in the point *A*. Suppose it therefore to be collected in *A*; and let an equal quantity of matter be collected in *F*; take *FG* constantly equal to *AD*; and let *AD* and *FG* flow: the fluxion of *CD* is to the fluxion of *FG*, as *AD* to *CD*; and the repulsion of *A* on the point *D*, in the direction *CD*, is to the repulsion of *F* on *G*, as *CD* to *AD*; and therefore the fluxion of the repulsion of *A* on the column *CD*, in the direction *CD*, is equal to the fluxion of the repulsion of *F* on *CG*; and when *AD* equals *AC*, the repulsion of both *A* and *F* on their respective columns vanishes; and therefore the repulsion of *A* on the whole column *CD* equals that of *F* on *CG*; and when *CD* and *CG* are both infinitely extended, they may be looked upon as the same column.

67] Prop. XVII. Let two similar bodies, of different sizes, and consisting of different sorts of matter, be both overcharged, or both undercharged, but in different degrees; and let the redundance or deficience of fluid in each be very small in respect of the whole quantity of fluid in them: it is impossible for the fluid to be disposed accurately in a similar manner in both of them*; as it has been shewn that there will be a space, close to the surface, which will either be as full of fluid as it can hold, or will be entirely deprived of fluid; but it will be disposed as nearly in a similar manner in both, as is possible. To explain this, let *BDE* and *bde* (Fig. 11) be the two similar bodies; and

* By the fluid being disposed in a similar manner in both bodies, I mean that the quantity of redundant or deficient fluid in any small part of one body, is to that in the corresponding small part of the other, as the whole quantity of redundant or deficient fluid in one body, to that in the other. By the quantity of deficient fluid in a body, I mean the quantity of fluid wanting to saturate it. Notwithstanding the impropriety of this expression, I must beg leave to make use of it, as it will frequently save a great deal of circumlocution. [See Note 1.]

Fig. 11.

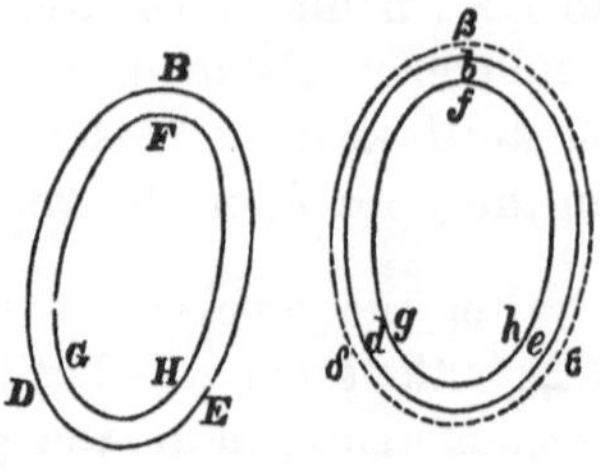

let the space comprehended between the surfaces *BDE* and *FGH* (or the space *BF* as I shall call it for shortness) be that part of *BDE*, which is either as full of fluid as it can hold, or entirely deprived of it: draw the surface *fgh*, such that the space *bf* shall be to the space *BF*, as the quantity of redundant or deficient fluid in *bde*, to that in *BDE*, and that the thickness of the space *bf* shall everywhere bear the same proportion to the corresponding thickness of *BF*: then will the space *bf* be either as full of fluid as it can hold, or entirely deprived of it; and the fluid within the space *fgh* will be disposed very nearly similarly to that in the space *FGH*.

For it is plain, that if the fluid could be disposed accurately in a similar manner in both bodies, the fluid would be in equilibrio in one body, if it was in the other: therefore draw the surface $\beta\delta\epsilon$, such that the thickness of the space βf shall be everywhere to the corresponding thickness of *BF*, as the diameter of *bde* to the diameter of *BDE*; and let the redundant fluid or matter in *bf* be spread uniformly over the space βf; then if the fluid in the space *fgh* is disposed exactly similarly to that in *FGH*, it will be in equilibrio; as the fluid will then be disposed exactly similarly in the spaces $\beta\delta\epsilon$ and *BDE*: but as by the supposition, the thickness of the space βf is very small in respect of the diameter of *bde*, the fluid or matter in the space *bf* will exert very nearly the same force on the rest of the fluid, whether it is spread over the space βf, or whether it is collected in *bf*.

68] PROP. XVIII. Let two bodies, *B* and *b*, be connected to each other by a canal of any kind, and be either over or undercharged: it is plain that the quantity of redundant or deficient fluid in *B*, would bear exactly the same proportion to that in *b*, whatever sort of matter *B* consisted of, if it was possible for the redundant or deficient fluid in any body to be disposed accurately in the same manner, whatever sort of matter it consisted of. For suppose *B* to consist of any sort of matter; and let the fluid in the canal and two bodies be in equilibrio: let now *B* be made to consist of some other sort of matter, which requires a different quantity of fluid to saturate it; but let the quantity and disposition of the redundant or deficient fluid in it remain the same as before: it is plain that the fluid will still be in equilibrio; as the attraction or repulsion of any body depends only on the quantity and disposition of the redundant and deficient fluid in it. Therefore, by the preceding proposition, the quantity of redundant or deficient fluid in *B*, will actually bear very nearly the same proportion to that in *b*, whatever sort of matter *B* consists of; provided the quantity of redundant or deficient fluid in it is very small in respect of the whole. [See Exp. IV., Art. 269.]

69] PROP. XIX. Let two bodies *B* and *b* (Fig. 12) be connected together by a very slender canal *ADda*, either straight or crooked: let the canal be everywhere of the same breadth and thickness; so that all sections of this canal made by planes perpendicular to the direction of the canal in that part, shall be equal and similar: let the canal be composed of uniform matter; and let the electric fluid therein be supposed incompressible, and of such density as exactly to saturate the matter therein; and let it, nevertheless, be able to move readily along the canal; and let each particle of fluid in the canal be attracted and repelled by the matter and fluid in the canal and in the bodies *B* and *b*, just in the same manner that it would be if it was not incompressible*; and let the bodies *B* and *b* be either over or undercharged. I say that the force with which the whole quantity of fluid in the canal is impelled from *A* towards *D*, in the direction of the axis of the canal, by the united attractions and repulsions of the two bodies, must be nothing; as otherwise the fluid in the canal could not be at rest: observing that by the force with which the whole quantity of fluid is impelled in the direction of the axis of the canal, I mean the sum of the forces, with which the fluid in each part of the canal is impelled in the direction of the axis of the canal in that place,

Fig. 12.

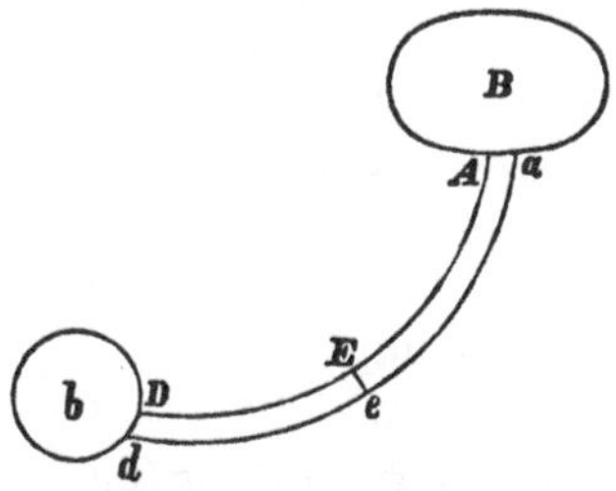

* This supposition of the fluid in the canal being incompressible, is not mentioned as a thing which can ever take place in nature, but is merely imaginary; the reason for making of which will be given hereafter.

from A towards D; and observing also, that an impulse in the contrary direction from D towards A must be looked upon as negative.

For as the canal is exactly saturated with fluid, the fluid therein is attracted or repelled only by the redundant matter or fluid in the two bodies. Suppose now that the fluid in any section of the canal, as *Ee*, is impelled with any given force in the direction of the canal at that place, the section *Dd* would, in consequence thereof, be impelled with exactly the same force in the direction of the canal at D, if the fluid between *Ee* and *Dd* was not at all attracted or repelled by the two bodies; and, consequently, the section *Dd* is impelled in the direction of the canal, with the sum of the forces, with which the fluid in each part of the canal is impelled by the attraction or repulsion of the two bodies in the direction of the axis in that part; and consequently, unless this sum was nothing, the fluid in *Dd* could not be at rest.

70] Cor. Therefore, the force with which the fluid in the canal is impelled one way in the direction of the axis, by the body B, must be equal to that with which it is impelled by b in the contrary direction.

71] Prop. XX. Let two similar bodies B and b (Fig. 13) be connected by the very slender cylindric or prismatic canal Aa, filled with incompressible fluid, in the same manner as described in the preceding proposition: let the bodies be overcharged; but let the quantity of redundant fluid in each bear so small a proportion to the whole, that the fluid may be considered as disposed in a similar manner in both; let the bodies also be similarly situated in respect of the canal Aa; and let them be placed at an infinite distance from each other, or at so great an one, that the repulsion of either body on the fluid in the canal shall not be sensibly less than if they were at an infinite distance: then, if the electric attraction and repulsion is inversely as the n power of the distance, n being greater than 1, and less than 3, the quantity of redundant fluid in the two bodies will be to each other as the $n-1$ power of their corresponding diameters AF and af.

Fig. 13.

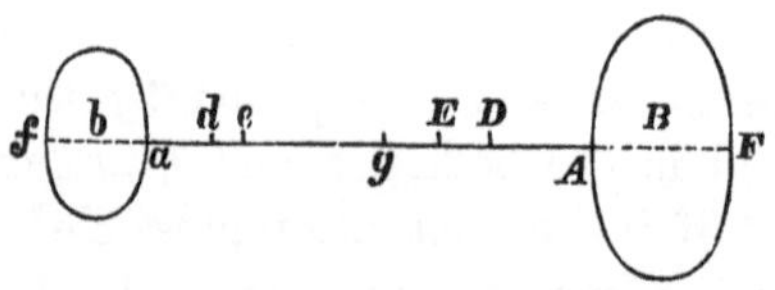

For if the quantity of redundant fluid in the two bodies is in this proportion, the repulsion of one body on the fluid in the canal will be equal to that of the other body on it in the contrary direction; and, consequently, the fluid will have no tendency to flow from one body to the other, as may thus be proved. Take the points D and E very near to each other; and take da to DA, and ea to EA, as af to AF; the repulsion of the body B on a particle at D, will be to the repulsion of b on a particle at d, as $\frac{1}{AF}$ to $\frac{1}{af}$; for, as the fluid is disposed similarly in both bodies, the quantity of fluid in any small part of B, is to the quantity in the corresponding part of b, as AF^{n-1} to af^{n-1}; and consequently the repulsion of that small part of B, on D, is to the repulsion of the corresponding part of b, on d, as $\frac{AF^{n-1}}{AF^n}$, or $\frac{1}{AF}$, to $\frac{1}{af}$. But the quantity of fluid in the small part DE of the canal, is to that in de, as DE to de, or as AF to af; therefore the repulsion of B on the fluid in DE, is equal to that of b on the fluid in de: therefore, taking ag to Aa, as af to AF, the repulsion of b on the fluid in ag, is equal to that of B on the fluid in Aa; but the repulsion of b on ag may be considered as the same as its repulsion on Aa; for, by the supposition, the repulsion of B on Aa may be considered as the same as if it was continued infinitely; and therefore, the repulsion of b on ag may be considered as the same as if it was continued infinitely.

N.B. If n was not greater than 1, it would be impossible for the length of Aa to be so great, that the repulsion of B on it might be considered as the same as if it was continued infinitely; which was my reason for requiring n to be greater than 1.

72] COR. By just the same method of reasoning it appears, that if the bodies are undercharged, the quantity of deficient fluid in b will be to that in B, as af^{n-1} to AF^{n-1}.

73] PROP. XXI. Let a thin flat plate be connected to any other body, as in the preceding proposition, by a canal of incompressible fluid, perpendicular to the plane of the plate; and let that body be overcharged, the quantity of redundant fluid in the plate will bear very nearly the same proportion to that in the other body, whatever the thickness of the plate may be, provided its thickness is very small in proportion to its breadth, or smallest diameter.

For there can be no doubt, but what, under that restriction, the fluid will be disposed very nearly in the same manner in the plate, whatever its thickness may be; and therefore its repulsion on the fluid in the canal will be very nearly the same, whatever its thickness may be. [See Exp. IV., Art. 272.]

74] Prop. XXII. Let *AB* and *DF* (Fig 14) represent two equal and parallel circular plates, whose centres are *C* and *E*; let

Fig. 14.

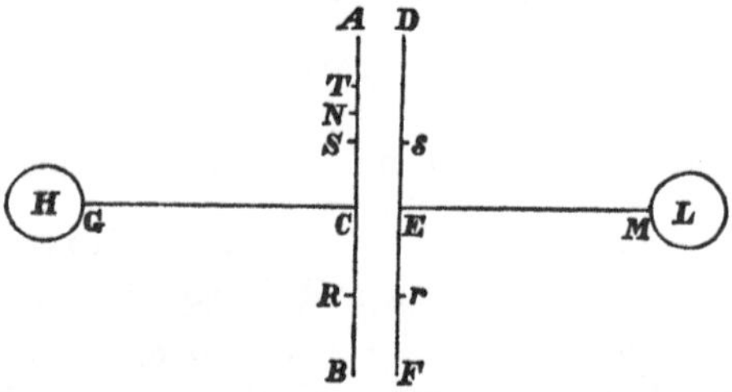

the plates be placed so, that a right line joining their centers shall be perpendicular to the plates; let the thickness of the plates be very small in respect of their distance *CE*; let the plate *AB* communicate with the body *H*, and the plate *DF* with the body *L*, by the canals *CG* and *EM* of incompressible fluid, such as are described in Prop. XIX; let these canals meet their respective plates in their centers *C* and *E*, and be perpendicular to the plane of the plates; and let their length be so great, that the repulsion of the plates on the fluid in them may be considered as the same as if they were continued infinitely; let the body *H* be overcharged, and let *L* be saturated. It is plain, from Prop. XII., that *DF* will be undercharged, and *AB* will be more overcharged than it would otherwise be. Suppose, now, that the redundant fluid in *AB* is disposed in the same manner as the deficient fluid is in *DF*; let *P* be to one as the force with which the plate *AB* would repel the fluid in *CE*, if the canal *ME* was continued to *C*, is to the force with which it would repel the fluid in *CM*; and let the force with which *AB* repels the fluid in *CG*, be to the force with which it would repel it, if the redundant fluid in it was spread uniformly, as π to 1; and let the force with which the body *H* repels the fluid in *CG*, be the same with which a quantity of redundant fluid, which we will call *B*, spread uniformly over *AB*, would repel it in the contrary direction. Then will the redundant fluid in *AB* be equal to $\frac{B}{2P\pi - P^2\pi}$, and therefore, if *P* is very small, will be very nearly equal to $\frac{B}{2P\pi}$; and the deficient fluid in *DF* will be to the redundant fluid in *AB*, as $1 - P$ to 1, and therefore, if *P* is very small, will be very nearly equal to the redundant fluid in *AB*.

For it is plain, that the force with which *AB* repels the fluid in *EM*, must be equal to that with which *DF* attracts it; for otherwise, some fluid would run out of *DF* into *L*, or out of *L* into *DF*: for the same reason, the excess of the repulsion of *AB* on the fluid in *CG*, above the attraction of *FD* thereon, must be equal to the force with which a quantity of redundant fluid equal to *B*, spread uniformly over *AB*, would repel it, or it must be equal to that with which a quantity equal to $\frac{B}{\pi}$, spread in the manner in which the redundant fluid is actually spread in *AB*, would repel it. By the supposition, the force with which *AB* repels the fluid in *EM*, is to the force with which it would repel the fluid in *CM*, supposing *EM* to be continued to *C*, as $1 - P$ to 1; but the force with which any quantity of fluid in *AB* would repel the fluid in *CM*, is the same with which an equal quantity similarly disposed in *DF*, would repel the fluid in *EM*; therefore the force with which the redundant fluid in *AB* repels the fluid in *EM*, is to that with which an equal quantity similarly disposed in *DF*, would repel it, as $1 - P$ to 1: therefore, if the redundant fluid in *AB* be called *A*, the deficient fluid in *DF* must be $A \times \overline{1 - P}$: for the same reason, the force with which *DF* attracts the fluid in *CG*, is to that with which *AB* repels it, as $A \times \overline{1 - P} \times \overline{1 - P}$, or $A \times (1 - P)^2$, to *A*; therefore, the excess of the force with which *AB* repels *CG* above that with which *DF* attracts it, is equal to that with which a quantity of redundant fluid equal to $A - A \times (1 - P)^2$, or $A \times (2P - P^2)$, spread over *AB*, in the manner in which the redundant fluid therein is actually spread, would repel it: therefore $A \times (2P - P^2)$ must be equal to $\frac{B}{\pi}$, or A must be equal to $\frac{B}{2P\pi - P^2\pi}$.

75] Cor. I. If the density of the redundant fluid near the middle of the plate AB, is less than the mean density, or the density which it would everywhere be of, if it was spread uniformly, in the ratio of δ to 1; and if the distance of the two plates is so small, that EC^{n-1} is very small in respect of AC^{n-1}, and that EC^{3-n} is very small in respect of AC^{3-n}, the quantity of redundant fluid in AB will be greater than $\frac{B}{2} \times \overline{\frac{AC}{EC}}\Big|^{3-n}$, and less than $\frac{B}{2\delta} \times \overline{\frac{AC}{EC}}\Big|^{3-n}$, but will approach much nearer to the latter value than the former. For, in this case, $P\pi$ is, by Lemma X. Corol. IV., less than $\overline{\frac{EC}{AC}}\Big|^{3-n}$, and greater than $\overline{\frac{EC}{AC}}\Big|^{3-n} \times \delta$, but approaches much nearer to the latter value than the former; and if EC^{3-n} is very small in respect of AC^{3-n}, P is very small.

76] Remarks. If DF was not undercharged, it is certain that AB would be considerably more overcharged near the circumference of the circle than near the center; for if the fluid was spread uniformly, a particle placed anywhere at a distance from the center, as at N, would be repelled with considerably more force towards the circumference than it would towards the center. If the plates are very near together, and, consequently, DF nearly as much undercharged as AB is overcharged, AB will still be more overcharged near the circumference than near the center, but the difference will not be near so great as in the former case: for, let NR be many times greater than CE, and NS less than CE; and take Er and Es equal to CR and CS, there can be no doubt, I think, but that the deficient fluid in DF will be lodged nearly in the same manner as the redundant fluid in AB; and therefore, the repulsion of the redundant fluid at R, on a particle at N, will be very nearly balanced by the attraction of the redundant matter at r, for R is not much nearer to N than r is; but the repulsion of S will not be near balanced by that of s; for the distance of S from N is much less than that of s. Let now a small circle, whose diameter is ST, be drawn round the center N, on the plane of the plate; as the density of the fluid is greater at T than at S, the repulsion of the redundant fluid within the small circle tends to impel the point N towards C; but as there is a much greater quantity of fluid between N and B, than between N and A, the repulsion of the fluid without the small circle tends to balance that; but the effect of the fluid within the small circle is not much less than it would be, if DF was not undercharged; whereas much the greater part of the effect of that part of the plate on the outside of the circle, is taken off by the effect of the corresponding part of DF: consequently, the difference of density between T and S will not be near so great as if DF was not undercharged. Hence I should imagine, that if the two plates are very near together, the density of the redundant fluid near the center will not be much less than the mean density, or δ will not be much less than 1; moreover, the less the distance of the plates, the nearer will δ approach to 1.

77] Cor. II. Let now the body H consist of a circular plate, of the same size as AB, placed so, that the canal CG shall pass through its center, and be perpendicular to its plane; by the supposition, the force with which H repels the fluid in the canal CG, is the same with which a quantity of fluid, equal to B, spread uniformly over AB, would repel it in the contrary direction: therefore, if the fluid in the plate H was spread uniformly, the quantity of redundant fluid therein would be B, and if it was all collected in the circumference, would be $\frac{2B}{3-n}$; and therefore the real quantity will be greater than B, and less than $\frac{2B}{3-n}$.

78] Cor. III. Therefore, if we suppose δ to be equal to 1, the quantity of redundant fluid in AB will exceed that in the plate H, in a greater ratio than that of $\left|\frac{AC}{CE}\right|^{3-n} \times \frac{3-n}{4}$ to 1, and less than that of $\left|\frac{AC}{CE}\right|^{3-n} \times \frac{1}{2}$ to 1; and from the preceding remarks it appears that the real quantity of redundant fluid in AB can hardly be much greater than it would if δ was equal to 1.

79] Cor. IV. Hence, if the electric attraction and repulsion is inversely as the square of the distance, the redundant fluid in AB, supposing δ to be equal to 1, will exceed that in the plate H, in a greater ratio than that of AC to $4CE$, and less than that of AC to $2CE$.

80] Cor. V. Let now the body H consist of a globe, whose diameter equals AB; the globe being situated in such a manner, that the canal CG, if continued, would pass through its center; and let the electric attraction and repulsion be inversely as the square of the distance, the quantity of redundant fluid in the globe will be $2B$: for the fluid will be spread uniformly over the surface of the globe, and its repulsion on the canal will be the same as if it was all collected in the center of the sphere, and will therefore be the same with which an equal quantity, disposed in the circumference of AB, would repel it in the contrary direction, or with which half that quantity, or B, would repel it, if spread uniformly over the plate. [See Art. 140.]

81] Cor. VI. Therefore, if δ was equal to 1, the redundant fluid in AB would exceed that in the globe, in the ratio of AC to $4CE$; and therefore, it will in reality exceed that in the globe, in a rather greater ratio than that of AC to $4CE$; but if the plates are very near together, it will approach very near thereto, and the nearer the plates are, the nearer it will approach thereto.

82] Cor. VII. Whether the electric repulsion is inversely as the square of the distance or not, if the body H is as much undercharged, as it was before overcharged, AB will be as much undercharged as it was before overcharged, and DF as much overcharged as it was before undercharged.

83] Cor. VIII. If the size and distance of the plates be altered, the quantity of redundant or deficient fluid in the body H remaining the same, it appears, by comparing this proposition with the 20th and 21st propositions, that the quantity of redundant and deficient fluid in AB will be as $AC^{n-1} \times \overline{\frac{AC}{EC}}\Big|^{3-n}$, or as $\frac{AC^2}{EC^{3-n}}$, supposing the value of δ to remain the same*.

[* Note 4.]

84] Prop. XXIII. Let AE (Fig. 15) be a cylindric canal, infinitely continued beyond E; and let AF be a bent canal, meeting the other at A, and infinitely continued beyond F: let the section of this canal, in all parts of it, be equal to that of the cylindric canal, and let both canals be filled with uniform fluid of the same density: the force with which a particle of fluid P, placed anywhere at pleasure, repels the whole quantity of fluid in AF, in the direction of the canal, is the same with which it repels the fluid in the canal AE, in the direction AE.

Fig. 15.

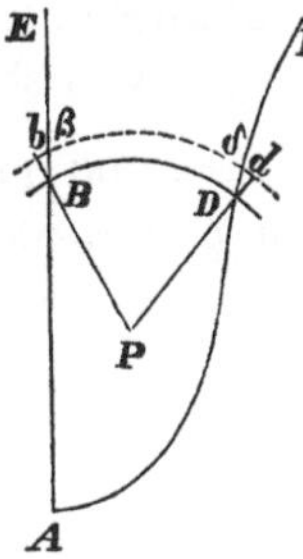

On the center P, draw two circular arches BD and bd, infinitely near to each other, cutting AE in B and β, and AF in D and δ, and draw the radii Pb and Pd. As $PB = PD$, the force with which P repels a particle at B, in the direction $B\beta$, is to that with which it repels an equal particle at D, in the direction $D\delta$, as $\frac{Bb}{B\beta}$ to $\frac{Dd}{D\delta}$, or as $\frac{1}{B\beta}$ to $\frac{1}{D\delta}$; and therefore, the force with which it repels the whole fluid in $B\beta$, in the direction $B\beta$, is the same with which it repels the whole fluid in $D\delta$, in the direction $D\delta$, that is in the direction of the canal; and therefore, the force with which it repels the whole fluid in AE, in the direction AE, is the same with which it repels the whole fluid in AF, in the direction of the canal.

85] Cor. If the bent canal ADF, instead of being infinitely continued, meets the cylindric canal in E, as in Fig. 16, the repulsion of P on the fluid in the bent canal ADE, in the direction of the canal, will still be equal to its repulsion on that in the cylindric canal AE, in the direction AE.

Fig. 16.

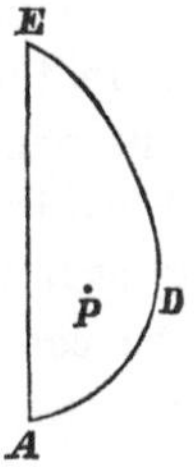

86] Prop. XXIV. If two bodies, for instance the plate AB, and the body H, of Prop. XXII. communicate with each other, by a canal filled with incompressible fluid, and are either over or undercharged, the quantity of redundant fluid in them will bear the same proportion to each other, whether the canal by which they communicate is straight or crooked, or into whatever part of the bodies the canal is inserted, or in whatever manner the two bodies are situated in respect of each other; provided that their distance is infinite, or so great that the repulsion of each body on the fluid in the canal shall not be sensibly less than if it was infinite.

Let the parallelograms *AB* and *DF* (Fig. 17) represent the two plates, and *H* and *L* the bodies communicating with them:

Fig. 17.

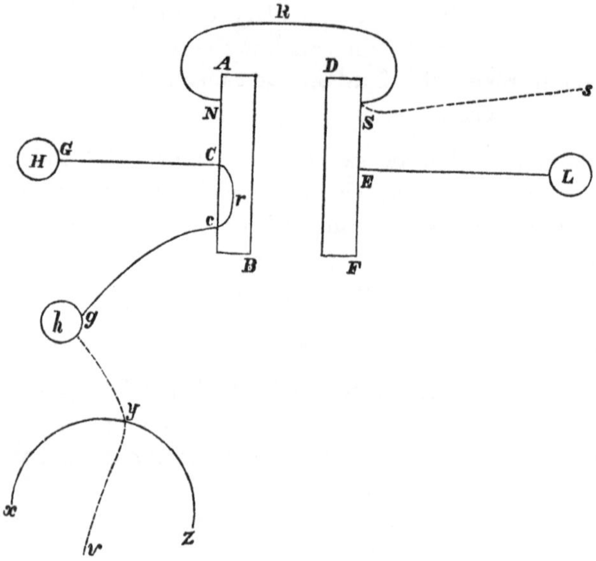

let now *H* be removed to *h*; and let it communicate with *AB* by the bent canal *gc*; the quantity of fluid in the plates and bodies remaining the same as before; and let us, for the sake of ease in the demonstration, suppose the canal *gc* to be everywhere of the same thickness as the canal *GC*; though the proposition will evidently hold good equally, whether it is or not: the fluid will still be in equilibrio. For let us first suppose the canal *gc* to be continued through the substance of the plate *AB*, to *C*, along the line *crC*; the part *crC* being of the same thickness as the rest of the canal, and the fluid in it of the same density: by the preceding proposition, the repulsion or attraction of each particle of fluid or matter in the plates *AB* and *DF*, on the fluid in the whole canal *Crcg*, in the direction of that canal, is equal to its repulsion or attraction on the fluid in the canal *CG*, in the direction CG; and therefore the whole repulsion or attraction of the two plates on the canal $Crcg$, is equal to their repulsion or attraction on CG: but as the fluid in the plate AB is in equilibrio, each particle of fluid in the part Crc of the canal is impelled by the plates with as much force in one direction as the other; and consequently the plates impel the fluid in the canal cg with as much force as they do that in the whole canal $Crcg$, that is, with the same force that they impel the fluid in CG. In like manner the body h impels the fluid in cg with the same force that H does the fluid in CG; and consequently h impels the fluid in cg one way in the direction of the canal, with the same force that the two plates impel it the contrary way; and therefore the fluid in cg has no tendency to flow from one body to the other.

87] COR. By the same method of reasoning, with the help of the corollary to the 23rd proposition, it appears, that if AB and H each communicate with a third body by canals of incompressible fluid, and a communication is made between AB and H by another canal of incompressible fluid, the fluid will have no tendency to flow from one to the other through this canal; supposing that the fluid was in equilibrio before this communication was made. In like manner if AB and H communicate with each other, or each communicate with a third body, by canals of real fluid, instead of the imaginary canals of incompressible fluid used in these propositions, and a communication is also made between them by a canal of incompressible fluid, the fluid can have no tendency to flow from one to the other. The truth of the latter part of this corollary will appear by supposing an imaginary canal of incompressible fluid to be continued through the whole length of the real one.

88] PROP. XXV. Let now a communication be made between the two plates AB and DF, by the canal NRS of incompressible fluid, of any length; and let the body H and the plate AB be overcharged. It is plain that the fluid will flow through that canal from AB to DF. Now the whole force with which the fluid in the canal is impelled along it by the joint action of the two plates is the same with which the whole quantity of fluid in the canal CG or cg is impelled by them; supposing the canal NRS to be everywhere of the same breadth and thickness as CG or cg.

For suppose that the canal *NRS*, instead of communicating with the plate *DF*, is bent back just before it touches it, and continued infinitely along the line *Ss*; the force with which the two plates impel the fluid in *Ss*, is the same with which they impel that in *EL*, supposing *Ss* to be of the same breadth and thickness as *EL*; and is therefore nothing; therefore the force with which they impel the fluid in *NRS*, is the same with which they impel that in *NRSs*; which is the same with which they impel that in *CG*.

89] PROP. XXVI. Let now *xyz* [Fig. 17] be a body of an infinite size, containing just fluid enough to saturate it; and let a communication be made between *h* and *xyz*, by the canal *hy* of incompressible fluid, of the same breadth and thickness as *gc* or *GC*; the fluid will flow through it from *h* to *xyz*; and the force with which the fluid in that canal is impelled along it, is equal to that with which the fluid in *NRS* is impelled by the two plates.

If the canal *hy* is of so great a length, that the repulsion of *h* thereon is the same as if it was continued infinitely, then the thing is evident: but if it is not, let the canal *hy*, instead of communicating with *xyz*, so that the fluid can flow out of the canal into *xyz*, be continued infinitely through its substance, along the line *yv*: now it must be observed that a small part of the body *xyz*, namely, that which is turned towards *h*, will by the action of *h* upon it, be rendered undercharged; but all the rest of the body will be saturated; for the fluid driven out of the undercharged part will not make the remainder, which is supposed to be of an infinite size, sensibly overcharged: now the force with which the fluid in the infinite canal *hyv* is impelled by the body *h* and the undercharged part of *xyz*, is the same with which the fluid in *gc* is impelled by them; but as the fluid in all parts of *xyz* is in equilibrio, a particle in any part of *yv* cannot be impelled in any direction; and therefore the fluid in *hy* is impelled with as much force as that in *hyv*; and therefore the fluid in *hy* is impelled with as much force as that in *gc*; and is therefore impelled with as much force as the fluid in *NRS* is impelled by the two plates.

90] It perhaps may be asked, whether this method of demonstration would not equally tend to prove that the fluid in *hy* was impelled with the same force as that in *NRS*, though *xyz* did not contain just fluid enough to saturate it. I answer not; for this demonstration depends on the canal *yv* being continued, within the body *xyz*, to an infinite distance beyond any over or undercharged part; which could not be if *xyz* contained either more or less fluid than that*.

91] PROP. XXVII. Let two bodies *B* and *b* (Fig. 13) be joined by a cylindric or prismatic canal *Aa*, filled with real fluid; and not by any imaginary canal of incompressible fluid as in the 20th proposition; and let the fluid therein be in equilibrio: the force with which the whole or any given part of the fluid in the canal is impelled in the direction of its axis by the united repulsions and attractions of the redundant fluid or matter in the two bodies and the canal, must be nothing; or the force with which it is impelled one way in the direction of the axis of the canal, must be equal to that with which it is impelled the other way.

For as the canal is supposed cylindric or prismatic, no particle of fluid therein can be prevented from moving in the direction of the axis of it, by the sides of the canal; and therefore the force with which each particle is impelled either way in the direction of the axis, by the united attractions and repulsions of the two bodies and the canal, must be nothing, otherwise it could not be at rest; and therefore the force with which the whole, or any given part of the fluid in the canal, is impelled in the direction of the axis, must be nothing

92] COR. I. If the fluid in the canal is disposed in such manner, that the repulsion or attraction of the redundant fluid or matter in it, on the whole or any given part of the fluid in the canal, has no tendency to impel it either way in the direction of the axis; then the force with which that whole or given part is impelled by the two bodies must be nothing; or the force with which it is impelled one way in the direction of the axis, by the body *B*, must be equal to that with which it is impelled in the contrary direction by the other body; but not if the fluid in the canal is disposed in a different manner.

93] COR. II. If the bodies, and consequently the canal, is overcharged; then, in whatever manner the fluid in the canal is disposed, the force with which the whole quantity of redundant fluid in the canal is repelled by the body *B* in the direction *Aa*,

[* Note 5.]

must be equal to that with which it is repelled by *b* in the contrary direction. For the force with which the redundant fluid is impelled in the direction *Aa* by its own repulsion, is nothing; for the repulsions of the particles of any body on each other have no tendency to make the whole body move in any direction.

94] REMARKS. When I first thought of the 20th and 22nd propositions, I imagined that when two bodies were connected by a cylindric canal of real fluid, the repulsion of one body on the whole quantity of fluid in the canal, in one direction, would be equal to that of the other body on it in the contrary direction, in whatever manner the fluid was disposed in the canal; and that therefore those propositions would have held good very nearly, though the bodies were joined by cylindric canals of real fluid; provided the bodies were so little over or undercharged, that the quantity of redundant or deficient fluid in the canal should be very small in respect of the quantity required to saturate it; and consequently that the fluid therein should be very nearly of the same density in all parts. But from the foregoing proposition it appears that I was mistaken, and that the repulsion of one body on the fluid in the canal is not equal to that of the other body on it, unless the fluid in the canal is disposed in a particular manner: besides that, when two bodies are both joined by a real canal, the attraction or repulsion of the redundant matter or fluid in the canal has some tendency to alter the disposition of the fluid in the two bodies; and in the 22nd proposition, the canal *CG* exerts also some attraction or repulsion on the canal *EM*: on all which accounts the demonstration of those propositions is defective, when the bodies are joined by real canals. I have good reason however to think, that those propositions actually hold good very nearly when the bodies are joined by real canals; and that, whether the canals are straight or crooked, or in whatever direction the bodies are situated in respect of each other: though I am by no means able to prove that they do: I therefore chose still to retain those propositions, but to demonstrate them on this ideal supposition, in which they are certainly true, in hopes that some more skilful mathematician may be able to shew whether they really hold good or not. [See Note 3.]

95] What principally makes me think that this is the case, is that as far as I can judge from some experiments I have made*,

[* Exp. III., Art. 265.]

the quantity of fluid in different bodies agrees very well with those propositions, on a supposition that the electric repulsion is inversely as the square of the distance. It should also seem from those experiments, that the quantity of redundant or deficient fluid in two bodies bore very nearly the same proportion to each other, whatever is the shape of the canal by which they are joined, or in whatever direction they are situated in respect of each other.

96] Though the above propositions should be found not to hold good when the bodies are joined by real canals, still it is evident, that in the 22nd proposition, if the plates AB and DF are very near together, the quantity of redundant fluid in the plate AB will be many times greater than that in the body H, supposing H to consist of a circular plate of the same size as AB, and DF will be near as much undercharged as AB is overcharged.

97] Sir Isaac Newton supposes that air consists of particles which repel each other with a force inversely as the distance: but it appears plainly from the foregoing pages, that if the repulsion of the particles was in this ratio; and extended indefinitely to all distances, they would compose a fluid extremely different from common air. If the repulsion of the particles was inversely as the distance, but extended only to a given very small distance from their centers, they would compose a fluid of the same kind as air, in respect of elasticity, except that its density would not be in proportion to its compression: if the distance to which the repulsion extends, though very small, is yet many times greater than the distance of the particles from each other, it might be shewn, that the density of the fluid would be nearly as the square root of the compression. If the repulsion of the particles extended indefinitely, and was inversely as some higher power of the distance than the cube, the density of the fluid would be as some power of the compression less than $\frac{2}{3}$. The only law of repulsion, I can think of, which will agree with experiment, is one which seems not very likely; namely, that the particles repel each other with a force inversely as the distance; but that, whether the density of the fluid is great or small, the repulsion extends only to the nearest particles: or, what comes to the same thing, that the distance to which the repulsion extends, is very small, and also is not fixed, but varies in proportion to the distance of the particles *.

[* Note 6.]

PART II.

CONTAINING A COMPARISON OF THE FOREGOING THEORY WITH EXPERIMENT.

98] § 1. It appears from experiment, that some bodies suffer the electric fluid to pass with great readiness between their pores; while others will not suffer it to do so without great difficulty; and some hardly suffer it to do so at all. The first sort of bodies are called conductors, the others non-conductors. What this difference in bodies is owing to I do not pretend to explain.

It is evident that the electric fluid in conductors may be considered as moveable, or answers to the definition given of that term in page 6. As to the fluid contained in non-conducting substances, though it does not absolutely answer to the definition of immoveable, as it is not absolutely confined from moving, but only does so with great difficulty; yet it may in most cases be looked upon as such without sensible error.

99] Air does in some measure permit the electric fluid to pass through it; though, if it is dry, it lets it pass but very slowly, and not without difficulty; it is therefore to be called a non-conductor.

It appears that conductors would readily suffer the fluid to run in and out of them, were it not for the air which surrounds them: for if the end of a conductor is inserted into a vacuum, the fluid runs in and out of it with perfect readiness; but when it is surrounded on all sides by the air, as no fluid can run out of it without running into the air, the fluid will not do so without difficulty.

100] If any body is surrounded on all sides by the air, or other non-conducting substances, it is said to be insulated: if on the other hand it anywhere communicates with any conducting body, it is said to be not insulated. When I say that a body communicates with the ground, or any other body, I would be understood to mean that it does so by some conducting substance.

101] Though the terms positively and negatively electrified are much used, yet the precise sense in which they are to be understood seems not well ascertained; namely, whether they are to be understood in the same sense in which I have used the words over or undercharged, or whether, when any number of bodies, insulated and communicating with each other by conducting substances, are electrified by means of excited glass, they are all to be called positively electrified (supposing, according to the usual opinion, that excited glass contains more than its natural quantity of electricity); even though some of them, by the approach of a stronger electrified body, are made undercharged. I shall use the words in the latter sense; but as it will be proper to ascertain the sense in which I shall use them more accurately, I shall give the following definition.

102] In order to judge whether any body, as A, is positively or negatively electrified: suppose another body B, of a given shape and size, to be placed at an infinite distance from it, and from any other over or undercharged body; and let B contain the same quantity of electric fluid as if it communicated with A by a canal of incompressible fluid: then, if B is overcharged, I call A positively electrified; and if it is undercharged, I call A negatively electrified; and the greater the degree in which B is over or undercharged, the greater is the degree in which A is positively or negatively electrified.

103] It appears from the corollary to the 24th proposition, that if several bodies are insulated, and connected together by conducting substances, and one of these bodies is positively or negatively electrified, all the other bodies must be electrified in the same degree: for supposing a given body B to be placed at an infinite distance from any over or undercharged body, and to contain the same quantity of fluid as if it communicated with one of those bodies by a canal of incompressible fluid, all the rest of those bodies must by that corollary contain the same quantity of fluid as if they communicated with B by canals of incompressible fluid:

but yet it is possible that some of those bodies may be overcharged, and others undercharged: for suppose the bodies to be positively electrified, and let an overcharged body D be brought near one of them, that body will become undercharged, provided D is sufficiently overcharged; and yet by the definition it will still be positively electrified in the same degree as before.

Moreover, if several bodies are insulated and connected together by conducting substances, and one of these bodies is electrified by excited glass, there can be no doubt, I think, but what they will all be positively electrified; for if there is no other over or undercharged body placed near any of these bodies, the thing is evident; and though some of these bodies may, by the approach of a sufficiently overcharged body, be rendered undercharged; yet I do not see how it is possible to prevent a body placed at an infinite distance, and communicating with them by a canal of incompressible fluid, from being overcharged.

In like manner if one of these bodies is electrified by excited sealing wax, they will all be negatively electrified *.

104] It is impossible for any body communicating with the ground to be either positively or negatively electrified: for the earth, taking the whole together, contains just fluid enough to saturate it, and consists in general of conducting substances; and consequently though it is possible for small parts of the surface of the earth to be rendered over or undercharged, by the approach of electrified clouds or other causes; yet the bulk of the earth, and especially the interior parts, must be saturated with electricity. Therefore assume any part of the earth which is itself saturated, and is at a great distance from any over or undercharged part; any body communicating with the ground, contains as much electricity as if it communicated with this part by a canal of incompressible fluid, and therefore is not at all electrified.

105] If any body A, insulated and saturated with electricity, is placed at a great distance from any over or undercharged body, it is plain that it cannot be electrified; but if an overcharged body is brought near it, it will be positively electrified; for supposing A to communicate with any body B, at an infinite distance, by a canal of incompressible fluid, it is plain that unless B is overcharged, the fluid in the canal could not be in equilibrio, but would

[* Note 7.]

run from *A* to *B*. For the same reason a body insulated and saturated with fluid, will be negatively electrified if placed near an undercharged body.

106] § 2. The phænomena of the attraction and repulsion of electrified bodies seem to agree exactly with the theory; as will appear by considering the following cases.

107] CASE I. Let two bodies, *A* and *B*, both conductors of electricity, and both placed at a great distance from any other electrified bodies, be brought near each other. Let *A* be insulated, and contain just fluid enough to saturate it; and let *B* be positively electrified. They will attract each other; for as *B* is positively electrified, and at a great distance from any overcharged body, it will be overcharged; therefore, on approaching *A* and *B* to each other, some fluid will be driven from that part of *A* which is nearest to *B* to the further part: but when the fluid in *A* was spread uniformly, the repulsion of *B* on the fluid in *A* was equal to its attraction on the matter therein; therefore, when some fluid is removed from those parts where the repulsion of *B* is strongest to those where it is weaker, *B* will repel the fluid in *A* with less force than it attracts the matter; and consequently the bodies will attract each other.

108] CASE II. If we now suppose that the fluid is at liberty to escape from out of *A*, if it has any disposition to do so, the quantity of fluid in it before the approach of *B* being still sufficient to saturate it; that is, if *A* is not insulated and not electrified, *B* being still positively electrified, they will attract with more force than before: for in this case, not only some fluid will be driven from that part of *A* which is nearest to *B* to the opposite part, but also some fluid will be driven out of *A*.

It must be observed, that if the repulsion of *B* on a particle at *E*, (Fig. 19) the farthest part of *A*, is very small in respect of its repulsion on an equal particle placed at *D*, the nearest part of *A*, the two bodies will attract with very nearly the same force, whether *A* is insulated or not; but if the repulsion of *B*, on a particle at *E*, is very near as great as on one at *D*, they will attract with very little force if *A* is

Fig. 19.

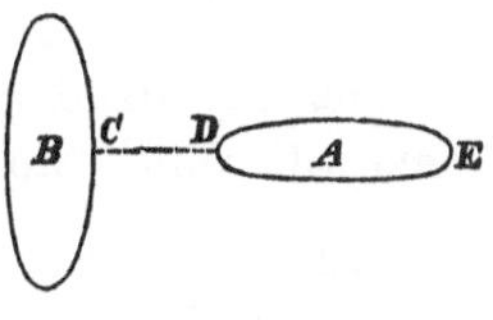

insulated. For instance, let a small overcharged ball be brought near one end of a long conductor not electrified; they will attract with very near the same force, whether the conductor be insulated or not; but if the conductor be overcharged, and brought near a small unelectrified ball, they will not attract with near so much force, if the ball is insulated, as if it is not.

109] CASE III. If we now suppose that *A* is negatively electrified, and not insulated, it is plain that they will attract with more force than in the last case; as *A* will be still more undercharged in this case, than in the last.

110] N.B. In these three cases, we have not as yet taken notice of the effect which the body *A* will have in altering the quantity and disposition of the fluid in *B*; but in reality this will make the bodies attract each other with more force than they would otherwise do; for in each of these cases the body *A* attracts the fluid in *B*; which will cause some fluid to flow from the farther parts of *B* to the nearer, and will also cause some fluid to flow into it, if it is not insulated, and will consequently cause *B* to act upon *A* with more force than it would otherwise do.

111] CASES IV. V. VI. Let us now suppose that *B* is negatively electrified; and let *A* be insulated, and contain just fluid enough to saturate it; they will attract each other; for *B* will be undercharged; it will therefore attract the fluid in *A*, and will cause some fluid to flow from the farthest part of *A*, where it is attracted with less force, to the nearer part, where it is attracted with more force; so that *B* will attract the fluid in *A* with more force than it repels the matter.

If *A* is now supposed to be not insulated and not electrified, *B* being still negatively electrified, it is plain that they will attract with more force than in the last case: and if *A* is positively electrified, they will attract with still more force.

In these three last cases also, the effect which *A* has in altering the quantity and disposition of the fluid in *B*, tends to increase the force with which the two bodies attract.

112] CASE VII. It is plain that a non-conducting body saturated with fluid, is not at all attracted or repelled by an over or undercharged body, until, by the action of the electrified body on it, it has either acquired some additional fluid from the air, or had some driven out of it, or till some fluid is driven from one part of the body to the other.

113] CASE VIII. Let us now suppose that the two bodies *A* and *B* are both positively electrified in the same degree. It is plain, that were it not for the action of one body on the other, they would both be overcharged, and would repel each other. But it may perhaps be said, that one of them as *A* may, by the action of the other on it, be either rendered undercharged on the whole, or at least may be rendered undercharged in that part nearest to *B*; and that the attraction of this undercharged part on a particle of the fluid in *B*, may be greater than the repulsion of the more distant overcharged part; so that on the whole the body *A* may attract a particle of fluid in *B*. If so, it must be affirmed that the body *B* repels the fluid in *A*; for otherwise, that part of *A* which is nearest to *B* could not be rendered undercharged. Therefore, to obviate this objection, let the bodies be joined by the straight canal *DC* of incompressible fluid (Fig. 19). The body *B* will repel the fluid in all parts of this canal; for as *A* is supposed to attract the fluid in *B*, *B* will not only be more overcharged than it would otherwise be, but it will also be more overcharged in that part nearest to *A* than in the opposite part. Moreover, as the near undercharged part of *A* is supposed to attract a particle of fluid in *B* with more force than the more distant overcharged part repels it; it must, *a fortiori*, attract a particle in the canal with more force than the other repels it; therefore the body *A* must attract the fluid in the canal; and consequently some fluid must flow from *B* to *A*, which is impossible; for as *A* and *B* are both electrified in the same degree, they contain the same quantity of fluid as if they both communicated with a third body at an infinite distance, by canals of incompressible fluid; and therefore, by the corollary to Prop. 24, if a communication is made between them by a canal of incompressible fluid, the fluid would have no disposition to flow from one to the other.

114] CASE IX. But if one of the bodies as *A* is positively electrified in a less degree than *B*, then it is possible for the bodies to attract each other; for in this case the force with which *B* repels the fluid in *A* may be so great, as to make the body *A* either intirely undercharged, or at least to make the nearest part of it so much undercharged, that *A* shall on the whole attract a particle of fluid in *B*.

It may be worth remarking with regard to this case, that when two bodies, both electrified positively but unequally, attract each other, you may by removing them to a greater distance from each other, cause them to repel; for as the stronger electrified body repels the fluid in the weaker with less force when removed to a greater distance, it will not be able to drive so much fluid out of it, or from the nearer to the further part, as when placed at a less distance.

115] CASES X. and XI. By the same reasoning it appears, that if the two bodies are both negatively electrified in the same degree, they must repel each other: but if they are both negatively electrified in different degrees, it is possible for them to attract each other.

All these cases are exactly conformable to experiment.

116] CASE XII. Let two cork balls be suspended by conducting threads from the same positively electrified body, in such manner that if they did not repel, they would hang close together: they will both be equally electrified, and will repel each other: let now an overcharged body, more strongly electrified than them, be brought under them; they will become less overcharged, and will separate less than before: on bringing the body still nearer, they will become not at all overcharged, and will not separate at all: and on bringing the body still nearer, they will become undercharged, and will separate again.

117] CASE XIII. Let all the air of a room be overcharged, and let two cork balls be suspended close to each other by conducting threads communicating with the wall. By Prop. 15, it is highly probable that the balls will be undercharged; and therefore they should repel each other.

These two last cases are experiments of Mr Canton's, and are described in *Philosophical Transactions* 1753, p. 350, where are other experiments of the same kind, all readily explicable by the foregoing theory.

I have now considered all the principal or fundamental cases of electric attractions and repulsions which I can think of; all of which appear to agree perfectly with the theory*.

118] § 3. On the cases in which bodies receive electricity from or part with it to the air.

LEMMA I. Let the body *A* (Fig. 6) either stand near some over or undercharged body, or at a distance from any. It seems highly probable, that if any part of its surface, as *MN*, is overcharged, the fluid will endeavour to run out through that part, provided the air adjacent thereto is not overcharged.

Fig. 6.

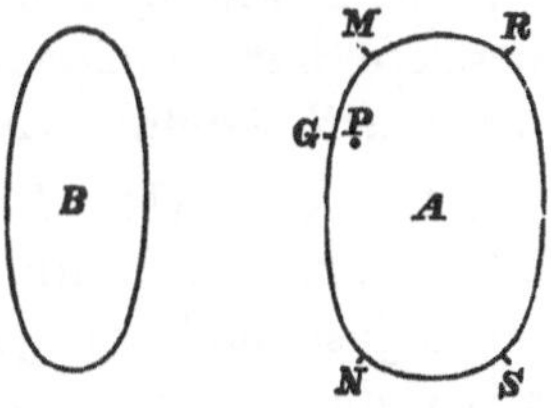

For let *G* be any point in that surface, and *P* a point within the body, extremely near to it; it is plain that a particle of fluid at *P* must be repelled with as much force in one direction as another (otherwise it could not be at rest) unless all the fluid between *P* and *G* is pressed close together, in which case it may be repelled with more force towards *G* than it is in the contrary direction: now a particle at *G* is repelled in the direction *PG*, *i.e.* from *P* to *G*, by all the redundant fluid between *P* and *G*; and a particle at *P* is repelled by the same fluid in the contrary direction; so that as the particle at *P* is repelled with not less force in the direction *PG* than in the contrary, I do not see how a particle at *G* can help being repelled with more force in that direction than the contrary, unless the air on the outside of the surface *MN* was more overcharged than the space between *P* and *G*.

In like manner, if any part of the surface is undercharged, the fluid will have a tendency to run in at that part from the air.

The truth of this is somewhat confirmed by the third problem; as in all the cases of that problem, the fluid was shewn to have a tendency to run out of the spaces *AD* and *EH*, at any surface which was overcharged, and to run in at any which was undercharged.

119] COR. I. If any body at a distance from other over or undercharged bodies be positively electrified, the fluid will gradually run out of it from all parts of its surface into the adjoining air; as it is plain that all parts of the surface of that body will be

[* Note 8.]

overcharged: and if the body is negatively electrified, the fluid will gradually run into it at all parts of its surface from the adjoining air.

120] COR. II. Let the body *A* (Fig. 6) insulated and containing just fluid enough to saturate it, be brought near the overcharged body *B*; that part of the surface of *A* which is turned towards *B* will by Prop. II. be rendered undercharged, and will therefore imbibe electricity from the air; and at the opposite surface *RS*, the fluid will run out of the body into the air.

121] COR. III. If we now suppose that *A* is not insulated, but communicates with the ground, and consequently that it contained just fluid enough to saturate it before the approach of *B*, it is plain that the surface *MN* will be more undercharged than before; and therefore the fluid will run in there with more force than before; but it can hardly have any disposition to run out at the opposite surface *RS*; for if the canal by which *A* communicates with the ground is placed opposite to *B*, as in figure 5, then the fluid will run out through that canal till it has no longer any tendency to run out at *RS*; and by the remarks at the end of Prop. 27, it seems probable, that the fluid in *A* will be nearly in the same quantity, and disposed nearly in the same manner, into whatever part of *A* the canal is inserted by which it communicates with the ground.

122] COR. IV. If *B* is undercharged the case will be reversed; that is, it will run out where it before run in, and will run in where it before run out.

As far as I can judge, these corollariés seem conformable to experiment: thus far is certain, that bodies at a distance from other electrified bodies receive electricity from the air, if negatively electrified, and part with some to it if positively electrified: and a body not electrified and not insulated receives electricity from the air if brought near an overcharged body, and loses some when brought near an undercharged body: and a body insulated and containing its natural quantity of fluid, in some cases, receives, and in others loses electricity, when brought near an over or undercharged body.

123] § 4. The well-known effects of points in causing a quick discharge of electricity seem to agree very well with this theory.

It appears from the 20th proposition, that if two similar bodies of different sizes are placed at a very great distance from each other, and connected by a slender canal, and overcharged, the force with which a particle of fluid placed close to corresponding parts of their surface is repelled from them, is inversely as the corresponding diameters of the bodies. If the distance of the two bodies is small, there is not so much difference in the force with which the particle is repelled by the two bodies; but still, if the diameters of the two bodies are very different, the particle will be repelled with much more force from the smaller body than from the larger. It is true indeed that a particle placed at a certain distance from the smaller body, will be repelled with less force than if it be placed at the same distance from the greater body; but this distance is, I believe, in most cases pretty considerable; if the bodies are spherical, and the repulsion inversely as the square of the distance, a particle placed at any distance from the surface of the smaller body less than a mean proportional between the radii of the two bodies, will be repelled from it with more force than if it be placed at the same distance from the larger body.

I think therefore that we may be well assured that if two similar bodies are connected together by a slender canal, and are overcharged, the fluid must escape faster from the smaller body than from an equal surface of the larger; but as the surface of the larger body is greatest, I do not know which body ought to lose most electricity in the same time; and indeed it seems impossible to determine positively from this theory which should, as it depends in great measure on the manner in which the air opposes the entrance of the electric fluid into it. Perhaps in some degrees of electrification the smaller body may lose most, and in others the larger.

124] Let now ACB (Fig. 18) be a conical point standing on any body DAB, C being the vertex of the cone; and let DAB be overcharged: I imagine that a particle of fluid placed close to the surface of the cone anywhere between b and C, must be repelled with at least as much, if not more, force than it would, if the part $AabB$ of the cone was taken away, and the part aCb connected to DAB by a slender canal; and con-

Fig. 18.

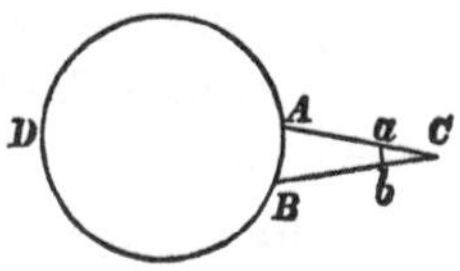

sequently, from what has been said before, it seems reasonable to suppose that the waste of electricity from the end of the cone must be very great in proportion to its surface; though it does not appear from this reasoning whether the waste of electricity from the whole cone should be greater or less than from a cylinder of the same base and altitude*.

All which has been here said relating to the flowing out of electricity from overcharged bodies, holds equally true with regard to the flowing in of electricity into undercharged bodies.

125] But a circumstance which I believe contributes as much as any thing to the quick discharge of electricity from points, is the swift current of air caused by them, and taken notice of by Mr Wilson and Dr Priestly (*vide* Priestly, p. 117 and 591); and which is produced in this manner.

If a globular body *ABD* is overcharged, the air close to it, all round its surface, is rendered overcharged by the electric fluid which flows into it from the body; it will therefore be repelled by the body; but as the air all round the body is repelled with the same force, it is in equilibrio, and has no tendency to fly off from it. If now the conical point *ACB* be made to stand out from the globe, as the fluid will escape much faster in proportion to the surface from the end of the point than from the rest of the body, the air close to it will be much more overcharged than that close to the rest of the body; it will therefore be repelled with much more force; and consequently a current of air will flow along the sides of the cone, from *B* towards *C*; by which means there is a continual supply of fresh air, not much overcharged, brought in contact with the point; whereas otherwise the air adjoining to it would be so much overcharged, that the electricity would have but little disposition to flow from the point into it.

The same current of air is produced in a less degree, without the help of the point, if the body, instead of being globular, is oblong or flat, or has knobs on it, or is otherwise formed in such manner as to make the electricity escape faster from some parts of it than the rest.

In like manner, if the body *ABD* be undercharged, the air adjoining to it will also be undercharged, and will therefore be

[* Note 9.]

repelled by it; but as the air close to the end of the point will be more undercharged than that close to the rest of the body, it will be repelled with much more force; which will cause exactly the same current of air, flowing the same way, as if the body was overcharged; and consequently the velocity with which the electric fluid flows into the body, will be very much increased. I believe indeed that it may be laid down as a constant rule, that the faster the electric fluid escapes from any body when overcharged, the faster will it run into that body when undercharged.

Points are not the only bodies which cause a quick discharge of electricity; in particular, it escapes very fast from the ends of long slender cylinders; and a swift current of air is caused to flow from the middle of the cylinder towards the end: this will easily appear by considering that the redundant fluid is collected in much greater quantity near the ends of the cylinders than near the middle. The same thing may be said, but I believe in a less degree, of the edges of thin plates.

What has been just said concerning the current of air, serves to explain the reason of the revolving motion of Dr Hamilton's and Mr Kinnersley's bent pointed wires, vide *Philosophical Trans.* Vol. LI., p. 905, and Vol. LIII., p. 86; also Priestly, p. 429: for the same repulsion which impels the air from the thick part of the wire towards the point, tends to impel the wire in the contrary direction.

126] It is well known, that if a body *B* is positively electrified, and another body *A*, communicating with the ground, be then brought near it, the electric fluid will escape faster from *B*, at that part of it which is turned towards *A*, than before. This is plainly conformable to theory; for as *A* is thereby rendered undercharged, *B* will in its turn be made more overcharged, in that part of it which is turned towards *A*, than it was before. But it is also well known that the fluid will escape faster from *B*, if *A* be pointed, than if it be blunt; though *B* will be less overcharged in this case than in the other; for the broader the surface of *A*, which is turned towards *B*, the more effect will it have in increasing the overcharge of *B*. The cause of this phænomenon is as follows:

If *A* is pointed, and the pointed end turned towards *B*, the air close to the point will be very much undercharged, and therefore will be strongly repelled by *A*, and attracted by *B*, which will cause a swift current of air to flow from it towards *B*; by which means a constant supply of undercharged air will be brought in contact with *B*, which will accelerate the discharge of electricity from it in a very great degree: and moreover, the more pointed *A* is, the swifter will be this current. If, on the other hand, that end of *A* which is turned towards *B* is so blunt, that the electricity is not disposed to run into *A* faster than it is to run out of *B*, the air adjoining to *B* may be as much overcharged as that adjoining to *A* is undercharged; and therefore may by the joint repulsion of *B* and attraction of *A*, be impelled from *B* to *A*, with as much or more force than the air adjoining to *A* is impelled in the contrary direction; so that what little current of air there is may flow in the contrary direction.

It is easy applying what has been here said to the case in which *B* is negatively electrified.

127] § 5. In the paper of Mr Canton's, quoted in the second section, and in a paper of Dr Franklin's *Philosophical Transactions* 1755, p. 300, and Franklin's letters p. 155, are some remarkable experiments, shewing that when an overcharged body is brought near another body, some fluid is driven to the further end of this body, and also some driven out of it, if it is not insulated. The experiments are all strictly conformable to the 11th, 12th, and 13th propositions: but it is needless to point out the agreement, as the explanation given by the authors does it sufficiently.

128] § 6. On the Leyden vial.

The shock produced by the Leyden vial seems owing only to the great quantity of redundant fluid collected on its positive side, and the great deficiency on its negative side; so that if a conductor was prepared of so great a size, as to be able to receive as much additional fluid by the same degree of electrification as the positive side of a Leyden vial, and was positively electrified in the same degree as the vial, I do not doubt but what as great a shock would be produced by making a communication between this conductor and the ground, as between the two surfaces of the Leyden vial, supposing both communications to be made by canals of the same length and same kind.

It appears plainly from the experiments which have been made on this subject, that the electric fluid is not able to pass through the glass; but yet it seems as if it was able to penetrate without much difficulty to a certain small depth, perhaps I might say an imperceptible depth within the glass; as Dr Franklin's analysis of the Leyden vial shews that its electricity is contained chiefly in the glass itself, and that the coating is not greatly over or undercharged.

It is well known that glass is not the only substance which can be charged in the manner of the Leyden vial; but that the same effect may be produced by any other body, which will not suffer the electricity to pass through it.

129] *Hence the phænomena of the vial seem easily explicable by means of the 22nd proposition. For let *ACGM*, Fig. 20, represent a flat plate of glass or any other substance which will not suffer the electric fluid to pass through it, seen edgeways; and let *BbdD*, and *EefF*, or *Bd* and *Ef*, as I shall call them for shortness, be two plates of conducting matter of the same size, placed in contact with the glass opposite to each other; and let *Bd* be positively electrified; and let *Ef* communicate with the ground; and let the fluid be supposed either able to enter a little way into the glass, but not to pass through it, or unable to enter it at all; and if it is able to enter a little way into it, let $b\beta\delta d$, or $b\delta$, as I shall call it, represent that part of the glass into which the fluid can enter from the plate *Bd*, and $e\phi$, that which the fluid from *Ef* can enter. By the abovementioned proposition, if *be*, the thickness of the glass, is very small in respect of *bd*, the diameter of the plates, the quantity of redundant fluid forced into the space *Bd*, or $B\delta$, (that is, into the plate *Bd*, if the fluid is unable to penetrate at all into the glass, or into the plate *Bd*, and the space $b\delta$ together, if the fluid is able to penetrate into the glass,)

Fig. 20.

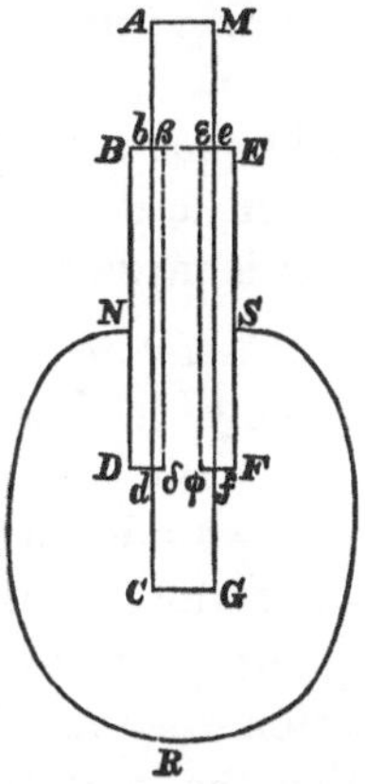

* The following explication is strictly applicable only to that sort of Leyden vial, which consists of a flat plate of glass or other matter. It is evident, however, that the result must be nearly of the same kind, though the glass is made into the shape of a bottle as usual, or into any other form; but I propose to consider those sort of Leyden vials more particularly in a future paper.

will be many times greater than what would be forced into it by the same degree of electrification if it had been placed by itself; and the quantity of fluid driven out of $E\phi$ will be nearly equal to the redundant fluid in $B\delta$.

If a communication be now made between $B\delta$ and $E\phi$, by the canal *NRS*, the redundant fluid will run from $B\delta$ to $E\phi$; and if in its way it passes through the body of any animal, it will by the rapidity of its motion produce in it that sensation called a shock.

130] It appears from the 26th proposition, that if a body of any size was electrified in the same degree as the plate *Bd*, and a communication was made between that body and the ground, by a canal of the same length, breadth and thickness as *NRS*; that then the fluid in that canal would be impelled with the same force as that in *NRS*, supposing the fluid in both canals to be incompressible; and consequently, as the quantity of fluid to be moved, and the resistance to its motion is the same in both canals, the fluid should move with the same rapidity in both: and I see no reason to think that the case will be different, if the communication is made by canals of real fluid.

Therefore what was said in the beginning of this section, namely, that as great a shock would be produced by making a communication between the conductor and the ground, as between the two sides of the Leyden vial, by canals of the same length and same kind, seems a necessary consequence of this theory; as the quantity of fluid which passes through the canal is, by the supposition, the same in both; and there is the greatest reason to think, that the rapidity with which it passes will be nearly if not quite the same in both. I hope soon to be able to say whether this agrees with experiment as well as theory.

131] It may be worth observing, that the longer the canal *NRS* is, by which the communication is made, the less will be the rapidity with which the fluid moves along it; for the longer the canal is, the greater is the resistance to the motion of the fluid in it; whereas the force with which the whole quantity of fluid in it is impelled, is the same whatever be the length of the canal. Accordingly, it is found in melting small wires, by directing a shock through them, that the longer the wire the greater charge it requires to melt it.

132] As the fluid in $B\delta$ is attracted with great force by the redundant matter in $E\phi$, it is plain that if the fluid is able to penetrate at all into the glass, great part of the redundant fluid will be lodged in $b\delta$, and in like manner there will be a great deficience of fluid in $e\phi$. But in order to form some estimate of the proportion of the redundant fluid which will be lodged in $b\delta$, let the communication between *Ef* and the ground be taken away, as well as that by which *Bd* is electrified; and let so much fluid be taken from $B\delta$, as to make the redundant fluid therein equal to the deficient fluid in $E\phi$. If we suppose that all the redundant fluid is collected in $b\delta$, and all the deficient in $e\phi$, so as to leave *Bd* and *Ef* saturated; then, if the electric repulsion is inversely as the square of the distance, a particle of fluid placed anywhere in the plane *bd*, except near the extremities *b* and *d*, will be attracted with very near as much force by the redundant matter in $e\phi$, as it is repelled by the redundant fluid in $b\delta$; but if the repulsion is inversely as some higher power than the square, it will be repelled with much more force by $b\delta$, than it is attracted by $e\phi$, provided the depth $b\beta$ is very small in respect of the thickness of the glass; and if the repulsion is inversely as some lower power than the square, it will be attracted with much more force by $e\phi$, than it is repelled by $b\delta$. Hence it follows, that if the depth to which the fluid can penetrate is very small in respect of the thickness of the glass, but yet is such that the quantity of fluid naturally contained in $b\delta$, or $e\phi$, is considerably more than the redundant fluid in $B\delta$; then, if the repulsion is inversely as the square of the distance, almost all the redundant fluid will be collected in $b\delta$, leaving the plate *Bd* not very much overcharged; and in like manner *Ef* will be not very much undercharged: if the repulsion is inversely as some higher power than the square, *Bd* will be very much overcharged, and *Ef* very much undercharged: and if the repulsion is inversely as some lower power than the square, *Bd* will be very much undercharged, and *Ef* very much overcharged.

133] Suppose, now, the plate *Bd* to be separated from the plate of glass, still keeping it parallel thereto, and opposite to the same part of it that it before was applied to; and let the repulsion of the particles be inversely as some higher power of the distance than the square. When the plate is in contact with the glass, the repulsion of the redundant fluid in that plate, on a particle in the plane *bd*, *id est*, the inner surface of the plate, must be equal to the excess of the repulsion of the redundant fluid in $b\delta$ on it, above the attraction of $E\phi$ on it; therefore, when the plate *Bd* is removed ever so small a distance from the glass, the repulsion of the redundant fluid in the plate, on a particle in the inner surface of that plate, will be greater than the excess of the repulsion of $b\delta$ on it, above the attraction of $E\phi$; for the repulsion of $b\delta$ will be much more diminished by the removal, than the attraction of $E\phi$: consequently, some fluid will fly from the plate to the glass, in the form of sparks: so that the plate will not be so much overcharged when removed from the glass, as it was when in contact with it. I should imagine, however, that it would still be considerably overcharged.

If one part of the plate is separated from the glass before the rest, as must necessarily be the case, if it consists of bending materials, I should guess it would be at least as much, if not more, overcharged, when separated, as if it is separated all at once.

In like manner, it should seem that the plate *Ef* will be considerably undercharged, when separated from the glass, but not so much so as when in contact with it.

From the same kind of reasoning I conclude, that if the repulsion is inversely as some lower power of the distance than the square, the plate *Bd* will be considerably undercharged, and *Ef* considerably overcharged, when separated from the glass, but not in so great a degree as when they are in contact with it.

134] § 7. There is an experiment of Mr. Wilke and Æpinus, related by Dr. Priestly, p. 258, called by them, electrifying a plate of air: it consisted in placing two large boards of wood, covered with tin plates, parallel to each other, and at some inches asunder. If a communication was made between one of these and the ground, and the other was positively electrified, the former was undercharged; the boards strongly attracted each other; and, on making a communication between them, a shock was felt like that of the Leyden vial.

I am uncertain whether in this experiment the air contained between the two boards is very much overcharged on one side, and very much undercharged on the other, as is the case with the plate of glass in the Leyden vial; or whether the case is, that the redundant or deficient fluid is lodged only in the two boards, and that the air between them serves only to prevent the electricity from running from one board to the other: but whichever of these is the case, the experiment is equally conformable to the theory *.

It must be observed, that a particle of fluid placed between the two plates is drawn towards the undercharged plate, with a force exceeding that with which it would be repelled from the overcharged plate, if it was electrified with the same force, the other plate being taken away, nearly in the ratio of twice the quantity of redundant fluid actually contained in the plate, to that which it would contain, if electrified with the same force by itself; so that, unless the plate is very weakly electrified, or their distance is very considerable, the fluid will be apt to fly from one to the other, in the form of sparks.

135] § 8. Whenever any conducting body as *A*, communicating with the ground, is brought sufficiently near an overcharged body *B*, the electric fluid is apt to fly through the air from *B* to *A*, in the form of a spark: the way by which this is brought about seems to be this. The fluid placed anywhere between the two bodies, is repelled from *B* towards *A*, and will consequently move slowly through the air from one to the other: now it seems as if this motion increased the elasticity of the air, and made it rarer: this will enable the fluid to flow in a swifter current, which will still further increase the elasticity of the air, till at last it is so much rarified, as to form very little opposition to the motion of the electric fluid, upon which it flies in an uninterrupted mass from one body to the other.

In the same manner may the electric fluid pass from one body to another, in the form of a spark, if the first body communicates with the ground, and the other body is negatively electrified, or in any other case in which one body is strongly disposed to part with its electricity to the air, and the other is strongly disposed to receive it.

136] In like manner, when the electric fluid is made to pass

[* See Articles 344, 345, 511, 516.]

through water, in the form of a spark, as in Signor Beccaria's* and Mr. Lane's† experiments, I imagine that the water, by the rapid motion of the electric fluid through it, is turned into an elastic fluid, and so much rarified as to make very little opposition to its motion: and when stones are burst or thrown out from buildings struck by lightning, in all probability that effect is caused by the moisture in the stone, or some of the stone itself, being turned into an elastic fluid.

137] It appears plainly, from the sudden rising of the water in Mr. Kinnersley's electrical air thermometer‡, that when the electric fluid passes through the air, in the form of a spark, the air in its passage is either very much rarified, or intirely displaced: and the bursting of the glass vessels, in Beccaria's and Lane's experiments, shews that the same thing happens with regard to the water, when the electric fluid passes through it in the form of a spark. Now, I see no means by which the displacing of the air or water can be brought about, but by supposing its elasticity to be increased, by the motion of the electric fluid through it, unless you suppose it to be actually pushed aside, by the force with which the electric fluid endeavours to issue from the overcharged body: but I can by no means think, that the force with which the fluid endeavours to issue, in the ordinary cases in which electric sparks are produced, is sufficient to overcome the pressure of the atmosphere, much less that it is sufficient to burst the glass vessels in Beccaria's and Lane's experiments.

138] The truth of this is confirmed by Prop. XVI. For, let an undercharged body be brought near to, and opposite to the end of a long cylindrical body communicating with the ground, by that proposition the pressure of the electric fluid against the base of the cylinder is scarcely greater than the force with which the two bodies attract each other, provided that no part of the cylinder is undercharged; which is very unlikely to be the case, if the electric repulsion is inversely as the square of the distance, as I have great reason to believe it is; and, consequently, if the spark was produced by the air being pushed aside by the force with which the fluid endeavours to issue from the cylinder, no sparks

* *Elettricismo artificiale e naturale*, p. 110. Priestly, p. 209.

† *Phil. Trans.* 1767, p. 451.

‡ *Phil. Trans.* 1763, p. 84. Priestly, p. 216.

should be produced, unless the electricity was so strong, that the force with which the bodies attracted each other was as great as the pressure of the atmosphere against the base of the cylinder: whereas it is well known, that a spark may be produced, when the force, with which the bodies attract, is very trifling in respect of that *.

139] One may frequently observe, in discharging a Leyden vial, that if the two knobs are approached together very slowly, a hissing noise will be perceived before the spark; which shews, that the fluid begins to flow from one knob to the other, before it passes in the form of a spark; and therefore serves to confirm the truth of the opinion, that the spark is brought about in the gradual manner here described.

[* Note 10.]

PRELIMINARY PROPOSITIONS*.

In this and all the following propositions and lemmata the electric attraction and repulsion is supposed to be inversely as the square of the distance.

140] PROP. XXIX. Let a thin circular plate be connected to a globe [of the same diameter] placed at an infinite distance from it by a straight canal of incompressible fluid such as is described in Pr. XIX., perpendicular to the plane of the plate and meeting it in its center, and let them be overcharged.

If we suppose that part of the redundant fluid in the plate is spread uniformly, and that the remainder is disposed in its circumference, and that the part which is spread uniformly is to that which is disposed in the circumference as p to one, the quantity of redundant fluid in the plate will be to that in the globe as $p+1$ to $2p+1$.

For by Prop. XXII., Cor. v., the force with which that part of the redundant fluid in the plate which is disposed in the circumference repels the fluid in the canal is the same with which an equal quantity placed in the globe repels it in the contrary direction, and the repulsion of that part which is spread uniformly is the same as that of twice that quantity placed in the globe, and therefore the repulsion of a quantity of fluid equal to $p+1$ disposed in the plate as expressed in the proposition is equal to that of the quantity $2p+1$ placed in the globe.

141] PROP. XXX. Fig. 1. Let two equal thin circular plates AB and ab communicate with each other, and also with a third circular plate

Fig. 1.

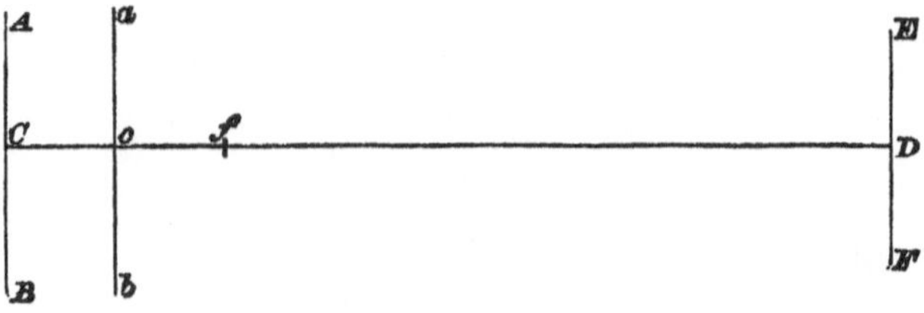

[* Hitherto unpublished.]

EF of the same size and shape and placed at an infinite distance from them, by the straight canal *CD* of incompressible fluid. Let the three plates be all parallel to each other and be placed so that *CD* shall pass through their centers and be perpendicular to their planes, and let the plates be overcharged. The quantity of redundant fluid in each of the plates *AB* and *ab* will be to that in *EF* as the repulsion of the plate *ab* on the canal *cD* to the sum of the repulsions on *cD* and *fD* (*cf* being taken equal to *cC*), supposing that the redundant fluid in all three plates is disposed in the same manner.

For first, as the plates *AB* and *ab* are at an infinite distance from any other over or undercharged body, the repulsion of *AB* on the canal *Cc* in one direction must be equal to that of *ab* on it in the contrary, and therefore the redundant fluid in *AB* must be equal to that in *ab*.

Secondly, the sum of the repulsions of *AB* and *ab* on the canal *cD* must be equal to that of *EF* on it in the contrary direction, as otherwise some fluid must flow from *ab* to *EF* or from *EF* to *ab*. But as all three plates are of the same size, and the fluid in them is disposed in the same manner, the repulsions of *EF* and *ab* on *cD* will be to each other as the quantity of redundant fluid in them, and therefore the quantity of redundant fluid in *ab* will be to that in *EF* as the repulsion of *ab* on *CD* to the sum of the repulsions of *AB* and *ab* on it, that is, as the repulsion of *ab* on *cD* to the sum of its repulsions on *fD* and *cD*, for the repulsion of *AB* on *cD* is equal to the repulsion of *ab* on *fD**.

142] Cor. I. If the fluid in these plates is disposed in the same manner as in Prop. XXIX. the quantity of redundant fluid in each of the plates *AB* and *ab* will be to that in *EF* as

$$AC\,(p+\tfrac{1}{2}) \text{ to } AC\,(p+\tfrac{1}{2}) + p\,(Ac - Cc) + \frac{AC^2}{2Ac}.$$

For by Lemma X. the repulsion of a given quantity of fluid spread uniformly over *ab* on the column *cD*; the repulsion of the same fluid on *cf*; the repulsion of the same quantity of fluid collected in the circumference of the plate *ab* on the column *cD*; and the repulsion of the same fluid on *cf* are to each other as ac; $ac + cf - af$; $\frac{ac}{2}$ and $\frac{ac}{2} - \frac{ac}{2af}$, and therefore the whole repulsion of the plate *ab* on *cD* is to its repulsion on *cf* as

$$p \times ac + \frac{ac}{2} : p \times (ac + cf - af) + \frac{ac}{2} - \frac{ac^2}{2af},$$

and therefore the repulsion of *ab* on *cD* is to the sum of its repulsions on *cD* and *fD* as

$$ac \times (p+\tfrac{1}{2}) : ac\,(2p+1) - p\,(ac + cf - af) - \frac{ac}{2} + \frac{ac^2}{2af},$$

or as
$$ac\,(p+\tfrac{1}{2}) : ac\,(p+\tfrac{1}{2}) + p\,(af - cf) + \frac{ac^2}{2af}.$$

[* Note 11].

143] COR. II. Therefore if all the redundant fluid in the plates is spread uniformly, the redundant fluid in each of the plates AB and ab will be to that in EF as $AC : AC + Ac - Cc$, and if it is all collected in the circumference, as $AC : AC + \frac{AC^2}{Ac}$.

144] COR. III. By Prop. XXIV. it appears that the redundant fluid in the plate AB or ab will bear the same proportion to that in EF though they communicate with EF by separate canals, and whether the canals by which they communicate with it are straight or crooked, or in whatever direction EF is placed in respect of them, provided the situation of AB and ab in respect of each other remains the same. Only it must be observed that if the fluid in the plates is not disposed so as to be in equilibrio, as will most likely be the case if it is disposed as in the two preceding corollaries, it is necessary that the canals should meet them in their centers, for if the fluid in a plate is not in equilibrio, its repulsion on a canal of infinite length will not be the same in whatever part the canal meets it, as it will if the fluid in the plate is in equilibrio.

145] LEMMA XII. Fig. 2. Let BA be an infinitely slender cylindric column of uniform matter infinitely continued beyond A: the

Fig. 2.

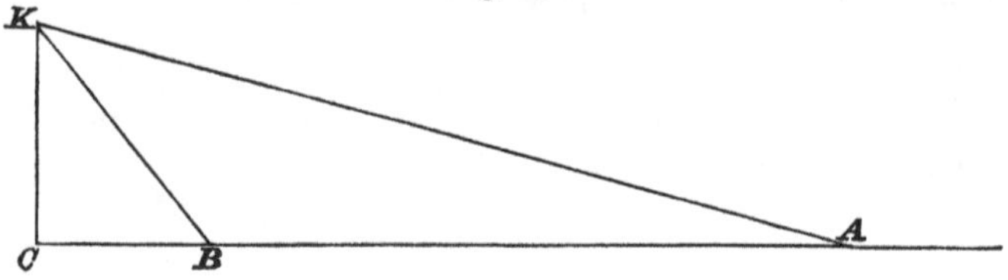

repulsion of a particle of matter K on this column in the direction BA is proportional to or may be represented by $\frac{1}{KB}$, supposing the size of the particle and [the] base of the column to be given.

For draw KC perpendicular to AB continued, and let the point B flow towards C, the fluxion of the repulsion of K on the column equals $\frac{-C\dot{B}}{KB^2} \times \frac{CB}{KB} = \frac{-K\dot{B}}{KB^2}$, the fluent of which, $\frac{1}{KB}$, is nothing when KB is infinite.

146] LEMMA XIII. Suppose now KC to represent an infinitely slender cylindric column of uniform matter: the repulsion of KC on the infinite column BA is to the repulsion of the same quantity of matter collected in the point C on the same column as the nat. log. of $\frac{KC + KB}{CB}$ to $\frac{KC}{CB}$.

For the repulsion of all the matter therein, when collected at C, on BA is proportional to $\frac{KC}{CB}$, and supposing the column CK to flow, the fluxion of its repulsion on BA is equal to $\frac{C\dot{K}}{KB}$, the fluent of which is the nat. log. of $\frac{KC + KB}{CB}$, and is nothing when CK is nothing.

147] **Lemma XIV.** The repulsion of CK on a particle at B, in the direction CB, is proportional to $\frac{CK}{KB \times CB}$, supposing the base of CK and the size of the particle B to be given.

For supposing CK to flow, the fluxion of its repulsion on B in the direction CB is proportional to $\frac{C\dot{K}}{KB^2} \times \frac{CB}{KB}$, the fluent of which is $\frac{CK}{KB \times CB}$, and is nothing when CK is nothing.

148] **Lemma XV.** Fig. 3. Let $GEFHMN$ be a cylinder whose bases are GEF and HMN and whose axis is CK. Let the convex

Fig. 3.

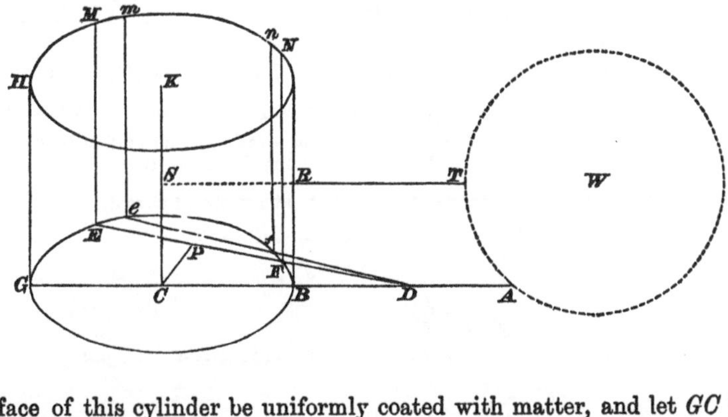

surface of this cylinder be uniformly coated with matter, and let GC be small in respect of CK. Let GA be a diameter of the base produced, and D any point therein. The repulsion of the convex surface of the cylinder on the point D in the direction CD is very nearly the same as if all the matter therein was collected in the axis CK and spread uniformly therein.

For let MED and med be two planes infinitely near to each other, parallel to CK and passing through D, and cutting the convex surface in ME and NF and in me and nf, which will consequently be right lines equal to each other and perpendicular to ED; and draw CP perpendicular to ED.

The repulsion of $NnfF$ on D in the direction CD is proportional to $\frac{Ff \times FN}{FD \times ND} \times \frac{PD}{CD}$, and that of $MmeE$ is proportional to $\frac{Ee \times EM}{ED \times MD} \times \frac{PD}{CD}$.

But Ff is to Ee as FD to ED, therefore $\frac{Ff}{FD}$ and $\frac{Ee}{ED}$ are each equal to $\frac{Ff + Ee}{FD + ED} = \frac{Ff + Ee}{2PD}$, therefore the sum of the repulsions of $MmeE$ and $NnfF$ is proportional to

$$\frac{(Ff + Ee)\, CK \times PD}{2PD \times CD} \times \left(\frac{1}{ND} + \frac{1}{MD}\right) = \frac{(Ff + Ee)\, CK}{2CD} \times \left(\frac{1}{ND} + \frac{1}{MD}\right).$$

But the repulsion of the same quantity of matter collected in CK is proportional to $\frac{(Ff + Ee) \times CK}{2CD} \times \frac{2}{KD}$, and, as CG is small in respect of CK, $\frac{1}{ND} + \frac{1}{MD}$ differs very little from $\frac{2}{KD}$ *, therefore the sum of the repulsions of $MmeE$ and $NnfF$ is very nearly the same as if all the matter in them was collected in CK, and consequently the repulsion of the whole convex surface of the cylinder will be very nearly the same as if all the matter in it was collected in CK.

149] Cor. Therefore if BA represents an infinitely thin cylindric column of uniform matter infinitely extended beyond A, the repulsion of the convex surface of the cylinder thereon in the direction BA is very

* As neither MD nor ND differ from KD by so much as CB, it is plain that $\frac{1}{MD} + \frac{1}{ND}$ cannot differ from $\frac{2}{KD}$ in so great a proportion as that of BC to KD, but in reality it does not differ from it in so great a ratio as that of CB^2 to KD^2, but as it is not material being so exact, I shall omit the demonstration. See A. 1.

[From MS. "A. 1"] Demonstration of note at bottom of page 8,

$$CB = r,\ CP = b,\ PF = d,\ PD = a,\ CR^2 + CD^2 = e^2,$$

$$b^2 - d^2 = f^2, \qquad e^2 - f^2 = g^2,$$

$$\frac{2}{g} = \frac{2}{e} + \frac{f^2}{e^3},$$

$$ND^2 = CR^2 + a^2 - 2ad + d^2 = e^2 - b^2 - 2ad + d^2 = e^2 - f^2 - 2ad$$

$$= g^2 - 2ad,$$

$$MD^2 = g^2 + 2ad,$$

$$\frac{1}{ND} + \frac{1}{MD} = \left\{\begin{matrix} \frac{1}{g} + \frac{ad}{g^3} + 3\frac{a^2d^2}{2g^5} \\ \frac{1}{g} - \frac{ad}{g^3} + 3\frac{a^2d^2}{2g^5} \end{matrix}\right\} = \frac{2}{g} + \frac{3a^2d^2}{g^5}$$

$$= \frac{2}{e} + \frac{f^2}{e^3} + 3\frac{a^2d^2}{g^5} = \frac{2}{e} + \frac{b^2}{e^3} + 3\frac{a^2d^2}{g^5} - \frac{d^2}{c^3},$$

which is less than

$$\frac{2}{e} + \frac{b^2}{e^3} + 2\frac{d^2}{e^3} = \frac{2}{e} + \frac{r^2 + d^2}{e^3}.$$

nearly the same as if all the matter therein was collected in CK, and therefore is to the repulsion of the same quantity of matter collected in the point C thereon very nearly as nat. log. $\frac{CK+KB}{CB}$ to $\frac{CK}{CB}$, that is very nearly as nat. log. $\frac{2CK}{CB}$ to $\frac{CK}{CB}$. In like manner the repulsion on the infinite column DA is to the repulsion of the same quantity of matter collected in C very nearly as nat. log. $\frac{CK+KD}{CD}$ to $\frac{CK}{CD}$.

150] PROP. XXXI. Fig. 3. Let the cylinder $GEFKMN$ be connected to the globe W, whose diameter is equal to GB and whose distance from it is infinite, by a canal TR of incompressible fluid of any shape, and meeting the cylinder in any part, and let them be overcharged: the quantity of redundant fluid in the cylinder will be to that in the globe in a less ratio than that of CK to nat. log. $\frac{2CK}{CB}$, and in a greater ratio than that of $\frac{CK}{2CB}$ to nat. log. $\frac{CK}{CB}$, provided CB is small in respect of CK.

By Prop. XXIV. the quantity of redundant fluid in the cylinder will bear the same proportion to that in the globe in whatever part the canal meets the cylinder, therefore first I say the redundant fluid in the cylinder will bear a greater proportion to that in the globe than that of $\frac{CK}{2CB}$ to nat. log. $\frac{CK}{CB}$.

For let the canal TR be straight and perpendicular to BL, and let it meet the cylinder in R, the middle point of the line BL, and let it, if produced, meet the axis in S, which will consequently be the middle point of CK; then, if the redundant fluid in the cylinder was spread uniformly on its convex surface, the quantity of redundant fluid therein would be to that in the globe very nearly as $\frac{CK}{2CB}$ to nat. log. $\frac{CK}{CB}$.

For in that case the repulsion of the cylinder on the canal RT would be to the repulsion of the same quantity of redundant fluid collected in C very nearly as nat. log. $\frac{2SK}{SR}$ to $\frac{SK}{SR}$ or as nat. log. $\frac{CK}{CB}$ to $\frac{CK}{2CB}$, and the force with which the globe repels the canal in the direction TR is the same with which a quantity of redundant fluid equal to that in the globe placed at S would repel it in the contrary direction.

But there can be no doubt but that almost all the redundant fluid in the cylinder will be collected on its surface, and also will be collected in greater quantity near the ends than near the middle, consequently the repulsion of the cylinder on RT will be less than if the redundant fluid was spread uniformly on its convex surface, and therefore the quantity of redundant fluid in it will bear a greater proportion to that in the globe than it would on that supposition.

Secondly, the quantity of fluid in the cylinder will bear a less proportion to that in the globe than that of $\frac{CK}{CB}$ to nat. log. $\frac{2CK}{CB}$.

For suppose the canal to meet the cylinder in B and to coincide with BA. Then, if the redundant fluid was spread uniformly on the convex surface, the quantity therein would be to that in the globe very nearly as $\frac{CK}{CB}$ to nat. log. $\frac{2CK}{CB}$, and the real quantity of redundant fluid in it will bear a less proportion to that in the globe than if it was spread uniformly on the convex surface.

151] Cor. Therefore the quantity of redundant fluid in the cylinder is to that in a globe whose diameter equals CK in a ratio between that of 2 to nat. log. $\frac{2CK}{CB}$ and that of 1 to nat. log. $\frac{CK}{CB}$*.

152] Prop. XXXII. Fig. 4. Let $ADFB$ and $adfb$ be two equal cylinders whose axes are EC and ec, let them be parallel to each other

Fig. 4.

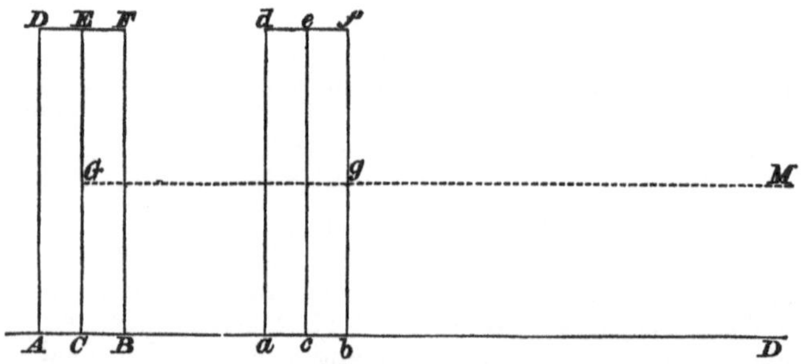

and placed so that Cc, the line joining the ends of the axes, shall be perpendicular to the axes, and let the lines EC and Fb be bisected in G and g, and let them be connected by canals of incompressible fluid of any shape to a third cylinder of the same size and shape placed at an infinite distance from them, and let them be overcharged: the quantity of redundant fluid in each of them will be to that in the third cylinder in a ratio between that of $\log\frac{EC}{CB}$ to $\log\frac{EC}{CB}+\log\frac{EG+eg}{Cb}$ and that of $\log\frac{2EC}{CB}$ to $\log\frac{2EC}{CB}+\log\frac{EC+Eb}{Cb}$, provided the redundant fluid in the third cylinder is disposed in the same manner as in the other two.

For let us suppose that $ADFB$ and $adfb$ are connected to the third cylinder by the canal GM, then, if the redundant fluid in each cylinder is disposed uniformly on its convex surface, the sum of the repulsions of $ADFB$ and $adfb$ on the canal gM will be to the repulsion of the third

[* Note 12.]

cylinder thereon (supposing the quantity of redundant fluid in it to be equal to that in each of the two others) as $\log\frac{2EG}{CB}+\log\frac{EG+Eg}{Gg}$ to $\log\frac{2EG}{CB}$.

Let us now suppose the fluid in the first two cylinders to be disposed so as to be in equilibrio, and consequently to be disposed in greater quantity near their extremities than near their middles, and let the fluid in the third cylinder be disposed in the same manner, and be the same in quantity as before. The repulsion of *ADFB* on *Gg* will be diminished in a greater ratio, and consequently its repulsion on *gM* will be diminished in a less ratio than that of *adfb* on *gM*, consequently the sum of the repulsions of *ADFB* and *adfb* on *gM* will be diminished in a less ratio than that of the third cylinder thereon, and therefore the sum of the repulsions of *ADFB* and *adfb* on *gM* will be to that of the third cylinder thereon in a greater ratio than that of

$$\log\frac{2EG}{CB}+\log\frac{EG+Eg}{Gg} \text{ to } \log\frac{2EG}{CB}.$$

Therefore the real quantity of redundant fluid in each of the first two cylinders will be to that in the third cylinder in a less ratio than that of $\log\frac{2EG}{CB}$ to $\log\frac{2EG}{CB}+\log\frac{EG+Eg}{Gg}$.

In like manner, by supposing them to be connected to the third cylinder by the canal *bD*, it may be shown that the quantity of redundant fluid in either of the first two cylinders is to that in the third in a greater ratio than that of $\log\frac{2EC}{CB}$ to $\log\frac{2EC}{CB}+\log\frac{EC+Eb}{Cb}$*.

153] Prop. XXXIII. If two bodies *B* and *b* are successively connected by canals of incompressible fluid to a third body *C* placed at an infinite distance from them, and are overcharged, that is, if one of them, as *B*, is first connected to *C* and afterwards *B* is removed and *b* put in its room, the quantity of redundant fluid in *C* being the same in both cases, it is plain that the quantity of redundant fluid in *B* will bear the same proportion to that in *b* that it would if *B* and *b* were placed at an infinite distance from each other, and connected by canals of incompressible fluid.

154] Lemma XV. Fig. 5. Let *AB* be a thin flat plate of any shape whatsoever, of uniform thickness and composed of uniform matter. Let *CG* be an infinitely slender cylindric column of uniform matter perpendicular to the plane of *AB* and meeting it in *C* and extended infinitely beyond *G*. Let *ab* be a thin circular plate perpendicular to *cG* whose center is *C*. Let the area of *ab* be equal to that of *AB*, and let the quantity of matter in it be the same, and let it be disposed uniformly.

[* Note 13.]

Fig. 5.

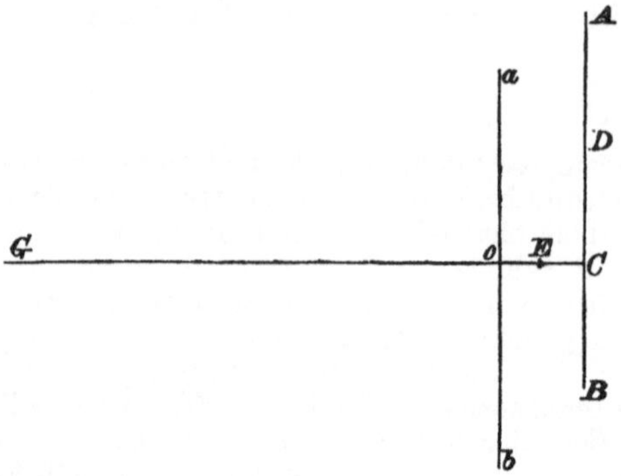

Let B be that point of the circumference of AB which is nearest to C. If EC is small in respect of CB, the repulsion of the plate AB on the short column EC is to the repulsion of ab on the infinite column cG nearly as EC to cb.

For let BD be a circle drawn through B with center C, as EC is very small in respect of CB, the repulsion of the circle BD on EC is to its repulsion on CG very nearly as EC to CB, and therefore is to the repulsion of ab on cG very nearly as EC to cb. But the repulsion of AB on EC is very little greater than that of DB, for the repulsion of DB is very near as great as it would be if its size was infinite.

155] Lemma XVI. Let ACB and DEF be two thin plates, not flat but concave on one side, let their distance be everywhere the same,

Fig. 6.

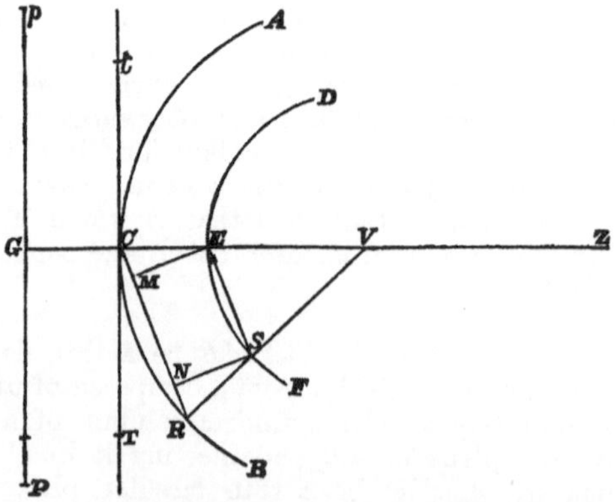

and let it be very small in respect of the radius of curvature of all parts of their surface. Let C be any point of the surface of AB, and let CE be perpendicular to the surface in that point. Let Tt be a flat plate perpendicular to CE.

Let R be any point in AB and S the corresponding point in DF, and let T be the corresponding point in Tt*: the sum of the repulsions of R on the column CE in the direction CE and of S on the same column in the opposite direction EC is very nearly equal to the force with which they would repel the same column in the direction CE if they were both transferred to T, provided CR^2 is very small in respect of the square of the least radius of curvature of the surface of AB.

Let RS be continued till it meets CE continued in V, draw EM and SN perpendicular to CR.

Let $CM = C$, $RE - RM = E$, $SC - NC = S$, and $SE - NM = D$.

As CE is very small in respect of the least radius of curvature of AB, and CV is not less than the least radius of curvature, CM and NR are each very small in respect of CR, and therefore CN, MR, and ES differ from CR in a very small ratio. Moreover as CR^2 is very small in respect of CV^2, CM^2 and RN^2 are very small in respect of CE^2, and therefore ME and NS differ in a very small ratio from CE; and, moreover, $2 \times (TE - TC)$ is greater than $\frac{CE^2}{TE}$.

Now the repulsion of the point R on the column CE in the direction CE is $\frac{1}{RC} - \frac{1}{RE}$† $= \frac{RE - RC}{RC \times RE}$, and the repulsion of the point S on the same column in the opposite direction is $\frac{SC - SE}{SC \times SE}$, and the sum of the repulsions of R and S is

$$\frac{RE - RC}{RC \times RE} + \frac{SC - SE}{SC \times SE} = \frac{E - C}{RC \times RE} + \frac{S + C - D}{SC \times SE}$$

$$= \frac{E}{RC \times RE} + \frac{S}{SC \times SE} - \frac{D}{SC \times SE} - \frac{C}{RC \times RE} + \frac{C}{SC \times SE},$$

and the repulsion of the two particles when transferred to T on the column CE, or the repulsion of T, as I shall call it for shortness, is $2\frac{TE - TC}{TE \times TC}$.

But as ME differs in a very small ratio from CE, and RM differs in a very small ratio from RC, $RE - RM$ or E differs in a very small ratio from $TE - TC$. In like manner $SC - NC$ or S differs in a very small ratio from $TE - TC$, and ER and CS both differ in a very small ratio from TE, and SE differs in a small ratio from TC.

Therefore $\frac{E}{RC \times RE} + \frac{S}{SC + SE}$ differs very little from $2 \times \frac{TE - TC}{TE \times TC}$, that is, from the repulsion of T.

* If RS is drawn perpendicular to the surface of AB at the point R cutting DF in S, I call S the corresponding point of the plate DF, and if CT is taken in the intersection of the plane RCE with that of the plate Tt equal to the right line CR, I call T the corresponding point of Tt.

† Lemma XII. [Art. 146].

Moreover, as EM and SN differ very little from each other, D is very small in respect of $TE - TC$, and $\frac{D}{SC \times SE}$ is very small in respect of the repulsion of T.

Moreover, $\frac{RC - SE}{RC}$ is less than $\frac{CM + RN}{RC}$ or than $\frac{CE}{CV}$, and $\frac{RE - SC}{RE}$ is hardly greater than $\frac{RM - CN}{RE}$, and is therefore still less than $\frac{RC - SE}{RC}$; therefore $\frac{RC}{SE}$ and $\frac{RE}{SC}$ each differ from one in a less ratio than that of CE to CV, and therefore $\frac{RC \times RE}{SE \times SC}$ differs from one in a less ratio than that of $2CE$ to CV.

Consequently, $-\frac{C}{RC \times RE} + \frac{C}{SE \times SC}$ or $\frac{-C}{RC \times RE} \times \left(1 - \frac{RC \times RE}{SE \times SC}\right)$ is less than $\frac{-C}{RC \times RE} \times \frac{2CE}{CV}$, which is less than

$$\frac{CE \times RC}{CV \times RC \times RE} \times \frac{2CE}{CV} = \frac{2CE^2}{CV^2 \times RE},$$

which is very small in respect of $\frac{2CE^2}{TE \times TC \times RE}$, that is, of the repulsion of T.

Therefore the sum of the repulsions of R and S differs very little from the repulsion of T.

N.B. Though the distance CR is ever so great, it may be shown that the sum of the repulsions of R and S cannot be more than double that of T *.

156] Cor. I. Let the edges of the plates ACB and DEF correspond, that is, let them be such that if a line is erected on any part of the circumference of one plate perpendicular to the [tangent] plane of the plate in that part, that line shall meet the other plate in its circumference. Let the two plates be of an uniform thickness, and let the thickness of DF bear such a proportion to that of AB that the quantity of matter shall be the same in both. Consequently the quantity of matter in each part of DF will be very nearly equal to that in the corresponding part of AB. Also let the size of the plates be such that CE shall be very small in respect of the distance of C from the nearest part of the circumference of AB, and let the least radius of curvature of the surface of AB be so great in respect of CE that a point R may be taken such that CR shall be small in respect of that radius of curvature, and yet very great in respect of CE.

Let Pp be a flat circular plate whose center is G and whose plane is perpendicular to GZ, and let its area be equal to that of AB, and let the quantity of matter in it be also equal to that in AB, and let it be

[* Note 14.]

disposed uniformly: the sum of the repulsions of AB and DF on CE in the opposite directions CE and EC will be to the repulsion of Pp on the infinite column GZ very nearly as $2CE$ to GP.

For suppose each particle of matter in all that part of AB whose distance from C is not greater than CR and in the corresponding part of DF to be transferred to its corresponding point in Tt, so as to form a circular plate whose radius is CR.

If we suppose that the thickness of the plates Tt and Pp are both equal to that of AB, the matter in all parts of Tt will be very nearly twice as dense as that in AB or as that in Pp. Therefore the repulsion of Tt on CE will be very nearly twice the repulsion of Pp on Gg, supposing Gg to be equal to CE.

But from the foregoing lemma it appears that the sum of the repulsions which the above-mentioned part of AB and DF exerted on CE before the matter was transferred is very nearly equal to that which Tt exerts thereon after the matter is transferred, and the sum of the repulsions of the remaining part of AB and DF, or that whose distance from C is greater than CR, is very small in respect of that part whose distance is less, therefore the sum of the repulsions of the whole plates AB and DF on CE is to the repulsion of Pp on GZ very nearly as $2CE$ to GP.

It may perhaps be supposed from this demonstration that it would be necessary that CE should be excessively small in respect of CV, in order that the sum of the repulsions of the plates on CE should be very nearly equal to the repulsion of Pp on Gg, but in reality this seems not to be the case, for if the plates are segments of concentric spheres whose center is V, the sum of their repulsions will exceed twice the repulsion of Pp on Gg in a not much greater ratio than that of $1 + \frac{CE}{CV}$ to 1, and if the radius of curvature of their surfaces is in some places greater than CV, and nowhere less, I should think that the sum of their repulsion could hardly exceed twice the repulsion of Pp in so great a ratio as that.

157] Cor. II. If we now suppose that the matter of the plate AB is denser near the circumference than near the point C, and that the density at and near C is to the mean density (or the density which it would everywhere be of if the matter was spread uniformly) as δ to one, and that the quantity of matter in each part of DF is equal to that in the corresponding part of AB as before, the sum of the repulsions of the plates on CE will be less than if the matter was spread uniformly in a ratio approaching much nearer to that of δ to one than to that of equality.

For if any particle of matter is removed from that part of AB which is near C to that point which is at a distance from it, and an equal alteration is made in the plate DF, the sum of the repulsions of these particles will be much less after their removal than before.

158] LEMMA XVII. Fig. 7. Let ACB be a thin plate, not flat but concave on one side, let the radius of curvature of its surface be

Fig. 7.

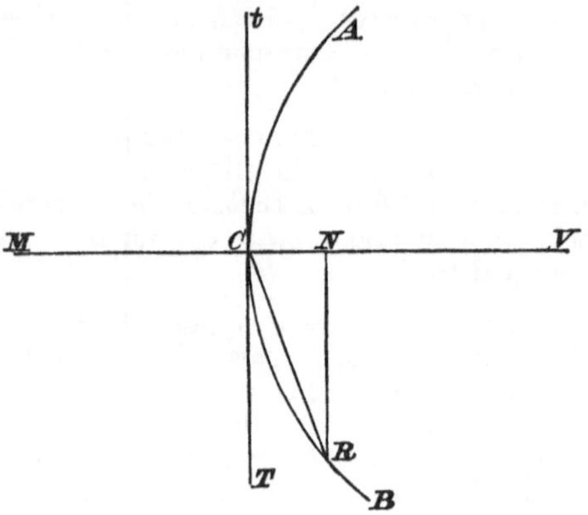

nowhere less than CV, and let MV be perpendicular to its surface at C; let MC be very small in respect of CV, and let Tt be a plane perpendicular to MC: the difference of the repulsion of any particle of matter as R in the plate ACB on the point M in the direction CM, and of its repulsion on the point C in the same direction, is very nearly the same as if the particle was transferred to T (CT being equal to the right line CR), provided CR is small in respect of CV.

Draw RN perpendicular to MC, the difference of the repulsions of R on the points M and $C = \frac{MN}{MR^3} - \frac{CN}{CR^3} = \frac{MC}{MR^3} + \frac{CN}{MR^3} - \frac{CN}{CR^3}$, and the difference of the repulsions of the same particle placed at T on the same points $= \frac{MC}{MT^3}$, but

$$\begin{aligned} MR^2 &= (MC + CN)^2 + RN^2 \\ &= MC^2 + CR^2 + 2MC \times CN \\ &= MT^2 + 2MC \times CN, \end{aligned}$$

and CN is not greater than $\frac{CR^2}{2CV}$, and therefore $2MC \times CN$ is not greater than $\frac{MC \times CR^2}{CV}$, and therefore is very small in respect of CR^2 or MT^2.

Therefore MR^2 differs very little from MT^2, and $\frac{1}{MR^3}$ from $\frac{1}{MT^3}$.

This being premised there are two cases to be considered.

First, if CR is considerably greater than MC, as

$$CR^2 = MR^2 - MC^2 - 2MC \times CN = MR^2 \times \left\{1 - \frac{MC\,(MC + 2CN)}{MR^2}\right\},$$

$\frac{1}{CR^3}$ differs not much from $\frac{1}{MR^3} \times \left\{1 - \frac{3}{2}\frac{MC(MC+2CN)}{MR^2}\right\}$,

and $\frac{CN}{MR^3} - \frac{CN}{CR^3}$ differs not much from $\frac{CN}{MR^3} \times \frac{-3MC(MC+2CN)}{2MR^2}$,

or from $\frac{MC}{MR^3} \times \frac{-3CN(MC+2CN)}{2MR^2}$, which is very small in respect of $\frac{MC}{MT^3}$, provided CR is small in respect of CV.

For as CN is less than $\frac{CR^2}{2CV}$, $\frac{3CN(MC+2CN)}{2MR^2}$ is less than $\frac{3CR^2 \times MC}{4MR^2 \times CV} + \frac{3CR^4}{4MR^2 \times CV^2}$, or than $\frac{3MC}{CV} + \frac{3CR^2}{4CV^2}$.

Therefore as $\frac{CN}{MR^3} - \frac{CN}{CR^3}$ is very small in respect of $\frac{MC}{MT^3}$, and as $\frac{MC}{MR^3}$ differs very little from $\frac{MC}{MT^3}$, $\frac{MC}{MR^3} + \frac{CN}{MR^3} - \frac{CN}{CR^3}$, or the difference of the repulsions of R on the points M and C differs very little from $\frac{MC}{MT^3}$, the difference of the repulsions of T on the same points.

Secondly, if CR is not considerably greater than MC, CN must be very small in respect of CR, and consequently must be very small in respect of MC. Therefore $\frac{CN}{CR^3} - \frac{CN}{MR^3}$ is very small in respect of $\frac{CN}{MR^3}$, and therefore the difference of the repulsions of R on C and M differs very little from $\frac{MC}{MT^3}$.

159] Cor. Therefore by the same method of reasoning as was used in Cor. to Lemma XVI., the difference of the repulsions of the whole plate ACB on the points M and C is very nearly the same as if each particle of matter in it was transferred to the plane Tt and placed at the same distance from C as before, and therefore its repulsion on M is very nearly equal to its repulsion on C, provided MC is very small in respect of the least distance of the circumference of the plate from C, and that the thickness of the plate is everywhere very nearly the same, except at such a distance from C as is very great in respect of MC.

160] Prop. XXXIV.* Fig. 8. Let $NnvV$ be a plate of glass or any other substance which does not conduct electricity, of uniform thickness, either flat, or concave on one side and convex on the other, and let the electric fluid be unable to penetrate at all into the glass or to move within it.

Let ACB and DEF be thin coatings of metal, or any substance which conducts electricity, applied to the glass.

* This proposition is nearly the same as Prop. XXII., only made more general.

Fig. 8.

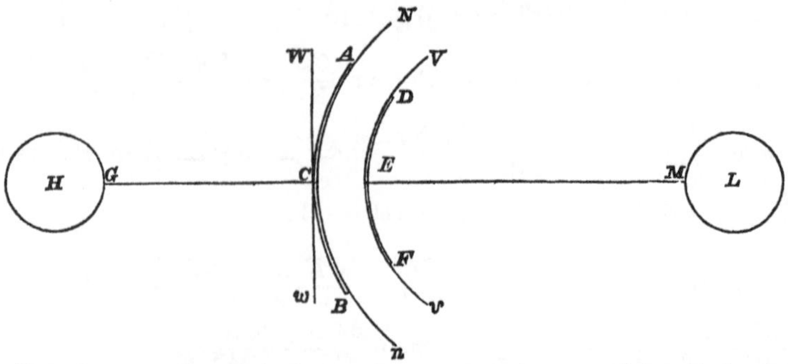

Let these coatings be of any shape whatsoever, and let their edges correspond as in Lemma XVI. Cor. I.

Let AB communicate with the body H, and DF with the body L, by the straight canals CG and EM of incompressible fluid.

Let the points C and E be so placed that the two canals shall form one right line perpendicular to AB at the point C, and let the lengths of these canals be so great that the repulsion of the coatings on the fluid in them shall be not sensibly less than if they were infinite, and let H be overcharged and let L be saturated.

It is plain from Prop. XII. that DF will be undercharged, and that AB will be more overcharged than it would otherwise be.

Let Ww be a thin flat circular plate whose center is C, perpendicular to CE, and whose area is equal to that of AB, let the force with which the redundant fluid in AB would repel the short column CE (if ME was continued to C) be called m, and let the force with which it would repel CM, or with which it repels CG (for they are both alike), be called M. Let the force with which the same quantity of redundant fluid disposed in DF, in the same manner in which the deficient fluid therein is actually disposed, would repel $\begin{Bmatrix} CE \\ EG \end{Bmatrix}$ be called $\dfrac{g}{G}$, let the force with which the same quantity of redundant fluid uniformly disposed on Ww would repel CG be called W, and let the force with which H repels CG be the same with which a quantity of fluid, which we will call B, uniformly distributed on Ww would repel it in the contrary direction: then will the quantity of redundant fluid in AB be $B \times \dfrac{GW}{Mg + Gm - mg}$, which, if M and G are very nearly alike, and m and g are very small in respect of G, differs very little from $\dfrac{BW}{g + m}$, and the deficient fluid in DF will be to the redundant fluid in AB as $M - m$ to G, and therefore on the same supposition will be very nearly equal to it.

For the force with which AB repels the fluid in EM must be equal to that with which DF attracts it, for otherwise some fluid would run out of DF into L, or out of L into DF. For the same reason the excess of the repulsion of AB on CG above the attraction of DF thereon must be equal to the force with which a quantity of redundant fluid equal to B spread uniformly on Ww would repel it.

By the supposition the force with which AB repels the canal EM is $M-m$, and the force with which the same quantity of redundant fluid, spread on DF in the same manner in which the deficient fluid therein is actually disposed, repels it is G, therefore if the redundant fluid in AB is called A, the deficient fluid in DF will be $A \times \frac{M-m}{G}$; therefore the force with which DF attracts CG is $(G-g)\frac{M-m}{G}$, and the excess of the force with which AB repels CG above that with which DF attracts it is

$$M - \frac{(G-g)(M-m)}{G} = \frac{Mg+Gm-mg}{G},$$

which must be equal to the force with which a quantity of fluid equal to B spread uniformly over Ww would repel it, that is, it must be equal to $W\frac{B}{A}$; therefore A equals $\frac{BGW}{Mg+Gm-mg}$.

161] Cor. I. If the plate of glass is flat, and its thickness is very small in respect of the least distance of the point C from the circumference of AB, and the fluid in AB and DF is spread uniformly, the quantity of redundant fluid in DF will differ very little from $\frac{B \times CW}{2CE}$, and the deficient fluid in DF will be very nearly equal to the redundant fluid in AB.

For as the plate of glass is flat, the two coatings will be equal to each other, and therefore M and G are equal to each other, and so are m and g, and $\frac{g}{W}$ differs very little from $\frac{CE}{CW}$ *, and moreover g is very small in respect of G.

162] Cor. II. If the plate is flat and the two coatings are circular, their centers being in C and E, the quantity of redundant fluid in AB will be more accurately equal to $\frac{B \times CW}{2CE} \times \frac{CW}{CW-CE}$, CW being in this case equal to the semi-diameter of the coatings, and the deficient fluid in DF will be to the redundant in AB nearly as $CW-CE$ to CW.

For in this case $\frac{m}{W}$ is accurately equal to

$$\frac{CE+CW-\sqrt{CE^2+CW^2}}{CW},$$

and therefore

$$\frac{2m}{W} - \frac{m^2}{W^2} = \frac{2CE\sqrt{CW^2+CE^2}-2CE^2}{CW^2},$$

* Lemma XV. [Art. 148].

which, if CE is small in respect of CW, differs very little from

$$\frac{2CE(CW-CE)}{CW^2}.$$

163] Cor. III. If the plate of glass is not flat, and its thickness is very small in respect of the radius of curvature of its surface at and near C, everything else being as in Cor. I., the quantity of redundant fluid in AB will still be very nearly equal to $\frac{B \times CW}{2CE}$.

For as CE is very small in respect of the radius of curvature, the two coatings will be very nearly of the same size, and therefore G differs very little from M, and $m+g$ is to W very nearly as CE to CW*, and moreover m and g are both very small in respect of M and G†.

164] Cor. IV. If we now suppose that the density of the redundant fluid in AB is greater at its circumference than it is near the point C, and that its density at and near C is less than the mean density, or the density which it would everywhere be of if it was spread uniformly, in the ratio of δ to one, and that the deficient fluid in DF is spread nearly in the same manner as the redundant in AB, the quantity of redundant fluid in AB will be greater than before in a ratio approaching much nearer that of one to δ than to that of equality, and that whether the glass is flat or otherwise.

For by Lemma [XVI. Cor. II.], m and g will each be less than before in the above-mentioned ratio.

165] Cor. V. Whether the plate of glass is flat or concave, or whatever shape the coatings are of, or whatever shape the canals CG and EM are of, or in whatever part they meet the coatings, provided the thickness of the plate is very small in respect of the smallest diameter of the coatings, and is also sufficiently small in respect of the radius of curvature of its surface in case it is concave, the quantity of redundant fluid in AB will differ very little from $\frac{B \times CW}{2CE}$.

For suppose that the canal GC meets the coating AB in the middle of its shortest diameter, and that the point in which ME meets DF is opposite to L, as in Prop. [XXII. Art. 74], the thickness of the glass will then be very small in respect of the distance of the point C from the nearest part of the circumference of AB, and moreover, by just the same reasoning as was used in the Remarks to Prop. XXII., it may be shown that δ will in all probability differ very little from one, and consequently by Cors. I. and III. the redundant fluid in AB will be as above assigned. But by Prop. XXIV. the quantity of fluid in the coatings will be just the same in whatever part the canals meet them, or whatever shape the canals are of.

* Lemma XVI. Cor.

† As the demonstration of the sixteenth Lemma and its corollary is rather intricate, I chose to consider the case of the flat plate of glass separately in Cor. I., and to demonstrate it by means of Lemma XV.

166] Cor. VI. On the same supposition, if the body H is a globe whose diameter equals Ww, *id est* the diameter of a circle whose area equals that of the coating AB, the redundant fluid in AB will be to that in H very nearly as CW to $4CE$.

For the quantity of redundant fluid in H will be $2B$.

167] Cor. VII. On the same supposition the redundant fluid in AB will be very nearly the same whether the glass is flat or otherwise, or whatever shape the coatings are of.

168] Cor. VIII. On the same supposition, if the size and shape of the coatings and also the thickness of the glass is varied, the size and quantity of redundant fluid in H remaining the same, the quantity of redundant fluid in AB will be very nearly directly as its surface, and inversely as the thickness of the glass.

169] Prop. XXXV. (Fig. 9). Let Pp, Rr, Ss, Tt represent any number of surfaces whose distance from Nn, and consequently from

Fig. 9.

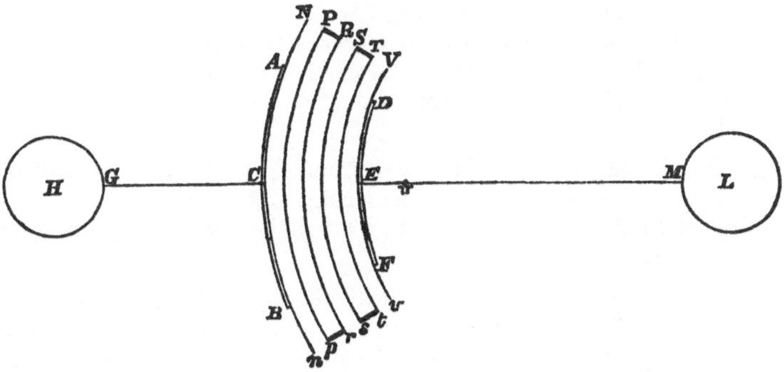

each other, is the same in all parts, and let everything be as in the preceding proposition, except that the fluid in the spaces $PprR$, $SstT$, &c., that is, in the spaces comprehended between the surfaces Pp and Rr, and between Ss and Tt, &c. is moveable*, in such manner, however, that though the fluid in any of these spaces as $PprR$ is able to move freely from Pp to Rr or from Rr to Pp, in a direction perpendicular to the surface Pp or Rr, yet it is not able to move sideways, or in a direction parallel to those surfaces†, and let the fluid in the remaining spaces $NnpP$, $RrsS$, $TtvV$, &c. be immoveable: the quantity of redundant fluid in AB and the deficient fluid in DF will be very nearly the same that they would be if the whole fluid within the glass was immoveable, and its thickness was only equal to $NP + RS + TV$, &c., that is, to the sum of the thicknesses of those spaces in which the fluid is immoveable, provided that NV, the thickness of the glass, is very small

* To avoid confusion I have drawn in the figure only two spaces in which the fluid is supposed to be moveable, but the case would be just the same if there were ever so many.

† [Note 15.]

in respect of the smallest diameter of AB, and also in respect of the radius of curvature of the surface of the glass.

Let the canals GC and EM be perpendicular to the plate of glass and opposite to each other, so as to form one right line, and let them meet AB and DF in the middle of their shortest diameters. The coating AB will be very much overcharged, and DF almost as much undercharged, in consequence of which some fluid will be driven from the surface Pp to Rr and from Ss to Tt. Moreover the quantity of fluid driven from any portion of the surface Pp near the line CE will be very nearly equal to the quantity of redundant fluid lodged in the corresponding part of AB, or more properly will be very nearly equal to a mean between that and the quantity of deficient fluid in the corresponding part of DF.

For a particle of fluid placed anywhere in the space $PprR$ near the line CE is impelled from Pp to Rr by the repulsion of AB and the attraction of DF, and it is not sensibly impelled either way by the spaces $SstT$, &c., as the attraction of the redundant matter in Ss is very nearly equal to the repulsion of the redundant fluid in Tt; and moreover the repulsion of AB on the particle and the attraction of DT are very nearly as great as if their distance from it was no greater than that of Pp and Rr, and therefore the particle could not be in equilibrio unless the quantity of fluid driven from Pp to Rr was such as we have a signed.

As to the quantity of fluid driven from Pp to Rr at a great distance from CE, it is hardly worth considering. It is plain, too, that the quantity of fluid driven from Ss to Tt will be very nearly the same as that driven from Pp to Rr.

Let now G, g, M, m and W signify the same things as in the preceding proposition, and let the quantity of redundant fluid in AB be called A as before, and let $NP + RS + TV +$ &c., *id est*, the sum of the thicknesses of those spaces in which the fluid is immoveable, be to NV, or the whole thickness of the glass as S to 1, and let $PR + ST +$ &c., or the sum of the thicknesses of those spaces in which the fluid is moveable be to NV as D to one.

Take $E\Pi$ equal to PR, the repulsion of the space $PprR$ on the infinite column EM is equal to the repulsion of the redundant fluid in Rr on $E\Pi$, and therefore is to the repulsion of AB on CE very nearly as $E\Pi$ or PR to CE. Therefore the repulsion of all the spaces $PprR$, $SstT$, &c. on EM is to the repulsion of AB on CE very nearly as D to one, or is equal to mD, and therefore the sum of the repulsions of AB and those spaces together on EM is very nearly equal to $M - m + mD$ or to $M - mS$.

But the attraction of DF on EM must be equal to the abovementioned sum of the repulsions, and therefore the deficient fluid in DF must be very nearly equal to $\frac{A(M - mS)}{G}$.

By the same way of reasoning it appears that the force with which

CG is repelled by AB, DF, and the spaces $PprR$ and $SstT$, &c. together is very nearly equal to

$$M - \frac{(M - mS)(G - g)}{G} - gD, \text{ or to } \frac{Mg}{G} + mS - \frac{mgS}{G} - gD,$$

which, as M differs very little from G, and $\frac{mgS}{G}$ is very small in respect of mS or gS, is very nearly equal to $g + mS - gD$ or to $(g + m)S$, therefore the quantity of redundant fluid in AB will be very nearly equal to $\frac{BW}{(g+m)S}$, and will therefore be greater than if the fluid within the glass was immoveable very nearly in the ratio of one to S, or will be very nearly the same as if the thickness of the glass was equal to $CE \times S$, and the fluid within it was immoveable.

170] Prop. XXXVI. Fig. 10. Let every thing be as in the preceding proposition, except that the electric fluid is able to penetrate into the glass on the side Nn as far as to the surface Kk, and on the side Vv

Fig. 10.

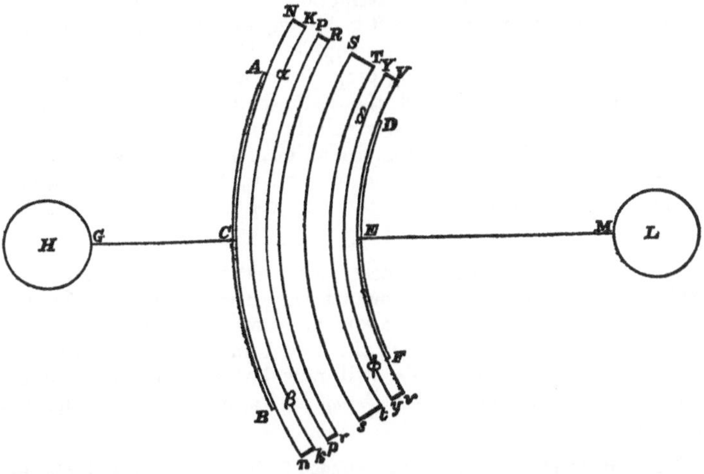

as far as to Yy; in such manner, however, that though the fluid can move freely from AB to $\alpha\beta$ or from $\alpha\beta$ to AB, and also from DF to $\delta\phi$ or from $\delta\phi$ to DF, in a direction perpendicular to those surfaces, yet it is unable to move sideways, or in a direction parallel to those surfaces: the quantity of redundant fluid on one side of the glass, and of deficient fluid on the other, will be very nearly the same as if the spaces $NnkK$ and $VvyY$ were taken away and the coatings AB and DF were applied to the surfaces Kk and Yy.

For by [Art. 132] of former Part, almost all the redundant and deficient fluid will be lodged on the surfaces $\alpha\beta$ and $\delta\phi$, and the coatings AB and DF will be not much over or undercharged. Now if the whole of the redundant and deficient fluid was lodged in $\alpha\beta$ and $\delta\phi$, it is evident that the quantity of redundant and deficient fluid would be exactly the same as if the spaces *NnkK* and *VvyY* were taken away, and therefore it will in reality be very nearly the same.

171] Cor. I. Therefore the quantity of redundant fluid on the positive side of the glass, that is, in the coating *AB*, and the space $A\alpha\beta B$ together, as well as the quantity of deficient fluid on the negative side of the glass, will be very nearly the same that they would be if the fluid was unable to penetrate into the glass or move within it, and that the thickness of the glass was equal only to the sum of the thicknesses of those spaces in which the fluid is immoveable.

172] Cor. II. Whether the electric fluid penetrates into the glass or not, it is evident that the quantity of redundant fluid on one side the glass, and of deficient fluid on the other, will be very nearly the same, whether the coatings are thick or thin.

173] Prop. XXXVII. It was shewn in the remarks on Prop. XXII. in the first Part, that when the plate of glass is flat, and the fluid within it is immoveable, the attraction of the deficient fluid in *DF* makes the redundant fluid in *AB* to be disposed more uniformly than it would otherwise be. Now if we suppose the fluid within the glass to be moveable as in the preceding proposition, and that the deficient fluid in the planes *Pp*, *Ss*, &c. and the redundant fluid in the planes *Rr*, *Tt*, &c. is equal to, and disposed similarly to that in *DF*, the redundant fluid in *AB* will be disposed more uniformly than it would be if the fluid within the glass was immoveable, and its thickness no greater than the sum of the thicknesses of those spaces in which the fluid is immoveable.

For let the intermediate spaces be moved so that *Tt* shall coincide with *Vv* and *Rr* with *Ss*, &c., but let the distance between *Tt* and *Ss* and between *Rr* and *Pp*, &c. remain the same as before, that is, let the thickness of the spaces in which the fluid is moveable remain unaltered. The distance of *Pp* from *Nn* will now be equal to the sum of the thicknesses of the spaces *TtVv*, *RrSs*, *NnPp*, &c. in which the fluid is immoveable.

Now, after this removal, the effect of the planes *Tt* and *DF* and of *Rr* and *Ss*, &c. will destroy each other, so that the intermediate spaces and *DF* together will have just the same effect in rendering the redundant fluid in *AB* more uniform than the plane *Pp* alone will have, that is, the fluid in *AB* will be disposed in just the same manner as if the thickness of the glass was no greater than the sum of the thicknesses of the spaces in which the fluid is immoveable, and the whole fluid within the glass was immoveable.

But the effect of the intermediate spaces in making the fluid in *AB* more uniform was greater before their removal than after, for the effect of the two planes *Pp* and *Rr* together, and also that of *Ss* and *Tt* together, &c. is the greater the nearer they are to *AB*.

174] Cor. The redundant and deficient fluid in the intermediate spaces will in reality be not exactly equal and similarly disposed to that in *DF*, and in all probability the quantity of deficient fluid disposed near the extremity of *DF* will be greater than that in the corresponding parts of *Pp*, *Ss*, &c., or than the redundant fluid in the corresponding parts of *Rr*, *Tt*, &c., so that the redundant fluid in *AB* will perhaps be disposed rather less uniformly than it would be if the deficient and redundant fluid in those spaces was equal to and similarly disposed to that in *DF*; but on the whole there seems no reason to think that it will be much less, if at all less, uniformly disposed than it would be if the thickness of the glass was equal to the sum of the thicknesses of the spaces in which the fluid is immoveable, and the whole fluid within the glass was immoveable.

APPENDIX.

175] As the following propositions are not so necessary towards understanding the experiment as the former, I chose to place them here by way of appendix.

Prop. I. Let everything be as in Prop. XXXIV., except that the bodies *H* and *L* are not required to be at an infinite distance from the plates of glass; let now an overcharged body *N* be placed near the glass in such manner that the force with which it repels the column *CG* towards *G* shall be to that with which it repels the column *EM* towards *M* as the force with which the deficient fluid in *DF* attracts the column *CG* is to that with which it attracts *EM*: it will make no alteration in the quantity of redundant fluid in *AB*, provided the repulsion of *N* makes no alteration in the manner in which the fluid is disposed in each plate.

For increase the deficience of fluid in *DF* so much as that that coating and *N* together shall exert the same attraction on *EM* as *DF* alone did before, they will also exert the same attraction on *CG* as *DF* alone did before, and consequently the fluid in the two canals will be in equilibrio.

176] Cor. In like manner, if the forces with which the body *N* repels the columns *CG* and *EM* bear the same proportion to each other as those with which the plate *AB* repels those columns, and therefore bear very nearly the same proportion to each other as those with which *EM* repels those columns, the quantity of deficient fluid in *DF* will be just the same as before *N* was brought near, and the redundant fluid in *AB* will be diminished by a quantity whose repulsion on *CG* is the same as that of *N* thereon.

Therefore, if the repulsion of N on CG is not greater than that of H thereon, the diminution of the quantity of redundant fluid in AB will bear but a very small proportion to the whole. For the quantity of redundant fluid in AB is many times greater than that which would be contained in it if DF was away, *id est*, than that whose repulsion on CG is equal to the repulsion of H thereon in the contrary direction.

177] PROP. II. From the preceding proposition and corollary we may conclude that if the force with which N repels the columns CG and EM bears very nearly the same proportion to each other as the force with which DF attracts those columns, the quantity of redundant fluid in AB will be altered by a quantity which will bear but a very small proportion to the whole, unless the repulsion of N on CG is much greater than that of H thereon.

If the reader wishes to see a stricter demonstration of this proposition, as well as to see it applied to the case in which the fluid is supposed moveable in the intermediate spaces, as in Prop. XXXV., he may read the following :

Fig. 10.

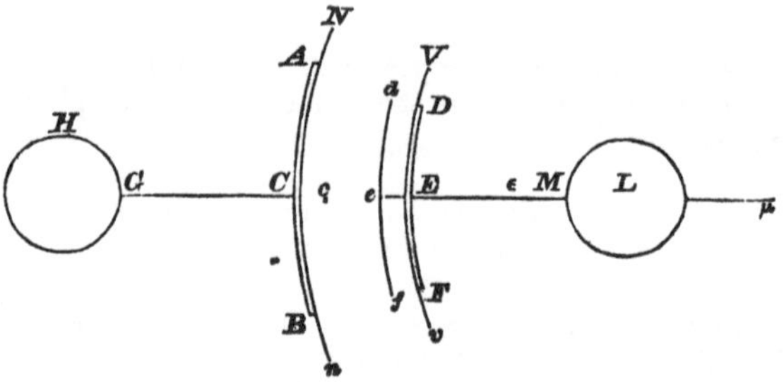

178] PART 1. Take $Ee = \frac{1}{2}$ thickness of those spaces in which the fluid is moveable, draw def equal and similar to DEF, and let the deficient fluid therein be equal to that in DF: the repulsion of the intermediate spaces on EM is to the difference of the attractions of DF on ϵM and $\epsilon\mu$ (supposing $E\epsilon$ and $M\mu$ to be equal to CE) very nearly as twice Ee to CE, and is therefore very nearly equal to twice the difference of the attraction of df and DF on EM.

In like manner the attraction of the intermediate spaces on CG is very nearly equal to twice the difference of the attraction of DF and df thereon.

Suppose now the quantity of deficient fluid in DF to be increased in the ratio of $1 + f$ to 1, the redundant fluid in AB remaining the same as before, a new attraction is produced on EM, very nearly equal to

$$f \times (\text{attraction of } DF \text{ on } EM) - \frac{f}{2} \times 2\,(\text{diff. attr. of } df \text{ and } DF \text{ on } EM),$$

that is, very nearly equal to $f \times$ (attraction of df on EM).

In like manner a new attraction is produced on CG, very nearly equal to $f \times$ (attraction of df on CG), therefore, the new attraction produced on EM is to that produced on CG very nearly as the attraction of df on EM is to its attraction on CG, and therefore in order that the quantity of redundant fluid in AB shall not be altered by the approach of N, the repulsion of N on EM must be to its repulsion on CG very nearly as the attraction of df on EM to its attraction on CG.

179] PART 2. Let the fluid within the glass be either moveable, as in Prop. [XXXV. Art. 169], or let it be immoveable, and let the distance of H and L from the glass be either great or not.

Let the repulsion of H on $\begin{cases} GC \\ EM \end{cases}$ in direction GC be $\begin{cases} H \\ Hp \end{cases}$, and let the sum of these repulsions $= S$.

Let the repulsion of N on $\begin{cases} GC \text{ in direction } CG = A \\ EM \text{ in direction } EM = B \end{cases}$, and let the repulsion which N should exert on CG in order that the redundant fluid in AB should remain unaltered be to that which it should exert on EM :: 1 : P.

The quantity of redundant fluid in AB will be increased in the ratio of $1 + \dfrac{B - PA}{S} \dfrac{1+p}{P+p}$ to 1, which, if P differs very little from 1, differs very little from that of

$$1 + \frac{B - PA}{S} \text{ to } 1.$$

For the force $\begin{cases} A \\ B \end{cases}$ may be divided into two parts, namely

$$\begin{cases} \dfrac{PA - B}{P+p} + \dfrac{pA + B}{P+p} \\[2ex] \dfrac{-PpA + pB}{P+p} + \dfrac{PpA + PB}{P+p}, \end{cases}$$

but the latter part of these two repulsions, or the force $\begin{Bmatrix} \dfrac{pA + B}{P+p} \\[2ex] \dfrac{PpA + PB}{P+p} \end{Bmatrix}$

has no tendency to alter the redundant fluid in AB, but the first part, or the force

$$\begin{cases} \dfrac{PA - B}{P+p} \\[2ex] \dfrac{-PpA + pB}{P+p} \end{cases} \text{acting on } \begin{cases} CG \text{ in direction } CG \\ EM \text{ in direction } EM \end{cases}$$

$$\text{or} \begin{cases} \dfrac{-PA+B}{P+p} \\ \dfrac{-PpA+pB}{P+p} \end{cases} \text{acting on} \begin{cases} CG \text{ in direction } GC \\ EM \text{ in direction } EM \end{cases}$$

as they are to the repulsion of H on $\begin{cases} GC \\ EM \end{cases}$ as $\dfrac{-PA+B}{P+p}$ to H,

$$\text{or as } (-PA+B)\,\frac{1+p}{P+p} \text{ to } S,$$

increases the redundant fluid in the ratio of

$$1+\frac{B-PA}{S}\,\frac{1+p}{P+p} \text{ to } 1.$$

180] COR. I. If the lengths of the columns CG and EM are such that the repulsion and attraction of AB and DF on them are not sensibly less than if they were of an infinite length, the attraction of DF on CG will be very nearly equal to its attraction on EM, and therefore, if the forces with which N repels the columns CG and EM are very nearly equal to each other, the quantity of redundant fluid in AB will be very little altered thereby.

N.B. If the size of H is much greater than that of AB, it is possible that its distance from the glass may be such as to exert a very considerable repulsion on EM, and yet that the action of AB and DF on CG shall be not sensibly less than if it was of an [infinite length].

181] COR. II. Let the bodies H and L be of the same size and shape and at an infinite distance from the glass, and let the fluid be in equilibrio. Let now an equal quantity of fluid be taken from H and L, the quantity of redundant fluid in AB will be very little altered thereby.

For the repulsion of the whole quantity of fluid in L on the canal EM will be as much diminished as that of H on CG, so that it comes to the same thing as placing an overcharged body N in such manner that its repulsion on CG shall be equal to that on EM, which by the preceding proposition will make very little alteration in the quantity of redundant fluid in AB.

182] COR. III. Let the bodies H and L be at an infinite distance, and either of the same or different size, and let the fluid be in equilibrio. Let now the body H be brought so near to AB that its repulsion on GC shall be sensibly less than before. The quantity of redundant fluid in AB will be very little altered thereby, provided the repulsion of the two plates on the column CG is not sensibly diminished.

For whereas when H was at an infinite distance from AB it exerted no repulsion on EM, now it is brought nearer it does exert some, and its repulsion on EM is very nearly equal to the diminution of its repulsion on CG, so that it comes to the same thing as placing a body N in such manner as to repel *EM* with very nearly the same force that it does *CG* in the contrary direction.

183] Cor. IV. Let the body *H* be brought near *AB* as in the preceding corollary, and let the fluid be in equilibrio; let now an overcharged body *R* be placed near *H*, the quantity of redundant fluid in *H* must be so much diminished, in order that the fluid may remain in equilibrio, supposing the fluid in *AB* to remain unaltered, as that the diminution of its repulsion on the two columns *GC* and *EM* shall be equal to the repulsion of *R* on the same columns. Consequently, if the repulsion of *R* on them is to the repulsion which *H* exerted on them before the approach of *R* as n to 1, the quantity of redundant fluid in *H* will be diminished in the ratio of $1-n$ to 1.

For supposing the quantity of fluid in *H* to be thus diminished, I say, the quantity of fluid in *A* will remain very nearly the same as before. For the repulsion of *H* and *R* on the two columns will be the same as that of *H* was before, but it is possible that their repulsion on *GC* may be a little less, and their repulsion on *EM* as much greater than that of *H* was before, but this, by the preceding corollaries, will make very little alteration in the quantity of fluid in *AB*.

184] Cor. V. It appears from Prop. XXIII. that the repulsion of the body *R* on the two columns *GC* and *EM* will be the same in whatever direction it is placed in respect of *H* and the canal, provided its distance from the point *G* is given, and consequently the diminution of the quantity of fluid in the body *H* will be very nearly the same in whatever direction *R* is situated, provided its distance from *G* is given.

185] Cor. VI. Fig. 11. Suppose now that instead of the body *H* there is placed a plate of glass *KkiI*, coated as in Props. XXXIV. and XXXV.,

Fig. 11.

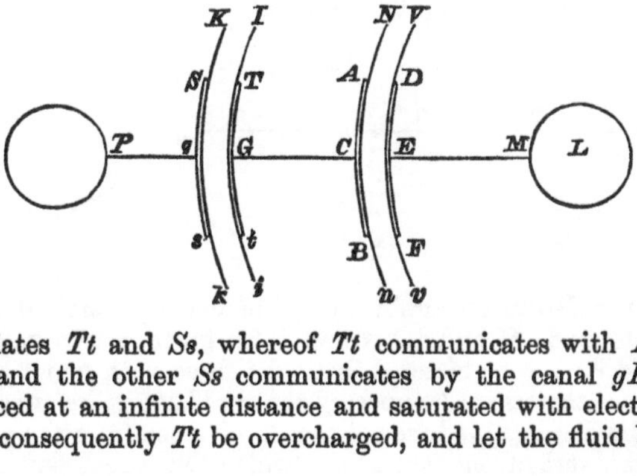

with the plates *Tt* and *Ss*, whereof *Tt* communicates with *AB* by the canal *GC*, and the other *Ss* communicates by the canal *gP* with the body *P*, placed at an infinite distance and saturated with electricity, and let *AB* and consequently *Tt* be overcharged, and let the fluid be in equilibrio.

Suppose now that an overcharged body *R* is brought near the glass *KkiI*, I say that the proportion which the redundant fluid in *Tt* bears to that in *AB* will be very little altered thereby, supposing the length of the canal CG to be such that the repulsion of the coatings AB and DF thereon shall be not sensibly less than if it was infinite, and that the thickness of the glass Gg is very small in respect of the distance of R from it, and that the repulsion of R does not sensibly alter the disposition of the fluid in Tt and Ss, and also that the repulsion of R on GC and EM together is not much less than if GM was infinite, and also not much greater than the repulsion of the glass $NnvV$ on CG.

For let the quantity of fluid in Tt and Ss be so much altered that the united repulsion of R and those two coatings on the two canals GC and EM together, and also their repulsion on gP, shall be the same as that of the two coatings alone before the approach of R.

By Prop. II. Cor. I. the quantity of fluid in Tt will be very little altered thereby, for the repulsion of R on the canal gP is very nearly the same as its repulsion on gC and EM together.

As the repulsion of Tt, Ss and R together on the two canals GC and EM together is the same as before the approach of R, it follows that if their repulsion on gC is less than before, their repulsion on EM will be as much increased.

Let now the quantity of fluid in AB and DF be so much altered that their repulsion on gC shall be as much diminished as that of $KkiI$ and R on the same column is diminished, and that their repulsion on EM shall be as much diminished as that of $KkiI$ and R on the same is increased, it is plain that the fluid in all three canals will be exactly in equilibrio, and by the preceding corollary the quantity of fluid in AB will be very little altered, and therefore the proportion of the redundant fluid in AB and Tt to each other will be very little altered*.

186] Cor. VII. By Prop. [XXIV. Art. 86] all which is said in this proposition and corollaries holds good equally whether the canals GC, EM and GP are straight or crooked.

Lemma.

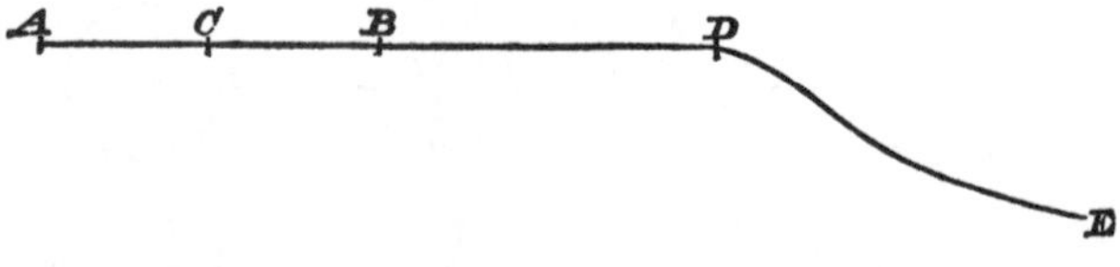

187] Let DE be an uniform canal of incompressible fluid infinitely continued towards E, and let A and B be given points in a right line with D, and let AB be bisected in C, the force with which any particle of fluid repels this canal (supposing the repulsion to be inversely as the square of the distance) is inversely as its distance from the point D, and therefore the sum of the forces with which two equal particles of fluid placed in A and B repels this canal is to the sum of the forces with which they would repel it if both collected in the point C,

* [Note 16.]

$$\text{as } \frac{1}{AD}+\frac{1}{BD} : \frac{2}{CD},$$

$$\text{or as } CD^2 : CD^2 - CB^2,$$

$$\text{or as } 1 : 1-\frac{CB^2}{CD^2}.$$

188] Let us now examine how far the proportion of the quantity of fluid in the large circle and the two small ones in Experiment v., [Art. 273] Fig. 18, bear to each other will be affected by the circumstances mentioned in [Art. 276], supposing the plates to be connected by canals of incompressible fluid.

First it appears from Cor. [VII. Art. 186], that the quantity of redundant fluid in the large circle, and also in the two small ones, will bear very nearly the same proportion to that in the jar A as it would if it had been placed at an infinite distance from A, for the distance of the plate from the jar was in neither experiment less than 63 inches, and neither the length nor the diameter of the coated part of the jar exceeded four inches, so that the repulsion of the jar on the canal connecting it to the plate could not differ by more than $\frac{1}{31}$ part from what it would be if the canal was infinitely continued, and would most probably differ from it by not more than $\frac{1}{2}$ or $\frac{1}{3}$ part of that quantity*; for the same reason the deficience of fluid in the trial plate will bear very nearly the same proportion to that in the jar, &c. as it would if it had been placed at an infinite distance from it.

Fig. 18.

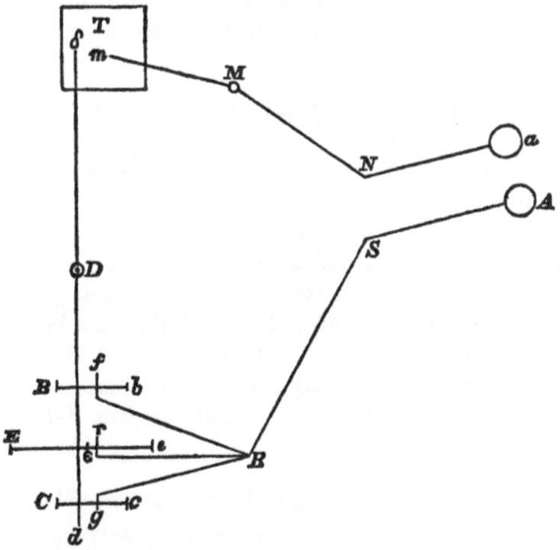

* The repulsion of a globe 4 inches diameter on a straight uniform canal of incompressible fluid extending 63 inches from it differs by only $\frac{1}{31}$ part from what it would be if the canal was infinitely continued, but the repulsion of a Leyden vial of that size on the same column differs probably not more than $\frac{1}{2}$ or $\frac{1}{3}$ of that quantity from what it would be if infinitely continued.

It is plain that if the plates had been placed at such a distance from the jar that the quantity of fluid in them had been considerably less than if they had been placed at an infinite distance, still the quantity in the large circle would bear very nearly the same proportion to that in the two small ones as it would if they had been placed at an infinite distance.

189] Secondly, it is plain that in trying the large circle, the repulsion of that circle increases the deficience of fluid in the trial plate, and the attraction of the trial plate increases the redundance in the circle. Now the repulsion of the plate *Ee* on the canal *mMNa*, and the attraction of the trial plate *T* on *rRSA* (supposing *mMNa* and *rRSA* to be infinitely continued beyond *a* and *A*) are by [Cor. IV. Art. 183] very nearly the same as if the redundant fluid in *Ee* and the deficient fluid in *T* were both collected in the centers of their respective plates, and the quantity of redundant fluid in *Ee* may be considered as equal to the deficient in *T*, and consequently the repulsion of *Ee* on *mMNa* is very nearly equal to the attraction of *T* on *rRSA*. Moreover, the repulsion of *Ee* on its own canal *rRSA* must be equal to the attraction of *T* on *mMNa*, as the jars with which they communicate are both equally electrified, and therefore, by Cor. [IV.], the quantity of redundant fluid in *Ee* will be increased in very nearly the same ratio as the deficient in *T*.

190] In like manner, in trying the two small circles, the quantity of redundant fluid in them is increased in very nearly the same ratio as the deficient in *T*, for as half the distance of the two circles never bore a greater proportion to ϵm than that of 18 to 72, the repulsion of the two circles on the canal *mMNa* will be very nearly the same, and the deficience of fluid in *T* will be increased in very nearly the same ratio as if all the redundant fluid in them were collected in ϵ, the middle point between them.

The quantity of redundant fluid in *Bb* indeed will be increased in a rather greater ratio, and that in *Cc* in a rather less ratio than if it was placed at ϵ, but the ratio in which the quantity of fluid in *Bb* is increased must very nearly as much exceed that in which it would be increased if it was placed at ϵ as that in which *Cc* is increased falls short of it, as the attraction of *T* on the canal *fRSA* exceeds that on *rRSA* by nearly the same quantity as its attraction *gRSA* falls short of it, and therefore the quantity of redundant fluid in both circles together is increased in very nearly the same proportion as that in a circle placed in ϵ would be, and consequently the redundance in the two circles is increased in very nearly the same ratio as the deficience in the trial plate*.

* Memorandum relating to the second article.

191] The attraction of the trial plate on the canals *fRSA* and *gRSA* and the repulsion of the circles *Bb* and *Cc* on the canal *mMNa* is very nearly the same as if the deficient or redundant fluid in the plates was collected in the centre of their respective plates, and therefore the repulsion of the circles *Bb* and *Cc* on the canal *mMNn* is inversely as the distances of their centres from *m*, and the increase of the quantity of redundant fluid in the circles *Bb* and *Cc* by the attraction of *T* is in the same proportion.

192] Consequently, in trying either the large circle or the two small ones, the trial plate must be opened to very nearly the same surface to contain the same charge as them as it must be if they were placed at an infinite distance from the trial plate, and consequently no sensible alteration can be produced in the phenomena of the experiment by the repulsion and attraction of the circles and trial plate on each other.

193] Thirdly, for the same reason it appears that as the circles and the trial plate are both at much the same distance from the ground and walls of the room, no sensible alteration can be produced in the experiment by the ground near the circles being rendered undercharged and that near the trial plate overcharged.

It must be observed, indeed, that the distance of the circles and trial plate from the ground is much less than their distance from each other, and consequently the alteration of the charge of the two circles and trial plate produced by this cause will not be so nearly alike as that caused by their attraction and repulsion on each other; but as, on the other hand, the whole alteration of their charge produced by this cause is, I imagine, much less than that produced by the other, I imagine that this cause can hardly have a more sensible effect in the experiment than the preceding.

194] Fourthly, we have not as yet taken notice that the canals by which the jars *Aa* communicate with the ground are but short, and meet the ground at no great distance from the jars.

But it may be shewn by the same kind of reasoning used in Prop. [II. Art. 178], with the help of the second corollary to the preceding proposition, that the quantity of redundant fluid in the circles will bear very nearly the same proportion to that in the positive side of the jar *A*, whether the canal by which *A* communicates with the ground is long or short.

Besides that, if it was possible for this circumstance to make much alteration in the proportion which the redundant fluid in the circles bears to that in *A*, it would in all probability have very nearly the same effect in trying the two small circles as in trying the large one, so that no sensible alteration can be produced in the experiment from this circumstance.

It appears, therefore, that none of the above-mentioned circumstances can cause any sensible alteration in this experiment*.

Therefore take the point *a* so that the repulsion of a particle at *a* on that canal shall be a mean between the repulsions of the same particle thereon when placed at *B* and *C*, the charge of *T* will be increased in the same proportion as it would be by the repulsion of a plate containing as much redundant fluid as the two plates together whose centre was *a*, and the charge of the two circles together will also be increased in the same proportion as that of the circle whose centre is *a* would be thereby.

* [Note 17.]

THOUGHTS CONCERNING ELECTRICITY.

195] Electricity seems to be owing to a certain elastic fluid interspersed between the particles of bodies, and perhaps also surrounding the bodies themselves in the form of an atmosphere.

196] This fluid, if it surrounds bodies in the form of an atmosphere, seems to extend only to an imperceptible distance from them*, but the attractive and repulsive power of this fluid extends to very considerable distances.

197] That the attraction and repulsion of electricity extend to considerable distances is evident, as corks are made to repel by an excited tube held out at a great distance from them. That the electric atmospheres themselves cannot extend to any perceptible distance, I think, appears from hence, that if two electric conductors be placed ever so near together so as not to touch, the electric fluid will not pass rapidly from one to the other except by jumping in the form of sparks, whereas if their electric atmospheres extended to such a distance as to be mixed with one another, it should seem as if the electricity might flow quietly from one to the other in like manner as it does through the pores of any conducting matter.

But the following seems a stronger reason for supposing that these atmospheres cannot extend to any perceptible distance from the body they surround, for if they did it should seem that two flat bodies whenever they were laid upon one another should always become electric thereby, for in that case there is no room for

* There are several circumstances which shew that two bodies, however smooth and strongly pressed together, do not actually touch each other. I imagine that the distance to which the electric atmospheres, if there are any, extend must be less than the smallest distance within which two bodies can be made to approach.

the electric atmosphere to extend to any sensible distance from those surfaces of the bodies which touch one another, so that the electric fluid which before surrounded those surfaces would be forced round to the opposite sides, which would thereby become overcharged with electricity, and consequently appear electrical, which is contrary to experience.

198] Many Electricians seem to have thought that electrified bodies were surrounded with atmospheres of electric matter extending to great distances from them. The reasons which may have induced them to think so may be first, that an electrified body affects other bodies at a considerable distance. But this may, with much more probability, be supposed owing to the attraction and repulsion of the electric matter within the body or close to its surface. And, secondly, because a body placed near a positively electrified body receives electricity itself, whence it is supposed to receive that electricity from the electrified body itself, and therefore to be within its atmosphere. But, in all probability, the body in this case receives its electricity from the contiguous air, and not immediately from the electrified body, as will be further explained in its place.

199] Let any number of bodies which conduct electricity with perfect freedom be connected together by substances which also conduct electricity. It is plain that the electric fluid must be equally compressed* in all these bodies, for if it was not, the electric fluid would move from those bodies in which it was more compressed to those in which it was less compressed till the compression became equal in all. But yet it is possible that some of these bodies may be made to contain more than their natural quantity of electricity, and others less. For instance, let some power be applied to some of these bodies which shall cause the electric fluid within their pores to expand and grow rarer†, those bodies will thereby be made to contain less electric matter than they would otherwise do, but yet the electric matter within them

* Note by Editor. [That is, must sustain an equal pressure. In modern scientific language the words compression, extension, distortion, are used to express *strain*, or change of form, while pressure, tension, torsion, are reserved to indicate the *stress* or internal force which accompanies this change of form. Cavendish uses the word compression to indicate stress. The idea is precisely that of *potential*.]

† [No such power has been discovered. There is nothing among electrical phenomena analogous to the expansion of air by heat.—Ed.]

will be just as much compressed as it would be if this power were not applied.

On the other hand, if some power were applied which shall diminish the elasticity of the electric fluid within them and thereby make it grow more dense, those bodies will be made to contain more electricity, but yet the compression will remain still the same.

200] To make what is here said more intelligible, let us suppose a long tube to be filled with air, and let part of this tube, and consequently the air within, be heated, the air will thereby expand, and consequently that part of the tube will contain less air than it did before, but yet the air in that part will be just as much compressed as in the rest of the tube.

In like manner, if you suppose the electric fluid to be not only confined within the pores of bodies, but also to surround them in the form of an atmosphere, let some power be applied to some of those bodies which shall prevent this atmosphere from extending to so great a distance from them, those bodies will thereby be made to contain less electricity than they would otherwise do, but yet the electric fluid that surrounds them will be just as much compressed as it would [be] if that power was not applied.

It will surely be needless to warn the reader here not to confound compression and condensation.

201] I now proceed to my hypothesis.

DEF. 1. When the electric fluid within any body is more compressed than in its natural state, I call that body positively electrified: when it is less compressed, I call the body negatively electrified.

It is plain from what has been here said that if any number of conducting bodies be joined by conductors, and one of the bodies be positively electrified, that all the others must be so too.

DEF. 2. When any body contains more of the electric fluid than it does in its natural state, I call it overcharged. When it contains less, I call it undercharged.

202] HYP 1st. Every body overcharged with electricity repels an overcharged body, and attracts an undercharged one.

HYP. 2nd. Every undercharged body attracts an overcharged body, and repels an undercharged one.

HYP. 3rd. Whenever any body overcharged with electricity is brought near any other body, it makes it less able to contain electricity than before.

HYP. 4th. Whenever an undercharged body is brought near another it makes it more able to contain electricity.

203] COR. I. Whenever any body at a distance from any other electrified body is positively electrified it will be overcharged, and if negatively electrified it will be undercharged.

COR. II. If two bodies, both perfectly insulated, so that no electricity can escape from them, be positively electrified and then brought near to each other, as they are both overcharged they will each, by the action of the other upon it, be rendered less capable of containing electricity, therefore, as no electricity can escape from them, the fluid within them will be rendered more compressed, just as air included within a bottle will become more compressed either by heating the air or by squeezing the bottle into less compass; but it is evident that the bodies will remain just as much overcharged as before.

204] COR. III. If two bodies be placed near together, and then equally positively electrified, they will each be overcharged, but less so than they would [be] if they had not been placed near together.

It may perhaps be said that this is owing to the electric atmosphere not having so much room to spread itself when the two bodies are brought near together as when they are at a distance; but I think it has already been sufficiently proved that these atmospheres cannot extend to any sensible distance from their respective bodies.

COR. IV. If two bodies are placed near together and then equally negatively electrified, they will each be undercharged, but less so (*id est*, they will contain more electricity) than if placed at a distance.

This phenomenon cannot be accounted for on the foregoing supposition.

205] COR. V. If a body overcharged with electricity be brought near a body not electrified and not insulated, part of the electric fluid will be driven out of this body, and it will become undercharged.

But if the body be insulated, as in that case the electric fluid cannot escape from it, it will not become undercharged, but the electric fluid within it will be more compressed than in its natural state, *id est*, the body will become positively electrified, and will remain so as long as the overcharged body remains near it, but will be restored to its natural state as soon as the overcharged body is taken away, provided no electricity has escaped during the mean time.

This is in effect the same case as that described in the 5th experiment of Mr Canton's paper in the 48th vol. of [the *Philosophical*] *Transactions*, p. 353, and is explained by him much in the same manner as is done here.

206] COR. VI. If a body positively electrified in such a manner that if it is by any means made more or less capable of containing electricity, the electric fluid shall run into it from without or shall run out of it, so as to keep it always equally electrified, be brought near another body not electrified and not insulated, the second body will thereby be rendered undercharged, whereby the first body will become more capable of containing electricity, and consequently will become more overcharged than it would otherwise be with the same degree of electrification. This again will make the second body more undercharged, which again will make the first body more overcharged, and so on.

It must be observed here, that if the two bodies are brought so near together that their action on one another shall be considerable, the electricity will jump from one to the other; otherwise if the two bodies were brought so near together that their distance should not be greater than the thickness of the glass in the Leyden bottle, it seems likely that the first body might receive many times as much additional electricity as it would otherwise receive by the same degree of electrification; and that the second body would lose many times as much electricity as it would by the same degree of negative electrification.

If the second body be negatively electrified, the same effect will be produced in a greater degree.

It may also happen that the second body shall be made undercharged though it is positively electrified, provided it be much less electrified than the first body, and that the two bodies be placed near enough to each other.

207] The shock produced by making a communication between the two surfaces of the Leyden vial seems owing only to the glass prepared in that manner containing vastly more electricity on its positive side than an equal surface of metal equally electrified, and vastly less on its negative side than the same surface of metal negatively electrified to the same degree, so that if two magazines of electricity were prepared, each able to receive as much additional electricity by the same degree of electrification as one of the surfaces of a Leyden vial, and one of the magazines was to be positively electrified and the other negatively, there is no doubt but what as great a shock would be produced by making a communication between the two magazines as between the two surfaces of the Leyden vial.

I think, therefore, that the phenomena of the Leyden vial may very well be accounted for on the principle of the 6th Corollary, for in the Leyden vial the two surfaces of the glass are so near together, that the electric matter on one surface may act with great force on that on the other, and yet the electricity cannot jump from one surface to the other, by which means perhaps the positive side may be made many times more overcharged, and the negative side many times more undercharged, than it would otherwise be.

208] Hyp. 5th. It seems reasonable to suppose that when the electric fluid within any body is more compressed than it is in the air surrounding it, it will run out of that body, and when it is less compressed it will run into the body.

Cor. I. Let the body *A*, not electrified, be perfectly insulated, and let an overcharged body be brought near it. The body *A* will thereby be rendered less capable of containing electricity, and therefore the electric fluid within it, as it cannot escape, will be rendered more compressed. But the electricity in the adjoining air will, for the same reason, be also compressed, and in all probability equally so, therefore the electricity will have no disposition either to run in or out of the body.

Cor. II. It is evidently the same thing whether *A* be insulated, or whether it be not insulated, but electrified in such manner that the fluid within it be as much compressed as it was before by virtue of the insulation. Therefore if the body *A* be now not insulated, but positively electrified, and an overcharged body be brought to such a distance from it that the electric fluid in the adjacent air be equally compressed with that in *A*, such a quantity of electricity will thereby be driven out of *A* that it will retain only its natural quantity. So that *A* will be neither overcharged nor undercharged, nor will the electricity have any disposition to run either in or out of it.

209] If the overcharged body be now brought nearer, *A* will become undercharged, and the electricity will run into it from the surrounding air. If the overcharged body be not brought so near *A* will be overcharged, and the electricity will run out of it. If an undercharged body be brought near *A* it will become more overcharged than before, and the electricity will run out stronger than before.

COR. III. If the body *A* be negatively electrified, and an undercharged body be brought near it till the electric fluid in the adjoining air is as much compressed as that in the body *A*, the electricity will have no disposition to run either in or out of *A*, nor will it be either overcharged or undercharged, as will appear from the same way of reasoning as was used with regard to the 2nd Corollary.

If the undercharged body be now brought nearer, *A* will become overcharged, and the electricity will also run out of it. If the undercharged body be removed farther off, *A* will become undercharged, and the electricity will also run into it. If an overcharged body be brought near to *A*, it will become more undercharged than before, and the electricity will also run in faster than before.

On the whole, therefore, it appears that whenever a body is undercharged the electricity will run into it, and whenever it is overcharged it will run out.

210] It has usually been supposed that two bodies, whenever the electricity either runs into or out of both of them, repel each other; but that when it runs into one and out of the other, they attract. In the beginning of this paper I laid down a different rule for the electric attraction and repulsion, namely, that when the two bodies are both overcharged or both undercharged they repel, but attract when one is overcharged and the other undercharged.

But by what has been just said it appears that these two rules agree together, or at least if they do differ, they differ so little that there is no reason to think my rule will agree less with experiment than the other.

The reasoning here used would have been more satisfactory if the bodies were capable of containing electricity only on one side, namely, on that which is turned towards the other body. But I do not imagine, however, that this will make much difference in the effect.

211] What has been here said holds good only in cases where the size of the body A is small in respect of the distance of the electrified body from it, so that the influence of the electrified body may be nearly the same on all parts of the body A as is the case in bits of cork held near an excited tube; but when the size of the body A is such that the influence of the electrified body may be much greater on that part of A which is directly under it than on that which is farther removed from it, as is the case in electrifying a prime conductor by an excited tube, then the case is very different, for then on approaching the electrified tube, part of the electric fluid will be driven away from that part of the prime conductor which is nearest the excited tube to the remoter parts where its influence is weaker, whereby that part of the conductor nearest the tube will be undercharged, and consequently the compression of the electric fluid in that part will be less than in the contiguous air, consequently some electric matter will flow into it from the adjoining air, whereby the conductor will be overcharged, and therefore on taking away the tube will be positively electrified.

Thus if the excited tube or other electrified body is not brought within a certain distance, the conductor receives its electricity only from the contiguous air, as was before said, and not immediately from the electrified body; but if the body be brought near enough, the electric matter jumps from the electrified body to the conductor in form of a spark.

212] The means by which this is brought about seems thus—When the part of the conductor nearest the excited tube has received any electricity from the contiguous air, that air will be undercharged, and will receive electricity from the adjacent air between it and the tube, by which means the electric matter will flow in gentle current between the particles of air from the excited tube to the conductor. It seems now as if the particles of air were by this means made to repel each other with more force, and thereby to become rarer, this will suffer the electric fluid to flow in a swifter current, which again will increase the repulsion of the particles of air, till at last a vacuum is made, upon which the electric fluid jumps in a continued body to the conductor.

213] That a vacuum is formed by the electric fluid when it passes in the form of a spark through air or water appears, I think, from the violent rising of the water in Mr Kinnersley's electrical air-thermometer (Priestley, p. 216), and still more strongly from the bursting the vial of water, in Mr Lane's experiment, by making the electrical fluid pass through the water in the form of a spark.

If I am not much mistaken I have frequently observed, in discharging a Leyden vial, that if the two knobs are approached together very gently, a hissing noise may be perceived before the spark, which shews that the electricity does begin to flow from one knob to the other before it moves in the form of the spark, and may therefore induce one to think that the spark is brought about in the gradual manner here described.

214] The attraction and repulsion of electrified bodies, according to the law I have laid down, may perhaps be accounted for in the following manner. Let a fluid consisting of particles mutually repelling each other, and whose repulsion extends to considerable distances, be spread uniformly all over the globe, except in the space A, which we will suppose to contain more than its proper quantity of the fluid. The fluid placed in any space B within reach of the repulsion of A will be repelled from A with more force than it will [be] in any other direction. But as it cannot recede from A without an equal quantity of the fluid coming into its room which will be equally repelled from A, it is plain that it will have no tendency to recede from A, any more than a body of the same specific gravity as water has any tendency to sink in water. Let now the space B be made to contain more than its natural quantity of this fluid, it will then really have a tendency to recede from *A*, or will appear to be repelled by it, just as a body heavier than water tends to descend in it, and, on the contrary, if *B* is made to contain less than its natural quantity of the fluid, it will have a tendency towards *A*, or will appear to be attracted by it.

215] Let now the space *A* be made to contain less than its natural quantity of the fluid (as the fluid in *B* is now repelled from *A* with less force than it is in any other direction, *id est*, apparently attracted towards it), if *B* also contain less than its natural quantity of the fluid it will tend to recede from *A*, *id est*, appear to be repelled by it; but if *B* contain more than its natural quantity, it will then tend to approach towards *A*, *id est*, appear to be attracted by it.

216] If the electric fluid is diffused uniformly through all bodies not appearing electrical and the repulsion of its particles extends to considerable distances, it is plain that the consequences are such as are here described; but how far that supposition will agree with experiment I am in doubt *.

* [Note 18.]

EXPERIMENTS ON ELECTRICITY.

EXPERIMENTAL DETERMINATION OF THE LAW OF ELECTRIC FORCE.

217] I now proceed to give an account of the experiments, in all of which I shall suppose, according to the received opinion, that the electricity of glass is positive, but it is not at all material to the purpose of this paper whether it is so or not, for if it was negative, all the experiments would agree equally well with the theory.

218] Experiment I. The intention of the following experiment was to find out whether, when a hollow globe is electrified,

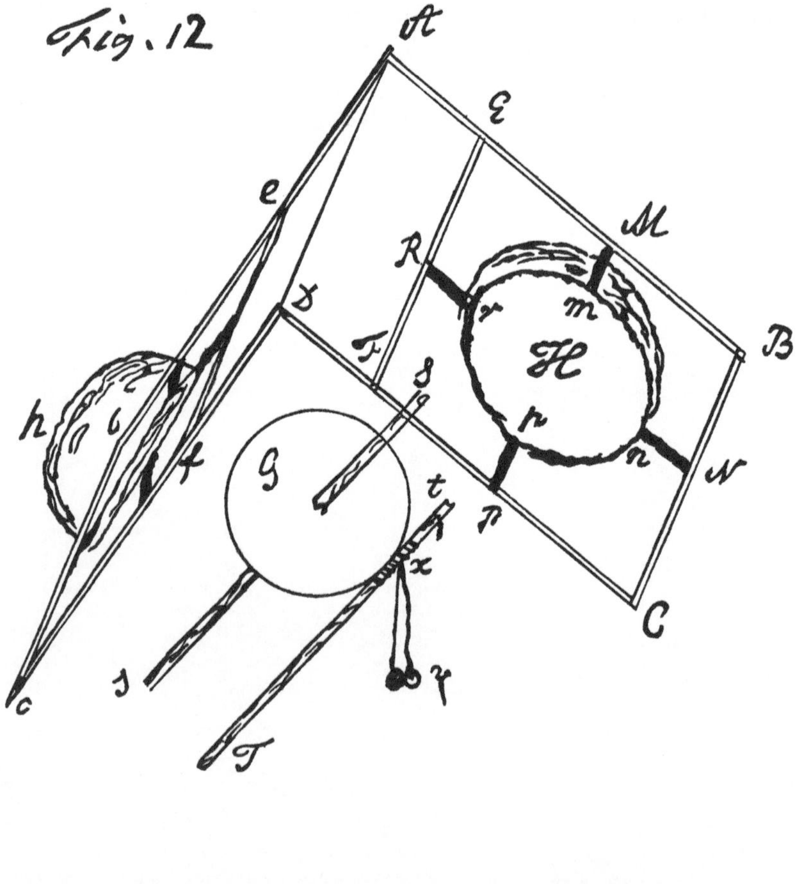

a smaller globe inclosed within it and communicating with the outer one by some conducting substance is rendered at all over or undercharged; and thereby to discover the law of the electric attraction and repulsion.

219] I took a globe 12·1 inches in diameter, and suspended it by a solid stick of glass run through the middle of it as an axis, and covered with sealing-wax to make it a more perfect non-conductor of electricity. I then inclosed this globe between two hollow pasteboard hemispheres, 13·3 inches in diameter, and about $\frac{1}{20}$ of an inch thick, in such manner that there could hardly be less than $\frac{4}{10}$ of an inch distance between the globe and the inner surface of the hemispheres in any part, the two hemispheres being applied to each other so as to form a complete sphere, and the edges made to fit as close as possible, notches being cut in each of them so as to form holes for the stick of glass to pass through.

By this means I had an inner globe included within an hollow globe in such manner that there was no communication by which the electricity could pass from one to the other.

I then made a communication between them by a piece of wire run through one of the hemispheres and touching the inner globe, a piece of silk string being fastened to the end of the wire, by which I could draw it out at pleasure.

220] Having done this I electrified the hemispheres by means of a wire communicating with the positive side of a Leyden vial, and then, having withdrawn this wire, immediately drew out the wire which made a communication between the inner globe and the outer one, which, as it was drawn away by a silk string, could not discharge the electricity either of the globe or hemispheres. I then instantly separated the two hemispheres, taking care in doing it that they should not touch the inner globe, and applied a pair of small pith balls, suspended by fine linen threads, to the inner globe, to see whether it was at all over or undercharged.

221] For the more convenient performing this operation, I made use of the following apparatus. It is more complicated, indeed, than was necessary, but as the experiment was of great importance to my purpose, I was willing to try it in the most accurate manner.

Fig. 12.

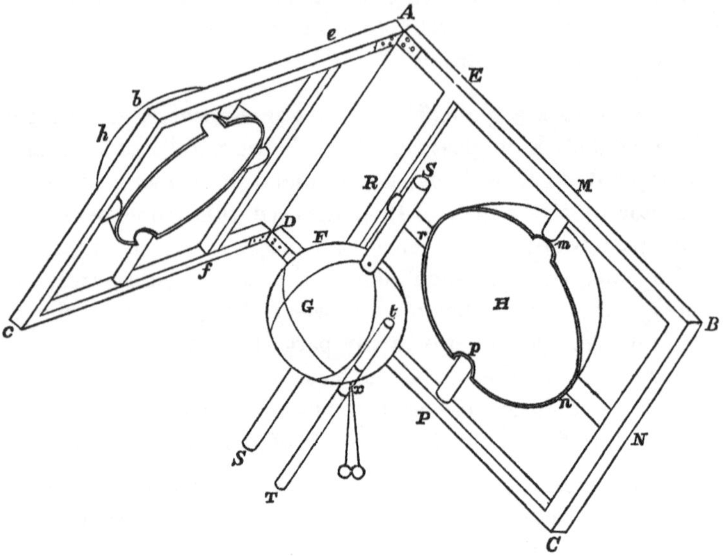

ABCDEF and *AbcDef* (Fig. 12) are two frames of wood of the same size and shape, supported by hinges at *A* and *D* in such manner that each frame is moveable on the horizontal line *AD* as an axis. *H* is one of the hemispheres, fastened to the frame *ABCD* by the four sticks of glass *Mm*, *Nn*, *Pp*, and *Rr*, covered with sealing-wax. *h* is the other hemisphere fastened in the same manner to the frame *AbcD*. *G* is the inner globe, suspended by the horizontal stick of glass *Ss*, the frame of wood by which *Ss* and the hinges at *A* and *D* are supported being not represented in the figure to avoid confusion.

Tt is a stick of glass with a slip of tinfoil bound round it at *x*, the place where it is intended to touch the globe, and the pith balls are suspended from the tinfoil.

The hemispheres were fixed within their frames in such manner that when the frames were brought near together the edges of the hemispheres touched each other all round as near as might be, so as to form a complete sphere, and so that the inner globe was inclosed within them without anywhere touching them, but on the contrary being at nearly the same distance from them in all parts.

222] It was also so contrived, by means of different strings, that the same motion of the hand which drew away the wire by which the hemispheres were electrified, immediately after that was done, drew out the wire which made the communication between the hemispheres and the inner globe, and immediately after that was drawn out, separated the hemispheres from each other and approached the stick of glass *Tt* to the inner globe. It was also contrived so that the electricity of the hemispheres and of the wire by which they were electrified was discharged as soon as they were separated from each other, as otherwise their repulsion might have made the pith balls to separate, though the inner globe was not at all overcharged.

The inner globe and hemispheres were also both coated with tinfoil to make them the more perfect conductors of electricity.

223] In trying the experiments a coated glass jar was connected to the wire by which the hemispheres were electrified, and this wire was withdrawn so as not to touch the hemispheres till the jar was sufficiently charged. It was then suffered to rest on them for two or three seconds and then withdrawn, and the hemispheres separated as above described.

224] An electrometer also was fastened to the prime conductor by which the coated jar was electrified, by which means the jar and consequently the hemispheres were always electrified in the same degree. This electrometer as well as the pith balls will be described in [Arts. 244 and 248]; the strength of the electricity was the same as was commonly used in the following experiments, and is described in [Arts. 263, 329, 359, 520].

225] My reason for using the glass jar was that without it it would have been difficult either to have known to what degree the hemispheres were electrified or to have kept the electricity of the same strength for a second or two together, and if the wire had been suffered to have rested on the hemispheres while the jar was charging, I was afraid that the electricity might have spread itself gradually on the sticks of glass which supported the globe and hemispheres, which might have made some error in the experiment.

226] From this manner of trying the experiment it appears: First, that at the time the hemispheres are electrified, there is a perfect communication by metal between them and the inner globe, so that the electricity has free liberty to enter the inner globe if it has any disposition to do so, and moreover that this communication is not taken away till after the wire by which the hemispheres are electrified is removed.

Secondly, before the hemispheres begin to be separated from each other, the wire which makes the communication between them and the globe is taken away, so that there is no longer any communication between them by any conducting substance.

Thirdly, from the manner in which the operation is performed, it is impossible for the hemispheres to touch the inner globe while they are removing, or even to come within $\frac{4}{10}$ths of an inch of it.

And Fourthly, the whole time of performing the operation is so short, that no sensible quantity of electricity can escape from the inner globe, between the time of taking away the communication between that and the hemispheres, and the approaching the pith balls to it, so that the quantity of electricity in the globe when the pith balls are approached to it cannot be sensibly different from what it is when it is inclosed within the hemispheres and communicating with them.

227] The result was, that though the experiment was repeated several times*, I could never perceive the pith balls to separate or show any signs of electricity.

228] That I might perceive a more minute degree of electricity in the inner globe, I tried the experiment in a different manner, namely, before the hemispheres were electrified, I electrified the pith balls positively, making them separate about one inch. When the hemispheres were then separated, and the tinfoil, x, brought in contact with the globe, and consequently the electricity of the pith balls communicated to the globe, they still continued to separate, though but just sensibly. I then repeated the experiment in the same manner, except that the pith balls were negatively electrified in the same degree that they before were positively. They still separated negatively after being brought in contact with the globe, and in the same degree that they before did positively.

[* Dec. 18—24, 1772, Arts. 512, 513, and April 4, 1773, Art. 562.]

229] It must be observed that if the globe was at all overcharged the pith balls should separate further when they were previously positively electrified than when negatively, as in the first case the pith balls must evidently separate further than they would do if the globe was not overcharged, and in the latter case less.

Moreover, a much smaller degree of electricity may be perceived in the globe by this manner of trying the experiment than the former, for when the pith balls have already got a sufficient quantity of electricity in them to make them separate, a sensible difference will be produced in their degree of divergence by the addition of a quantity of fluid several times less than what was necessary to make them separate at first. It is plain that this method of trying the experiment is not just, unless the hemispheres are electrified in nearly the same degree when the pith balls are previously electrified positively as when negatively, which was provided for by the electrometer.

230] In order to find how small a quantity of electricity in the inner globe might have been discovered by this experiment, I took away the hemispheres with their frames, leaving the globe and the pith balls as before. I then took a piece of glass, coated as a Leyden vial, which I knew by experiment contained not more than $\frac{1}{59}$th of the quantity of redundant fluid on its positive side that the jar by which the hemispheres were electrified did, when both were charged from the same conductor.

I then electrified this coated plate to the same degree, as shewn by the electrometer, that the jar was in the former experiment, and then separated it from the prime conductor, and communicated its electricity to the jar, which was not at all electrified. Consequently the jar contained only $\frac{1}{60}$th part of the redundant fluid in this experiment that it did in the former, for the coated plate and jar together contained only $\frac{1}{59}$th, and therefore the jar alone contained only $\frac{1}{60}$th.

By means of this jar, thus electrified, I electrified the globe in the same manner that the hemispheres were in the former experiment, and immediately after the electrifying wire was withdrawn, approached the pith balls. The result was that by previously electrifying the balls, as in the second way of trying the experiment, the electricity of the globe was very manifest, as the balls separated very sensibly more when they were previously electrified positively than when negatively, but the electricity of the globe was not sufficient to make the balls separate, unless they were previously electrified.

It is plain that the quantity of redundant fluid communicated to the globe in this experiment was less than $\frac{1}{60}$th part of that communicated to the hemispheres in the former experiment, for if the hemispheres themselves had been electrified they would have received only $\frac{1}{60}$th of the redundant fluid they did before, and the globe, as being less, received still less electricity.

231] It appears, therefore, that if a globe 12·1 inches in diameter is inclosed within a hollow globe 13·3 inches in diameter, and communicates with it by some conducting substance, and the whole is positively electrified, the quantity of redundant fluid lodged in the inner globe is certainly less than $\frac{1}{60}$th of that lodged in the outer globe, and that there is no reason to think from any circumstance of the experiment that the inner globe is at all overcharged.

232] Hence it follows that the electric attraction and repulsion must be inversely as the square of the distance, and that when a globe is positively electrified, the redundant fluid in it is lodged intirely on its surface.

For by Prop. V. [Art. 20], if it is according to this law, the whole redundant fluid ought to be lodged on the outer surface of the hemispheres, and the inner globe ought not to be at all over or undercharged, whereas, if it is inversely as some higher power of the distance than the square, the inner globe ought to be in some degree overcharged.

233] For let *ADB* (Fig. 13) be the hemispheres and *adb* the inner globe, and *Aa* the wire by which a communication is made between them. By Lemma IV. [Art. 18], if the electric attraction and repulsion is inversely as some higher power of the distance than the square, the redundant fluid in *ABD* repels a particle of fluid placed anywhere in the wire *Aa* towards the center, and consequently, unless the inner globe was sufficiently overcharged to prevent it, some fluid would flow from the hemispheres to the globe.

Fig. 13.

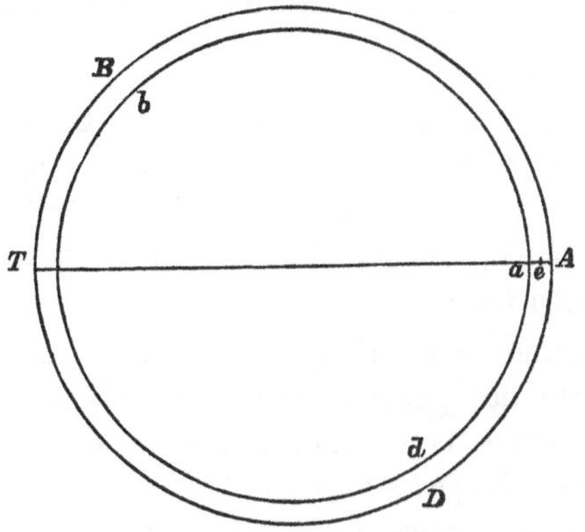

But if the electric attraction and repulsion is inversely as some lower power of the distance than the square, the redundant fluid in ABD impels the particle in the contrary direction, that is, from the center, and therefore the inner globe must be undercharged.

234] In order to form some estimate how much the law of the electric attraction and repulsion may differ from that of the inverse duplicate ratio of the distances without its having been perceived in this experiment, let AT be a diameter of the two concentric spheres ABD and abd, and let Aa be bisected in e. Ae in this experiment was about 35 of an inch and Te 13·1 inches, therefore if the electric attraction and repulsion is inversely as the $2+\frac{1}{50}$th power of the distance, it may be shewn that the force with which the redundant fluid in ABD repels a particle at e towards the center is to that with which the same quantity of fluid collected in the center would repel it in the contrary direction as 1 to 57.

But as the law of repulsion differs so little from the inverse duplicate ratio, the redundant fluid in the inner globe will repel the point e with very nearly the same force as if it was all collected in the center, and therefore if the redundant fluid in the inner globe is $\frac{1}{57}$th part of that in ABD the particle at e will be in equilibrio, and as e is placed in the middle between A and a, there is the utmost reason to think that the fluid in the whole wire Aa will be so too. We may therefore conclude that the electric attraction and repulsion must be inversely as some power of the distance between that of the $2+\frac{1}{50}$th and that of the $2-\frac{1}{50}$th, and there is no reason to think that it differs at all from the inverse duplicate ratio*.

235] EXPERIMENT II. A similar experiment was tried with a piece of wood 12 inches square and 2 inches thick, inclosed between two wooden drawers each 14 inches square and 2 inches deep on the outside, so as to form together a hollow box 14 inches square and 4 thick, the wood of which it was composed being ·5 to ·3 of an inch thick.

The experiment was tried in just the same manner as the former. I could not perceive the inner box to be at all over or undercharged, which is a confirmation of what was supposed at the end of Prop. IX. [Art. 41]—that when a body of any shape is overcharged, the redundant fluid is lodged entirely on the surface, supposing the electric attraction and repulsion to be inversely as the square of the distance †.

DEMONSTRATION OF COMPUTATIONS IN [ART. 234].

Fig. 13 *a*.

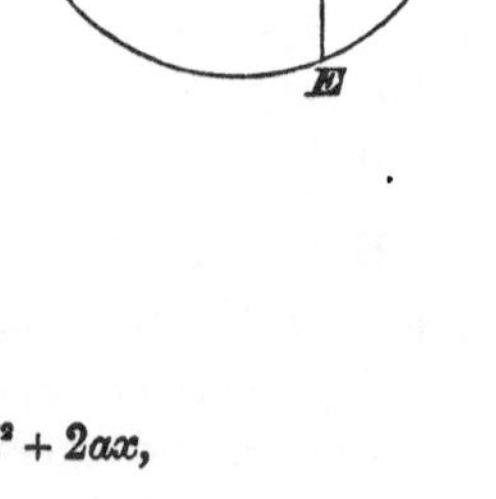

Let *aef* be a sphere, *c* its center, *b* any point within it, *af* a diameter, *Ee* any plane perpendicular to *af*.

Let $cb=a$, $ba=d$, $bf=s$ and $ad=x$, and let the repulsion be inversely as the n power of the distance. The convex surface of the segment *Eae* is to that of the whole globe as $ad : af$, and therefore if the point d is supposed to flow towards f, the fluxion of the surface *Eae* is proportional to $\dot{x}$, and the fluxion of its repulsion on b in the direction dc is proportional to

$$\frac{\dot{x}(d-x)}{be^{n+1}},$$

or may be represented thereby, but

$$be^2=(d-x)^2+x(2a+2d-x)=d^2+2ax,$$

therefore the fluxion of the repulsion is

$$\frac{\dot{x}(d-x)}{(d^2+2ax)^{\frac{n+1}{2}}},$$

* [Note 19.] † [Art. 561.]

the variable part of the fluent of which is

$$\frac{-2ad-d^2}{4a^2\frac{n-1}{2}(d^2+2ax)^{\frac{n-1}{2}}}-\frac{(d^2+2ax)^{\frac{3-n}{2}}}{4a^2\frac{3-n}{2}};$$

but when x is nothing, d^2+2ax, or $be^2=d^2$, and when $x=af$, or $s+d$, it $=s^2$, therefore the whole fluent generated while b moves from a to f is

$$\frac{2ad+d^2}{2a^2(n-1)}\left(\frac{1}{d^{n-1}}-\frac{1}{s^{n-1}}\right)+\frac{d^{3-n}-s^{3-n}}{2a^2(3-n)};$$

but the repulsion of all the fluid collected in the center on b

$$=\frac{s+d}{a^n},$$

and $$a=\frac{s-d}{2},$$

and $$2ad+d^2=ds,$$

therefore the repulsion of the surface of the globe is to that of the same quantity of fluid collected in the center as

$$\frac{ds}{n-1}\times\frac{s^{n-1}-d^{n-1}}{(ds)^{n-1}}+\frac{d^{3-n}-s^{3-n}}{3-n}:\frac{2(s+d)}{a^{n-2}},$$

or as $$\frac{s^{n-1}-d^{n-1}}{(n-1)(ds)^{n-2}}+\frac{d^{3-n}-s^{3-n}}{3-n}:\frac{(s+d)\,2^{n-1}}{(s-d)^{n-2}},$$

or dividing by s^{3-n}, as

$$\frac{s^{n-2}}{d^{n-2}(n-1)}-\frac{d}{s(n-1)}+\frac{d^{3-n}}{s^{3-n}(3-n)}-\frac{1}{3-n}:\frac{s+d}{s}\left(\frac{s-d}{s}\right)^{2-n}2^{n-1},$$

or as $$\frac{p^{2-n}-p}{n-1}+\frac{p^{3-n}-1}{3-n}:(1+p)\,2^{n-1}(1-p)^{2-n},$$

supposing $\frac{d}{s}$ to be called p.

[EXPERIMENTS ON THE CHARGES OF BODIES.]

236] The intention of the remaining experiments was to find out the proportion which the quantity of redundant fluid in bodies of several different shapes and sizes would bear to each other if placed at a considerable distance from each other and connected together by a slender wire, or, which comes to the same thing, to find the proportion which the quantity of redundant fluid in them would bear to each other if they were successively connected by a slender wire to a third body placed at a great distance from them, supposing the quantity of redundant fluid in the third body to be the same each time; and to examine how far that proportion agrees with what it should be by theory if the bodies were connected by canals of incompressible fluid.

237] To avoid circumlocution I shall frequently in the following pages make use of a term the meaning of which is given in the following definition.

DEF. When in relating any experiment in which two bodies B and b were successively connected to a third body and overcharged, I say that the charge of B was found to be to that of b as P to 1, I mean that the quantity of redundant fluid in B would have been to that in b in the above proportion, provided the quantity of redundant fluid in the third body was exactly the same each time, everything else being exactly the same as in the experiment, that is, the bodies being situated exactly as in the experiment. But when I say simply that the charge of one body is to that of another in any particular proportion, for instance, when I say that the charge of a thin circular plate is to that of a globe of the same diameter as 1 to 1·57, I would be understood to mean that if the circular plate and globe are successively connected to a third body by a thin wire the redundant fluid in the plate would be to that in the globe in that proportion, provided they were placed at a very great distance both from the third body and from any other over- or undercharged matter, and that the quantity of redundant fluid in the third body was exactly the same each time.

238] The method I took in making these experiments was by comparing each of the two bodies I wanted to examine, or B and b as I shall call them, one after another with a third body, which I shall call the trial plate, in this manner. I took two Leyden vials and charged both of them from the same conductor; I then electrified B positively by the inside of one of the vials, and at the same time electrified the trial plate negatively by the coating of the other vial. Having done this I tried whether the redundant fluid in B was more or less than sufficient to saturate the redundant matter in the trial plate, by making a communication between them by a piece of wire; for if the redundant fluid in B was more than sufficient to saturate the redundant matter in the trial plate, they would both be overcharged after the communication was made between them; if, on the other hand, the redundant fluid in B was not sufficient to saturate the redundant matter in the trial plate, they would be undercharged. Having by these means found what size the trial plate must be made so that the redundant matter in it should be just sufficient to saturate the redundant fluid in B, I tried the body b in the same manner, and if I found that it required the trial plate to be of the same size in order that the redundant matter in it should be just sufficient to saturate the redundant fluid in b, I was well assured that if B and b were successively made to communicate with a third body and positively electrified they would each of them contain the same quantity of fluid, supposing the quantity of redundant fluid in the third body to be the same each time; that is, that the charge of B was equal to that of b.

Having thus given a general idea of the method I used, I proceed to describe it more particularly.

239] The trial plates I made use of consisted of two flat tin plates *ABCD* and *abcd* (Fig. 15), made to slide one upon the other, so that by making the side *bc* of one plate extend more or less beyond the side *BC* of the other it formed a plate of a greater or less size, and which consequently contained more or less electricity*.

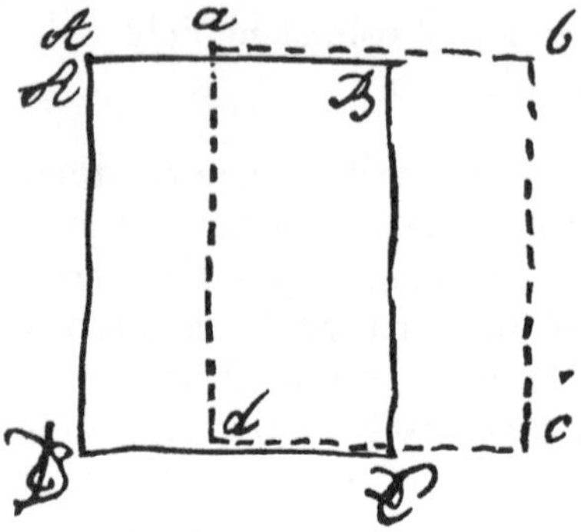

Fig. 15

240] The apparatus used in making these experiments is represented in Fig. 14, where the parallelogram *T* represents the trial plate and *B* one of the bodies to be compared together, each supported on non-conductors. *dDδ* is the wire for making a communication between them, having a joint in it at *D*, where it is supported by a non-conductor, and where are also hung two small pith balls to show whether *B* and *T* are over- or undercharged after the communication is made between them. *A* and *a* are the two vials; *Ee* is a wire communicating with the inside coating of *A*, *aCc* a wire communicating with the same coating of *a*; and *Ff* and *Gg* are wires fastened to the outside coating of *a*; *RrSs* is a wire for making a communication between *B* and the vial *A*, having joints in it at *R* and *S*, where it is supported by non-conductors, and *mMNn* is another wire of the same kind for making a communication between *T* and the vial *a*.†

241] In order to try the experiment I proceed in this manner: the wires *Dd* and *Dδ* are lifted off from the plates *B* and *T* so as not to touch them, and consequently so that there is no communication between *B* and *T*: the wires *Rr* and *Mm* are suffered to rest on *B* and *T*, and the wires *Ss* and *Nn* are lifted up so as not to touch *Ee* and *Ff*. The vials are then charged by means of the wire *bt* which rests on *Ee* and *Cc*, and communicates by the wire

* [See table for trial plate at Art. 468.]

† [See plan, Fig. 17, p. 128.]

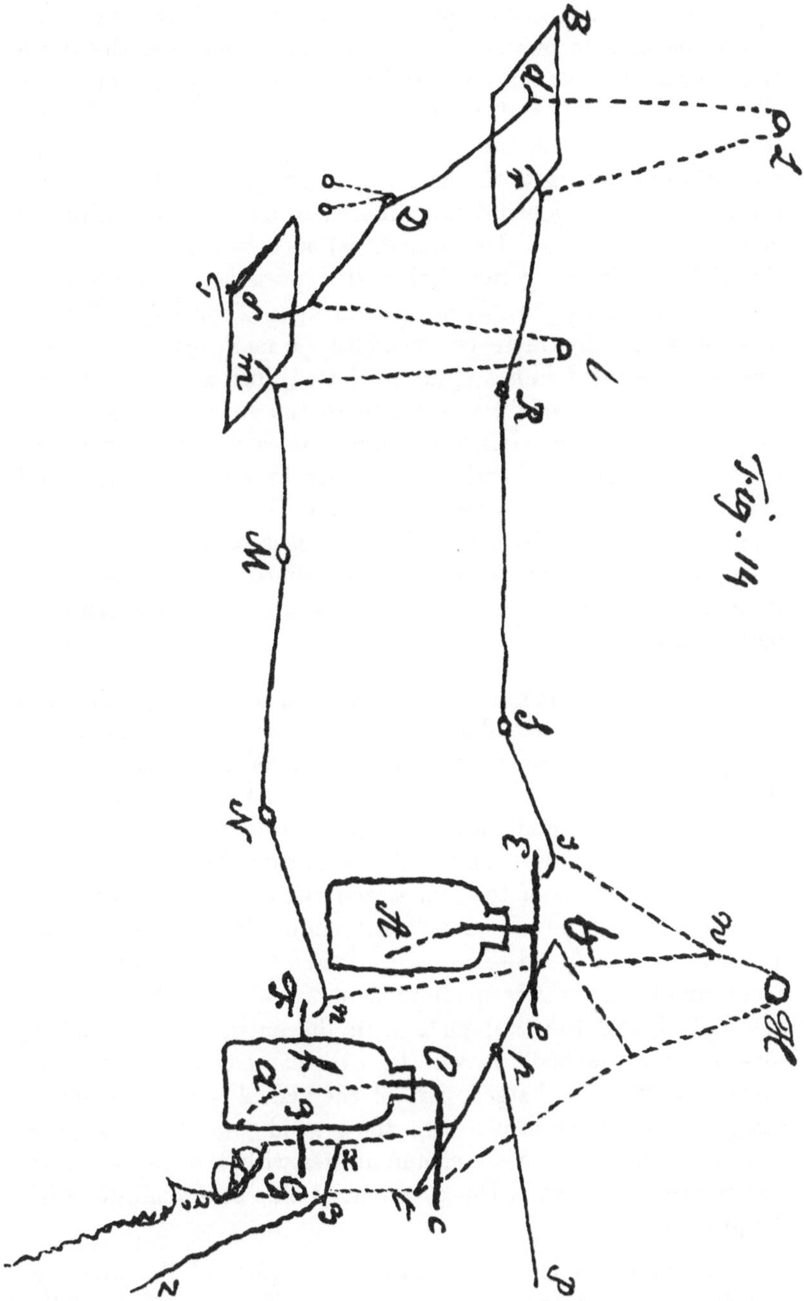

Fig. 14

Pp with the prime conductor, a communication being made between the outside of the vial *A* and the ground, and the vial *a* being made to communicate with the ground by the wire *xyz*, which rests on *Gg*, and is suspended from the wire *bt* by silk strings represented in the figure by dotted lines. When the vials are sufficiently charged, the wire *bt* is lifted up till *xy* bears against the bottom of *Cc*, *xy* being still suffered to communicate with the ground as before, and the communication between the outside of the vial *A* and the ground being still preserved. At the same time the wires *Ss* and *Nn* are let fall upon *Ee* and *Ff*. For the sake of doing this more commodiously I make use of the silk strings represented in the figure by dotted lines and passing over the pulley *H*. A weight is fastened to the string at *w*, which is supported while the vials are charging in such manner that the wires *Ss* and *Nn* are lifted up so as not to touch *Ee* and *Ff*, and the wire *bt* is suffered to rest on *Ee* and *Cc*, and the wire *xy* on *Gg*; and when the vials are sufficiently charged the weight is let down, by which means *Ss* and *Nn* are suffered to fall down upon *Ee* and *Ff*, and the wire *bt* is lifted up till *xy* bears against the bottom of *Cc*.

242] From what has been said it appears that whilst the vials were charging, the outsides of each of them communicated with the ground, and consequently the inside of each vial is overcharged and the outside undercharged. As soon as the vials are charged the communication of each of them with the prime conductor is taken away, and at the same time the communication between the outside of the vial *a* and the ground is taken away, so that it is intirely insulated, and, immediately after, a communication is made between its inside and the ground, and at the same time the body *B* is made to communicate with the inside of the vial *A*, and the trial plate with the outside of the vial *a*; consequently the body *B* will be overcharged as it communicates with the overcharged part of the vial *A*, while the undercharged side communicates with the ground; and the trial plate will be undercharged, as it communicates with the undercharged side of the vial *a*, while the overcharged side communicates with the ground.

Immediately after this operation is performed the wires *Rr* and *Mm* are lifted up, so as to cut off the communications of the bodies *B* and *T* with the vials, and, instantly after, the wires *Dd* and *Dδ* are let down, so as to make a communication between the body *B* and the trial plate. For the sake of expedition this operation was performed nearly in the same manner as the former, by means of the silk strings passing over the pullies *L* and *l*, and represented in the figure by dotted lines. I also employed an assistant to turn the electrical machine and to manage the silk strings passing over the pulley *H*, while I stood ready near *D* to perform the last mentioned operation as soon as the wires *Ss* and *Nn* were let down, and also to see whether the pith balls separated or not.

243] From the manner of performing the last mentioned operation it appears that the communication is not made between *B* and *T* till after their communication with the vials and all other bodies is cut off; consequently, if the quantity of redundant fluid communicated to *B* is more than sufficient to saturate the redundant matter in *T*, they will be overcharged after the communication is made between them, and the pith balls at *D* will separate positively, but if the redundant fluid in *B* is not sufficient to saturate the redundant matter in *T* they will be undercharged, and the pith balls will separate negatively.

244] The balls were made of pith of elder, turned round in a lathe, about one-fifth of an inch in diameter, and were suspended by the finest linen threads that could be procured, about 9 inches long.

245] In making these experiments I did not open the trial plate to such a surface that the pith balls should not separate at all on making the communication between *B* and *T*, and assume that for the size which must be given to the trial plate in order that the deficience of fluid in it should be equal to the redundance in *B* (or for the required surface of the trial plate, as I shall call it for shortness); but I first made the surface of the trial plate such that the deficient fluid therein should exceed the redundant in *B*, and that the pith balls should separate negatively, just enough for me to be sure they separated: I then diminished the surface of the trial plate till I found, on repeating the experiment, that the pith balls separated positively as much as they before separated negatively, and the mean between these I concluded to be the required surface of the trial plate.

246. This way of making the experiment I found much more accurate than the other, for supposing the required surface of the trial plate to be expressed by the number 16, I found that its surface must be increased to about 20 before I could be certain that the pith balls would separate negatively, and that it must be diminished to about 12 before they would separate positively; whereas I found that increasing its surface from 20 to 21 would make the balls separate sensibly further, and that diminishing its surface from 12 to 11 would have the same effect; so that I could determine the required surface of the trial plate at least four times more exactly by the latter method than by the former.

247] It will be shewn hereafter * that the quantity of deficient fluid in the trial plate is in proportion to the square root of its surface; consequently the redundant fluid in *B* must exceed, or fall short of, the deficient fluid in the trial plate by about $\frac{1}{8}$th part, in order that the balls should separate, and moreover the increasing or diminishing the deficience of fluid in the trial plate by about $\frac{1}{32}$ part will make a sensible difference in the separation of the balls.

248] It is plain that this way of finding the required surface of the trial plate is not just, unless the vials are charged equally in both trials, namely, that in which the balls separate positively and that in which they separate negatively; I therefore fastened an electrometer to the wire *Pp*, at a sufficient distance from the vials, consisting of two paper cylinders about three-quarters of an inch in diameter and one inch in height, suspended by linen threads about eight inches long, and in changing the vials took care always to turn the globe† till these cylinders just began to separate.

249] In all the later experiments, however, I made use of a more exact kind of electrometer, consisting of two wheaten straws, *Aa* and *Bb* (Fig. 30), eleven inches long, with cork balls *A* and *B* at the bottom, each one-third of an inch in diameter, and supported at *a* and *b* by fine steel pins bearing on notches in the brass plate *C*, and turning on these pins as centers. This electrometer was suspended by the piece of brass *C* from the prime conductor, and a piece of pasteboard, with two black lines drawn upon it, was placed six inches behind the electrometer on a level with the balls, in order to judge of the distance to which the balls separated, the eye being placed before the electrometer at thirty inches distance

* [Arts. 284, 479, 682.]

† [Of Nairne's electrical machine, see Art. 563.]

from them (a guide for the eye being placed for that purpose*), and the electrical machine was turned till the balls appeared even

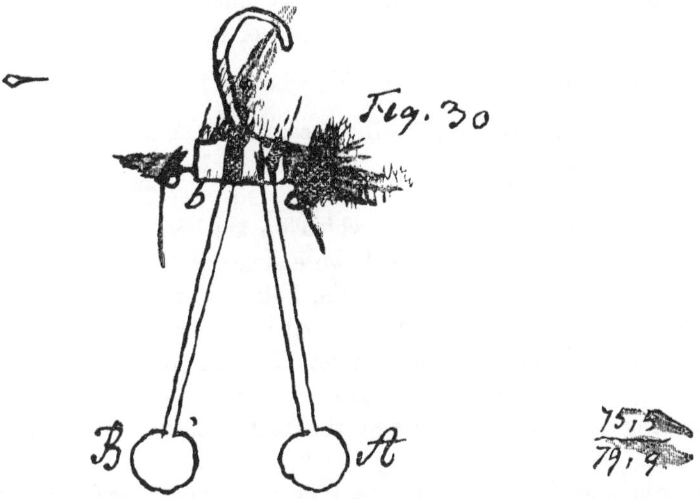

with those lines. By these means I could judge of the strength of the electricity to a considerable degree of exactness. In order to make the straws conduct the better they were gilt over.

250] In order to estimate what error may arise from the vials being not equally charged in both trials, let the required surface of the trial plate be called 16; then must the surface which must be given to it in order that the balls may separate negatively be 20, or $16+4$, supposing the vials to be charged with the usual degree of strength. Suppose now that in the next trial, in which the balls are to separate positively, the vials are charged stronger than before, in the ratio of x to 1, so that the quantity of redundant fluid in B shall be greater than before, in the ratio of x to 1, and that the deficience in the trial plate should be greater than before in the same ratio, provided its surface remained unaltered; then must the surface which must be given to the trial plate, in

* It is necessary that the eye should always be placed nearly at the same distance from the electrometer, as it is evident that the nearer the eye is placed the further the balls will appear to separate. But as the distance of the balls from the eye is so much greater than their distance from the pasteboard, a small alteration in the distance of the balls either from the eye or the pasteboard will make no sensible alteration in the distance to which the balls appear to separate.

order that the balls shall separate positively as much as they did negatively, be $16-\frac{4}{x}$; for, if this surface is given to it, it is plain that the redundant fluid in B will as much exceed the deficient in the trial plate as it before fell short of it. The mean between these two surfaces is $16+\frac{4(x-1)}{2x}$, whereas it ought to have been 16, so that the error which will proceed from thence in finding the required surface of the trial plate is $\frac{2(x-1)}{x}$, and, consequently, is less than half of the error which we are liable to in finding it the other way (or that in which we endeavour to find that surface of the trial plate with which the balls do not separate at all), though x is ever so great; for in that way it was before said that we were liable to an error of four. But if x is equal to $\frac{5}{4}$, which is as great an error of strength as I think can well arise in charging the vials, even when the first mentioned electrometer is used, the error in finding the required surface is only $\frac{1}{32}$ of the whole surface, or only $\frac{1}{8}$ part of what might arise the other way.

251] Having thus found what surface must be given to the trial plate, in order that the deficience of fluid in it shall be equal to the redundance in B, I take away the body B and put the other body b, which I want to compare with it, in its room, and if I find on repeating the experiment that the trial plate must be drawn out to the same surface as before, in order that the deficience of fluid in it shall be equal to the redundance in b, or, in other words, if the required surface of the trial plate is the same in trying b as in trying B, I am well assured that if B and b were successively made to communicate with one of the vials, or with any other third body, and were positively electrified, they would each of them contain the same quantity of redundant fluid, supposing the quantity of redundant fluid in the third body to remain the same each time. On the other hand, if I find that the required surface of the trial plate is greater in trying b than in trying B in the ratio of t^2 to T^2, I am well assured that the quantity of redundant fluid in b would exceed that in B in the ratio of t to T, supposing, as was said before, that the deficience of fluid in the trial plate is in proportion to the square root of its surface.

252] If the reader should think that this conclusion requires any proof it may be thus demonstrated:

Suppose that in trying *B* it was found that the required surface of the trial plate was T^2 and that in trying *b* it was t^2, and let us first suppose that the vials are charged in exactly the same degree in trying *b* as in trying *B*, then is the conclusion evident, for then are *B* and *b* successively made to communicate with the vial *A*, the charge of this vial being exactly the same each time, and the quantity of redundant fluid communicated to *b* is, actually, to that communicated to *B* as *t* to *T*. But it is plain that the conclusion is equally just, though the vials are charged higher in trying one than in trying the other. For though, in this case, the redundant fluid actually communicated to *b* will not be to that communicated to *B* in the ratio of *t* to *T*, yet we are sure that it would have been so if the vials had been charged in the same degree each time, for the required surfaces which must be given to the trial plate in trying *b* must evidently be the same whether the vials are charged to the same degree as they were in trying *B*, or to a different degree.

253] Though it is of no signification whether the vials are charged to the same degree in trying *b* as in trying *B*, yet it is necessary, as I said before, that in trying either *B* or *b* the vials should be charged nearly with the same strength when the balls are to separate positively as when they are to separate negatively, as otherwise a small error will arise in finding the required surface of the trial plate.

254] In all the following experiments I took care to proportion the size of the bodies *B* and *b* in such manner that the quantity of redundant fluid in one should not be very different from that in the other, so that, though the deficience of fluid in the trial plate should not be very nearly as the square root of its surface, it would make very little error in the conclusion.

255] The usual distance of the centers of *B* and *J* in these experiments was 83 inches, the distance of *B* from the vial *A* 106 inches, and that of *T* from *a* 86 inches, and the distance of the two vials about 10 inches*. The usual height of the body *B* and the trial plate above the ground was 50 inches; they were commonly supported upon pillars such as are represented in fig. 16,

* [See plan at Art. 265, details at Art. 466, and theory in Note 17.]

where *Ee*, *Bb* and *Dd* are three upright pillars of baked wood about 40 inches long, and *eε*, *bβ*, and *dδ* are sticks of glass 10 inches long and $\frac{1}{3}$ inch thick let into the wood, and covered with sealing-wax. *ACGF* is a piece of board which the pillars are fastened into. The points *M*, *N*, *R*, and *S* were each supported by a pillar of the same kind, and the point *D* was supported nearly in the same manner. In some experiments, however, the body *B* was suspended by silk strings. The wires *dDδ*, *rRSs*, and *mMNn* were about $\frac{1}{14}$ inch thick.

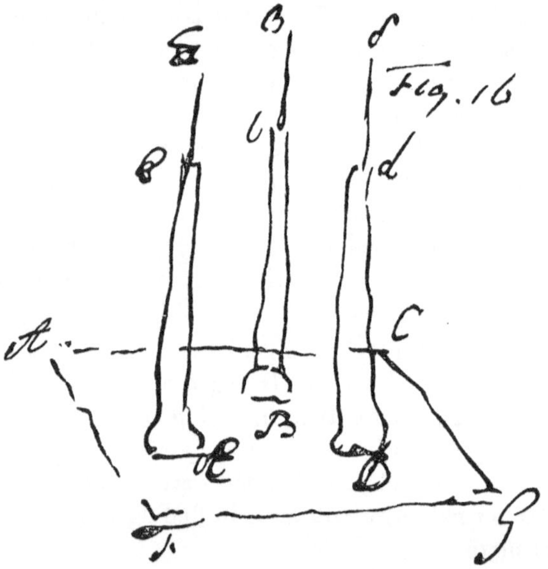

256] It is well known that the air of a room is easily rendered over- or undercharged, in particular if a wire such as *rRSs* [Fig. 14] is positively electrified, though even in no greater degree than in these experiments, and kept so for a second or two, and its electricity then destroyed, the air near it will be sensibly overcharged, as may be thus shewn. Take a pair of pith balls, like those hung at *D*, and suspend them within a few feet of the wire from some body communicating with the ground. The balls will instantly separate on electrifying the wire on account of the repulsion of the redundant fluid in it, but they will also continue to separate, though in a less degree, after the electricity of the wire is destroyed, which can be owing only to the air being rendered overcharged by it.

257] It may be suspected that this electrification of the air by the wires may affect the separation of the pith balls at *D* and thereby cause an irregularity in the experiments, but it must be considered that the wire *mMNn* is made as much undercharged as *rRSs* is overcharged, and the pith balls are placed about equally distant from both, so that the undercharged air near one wire will nearly balance the effect of the overcharged air near the other. Besides that, if it had any effect upon the separation of the balls, it would have much the same effect in trying *B* as in trying *b*, and therefore could hardly cause any error in the result of the experiment. However, still further to obviate any error from that cause, I had a contrivance by which the electricity of the wires *rRSs* and *mMNn*, as well as that of the vials, was destroyed as soon as the wires *rR* and *mM* were lifted up from *B* and *T*.

258] It is necessary that the outside of the bottle *A* and the wire *yz* should have as perfect a communication with the ground as possible, as otherwise it might happen that the body *B* and the trial plate might not receive their full degree of electrification before the wires *rR* and *mM* were lifted up. I therefore made them to communicate by a piece of wire with the outside wall of the house. This I found to be sufficient, for if I charged a vial, making the outside to communicate with the outside wall, and then made a communication by another wire between the inside of the vial and another portion of the outside wall of the house at several feet distance from the other, I found the vial to be discharged instantly; but if I made the wires to communicate only with the floor of the room instead of the wall of the house, I found it took up some time before the vial was discharged.

It must be observed that in this case, where you want to carry off the electricity very fast by an imperfect conductor, such as the wall, the best way is to apply a pretty broad piece of metal to the wall, so as to touch it in a considerable surface, and to fasten the wire to that, which was the way I last made use of, for if you only apply the wire against the wall, as it will touch the wall only in a few points, the electricity will not escape near so fast.

259] In dry weather the linen threads by which the pith balls are suspended are very imperfect conductors, so that the balls are apt not to separate or close immediately on giving or taking away the electricity. To remedy this inconvenience I moistened the threads with a solution of sea-salt, which I found answered the end perfectly well, for the threads after having been once moistened conveyed the electricity ever after very well, though the air was ever so dry.

260] As the charge of the vials *A* and *a* is continually diminishing from the time that the communication between them and the electrical machine is taken away, both by the electricity running along the surface of the vial from the inside to the outside, and by the waste of electricity from the wires *rRSs* and *mMNn* and their supports, it is necessary that the operation of electrifying *B* and *T* and lifting up the wires *rR* and *mM* should be performed as soon as possible, and, above all, it is necessary that the communication should be made between *B* and *T* as soon as possible after lifting up the wires *rR* and *mM*. This end was obtained very well by the manner, already described, of performing the operation.

261] Before I begin to relate the experiments, it will be proper to say something more about the accuracy that is to be expected in them. I before said that increasing or diminishing the surface of the trial plate by $\frac{1}{18}$ of what I called the required surface, *i.e.*, that surface in which the deficience was equal to the redundance in *B*, made a sensible alteration in the distance to which the pith balls separated. In reality I found that increasing or diminishing it by only $\frac{1}{24}$ part of the required surface would in general make a sensible alteration, but I could not be certain to nearly so small a quantity, for it would frequently happen that after having determined the surface of the trial plate at which the balls separated to a given degree, that on repeating the experiment a little after, the balls would separate differently from what they did before, and that I was obliged to alter the surface of the trial plate by $\frac{1}{12}$ and sometimes even $\frac{1}{8}$ of the required surface in order to make the balls separate in the same degree as before. Therefore, as increasing the surface of the trial plate by $\frac{1}{12}$ part increases the deficience of fluid therein by $\frac{1}{24}$ part, it appears that if the bodies *B* and *b* really contain the same quantity of redundant fluid, it might seem from the experiments as if *B* contained $\frac{1}{24}$ or even $\frac{1}{16}$ part more or less redundant fluid than *b*, so that I am liable to make an error of

$\frac{1}{24}$ or $\frac{1}{16}$ part in judging of the proportion of the quantity of redundant fluid in two bodies. I imagine, however, that it will not often happen that the error will amount to as much as $\frac{1}{32}$.

262] I do not very well know what this irregularity proceeded from. Part of it might arise from the difference in the strength with which the vials were charged, but I believe that part of it must arise from some other cause which I am not acquainted with. For greater security I always compared each body with the trial plate 6 or 7 times running.

263] It appears from the description of the electrometer fastened to the wire *Pp* that the vials were charged extremely weakly in these experiments, (they were indeed charged so weakly that if tried by Lane's electrometer they would not discharge themselves, if the distance of the knobs was more than $\frac{1}{25}$ of an inch,)* and it perhaps may be asked why I chose to charge them so weakly, as it is plain that the stronger the vials are charged the less alteration in the size of the trial plate would it have required to produce the same alteration in the separation of the pith balls.

264] My reason was this,—that the electricity seems to escape remarkably faster from any body, both by running into the air and by running along the surface of the non-conductor on which it is supported, when the body is electrified strongly than when it is weak, which made me afraid that if I had charged the vials much stronger the experiment might have been too much disturbed by the diminution of the quantity of redundant fluid in *B* and the deficience in the trial plate between the lifting up of the wires *Rr* and *Mm* and letting fall the wires *Dd* and *Dδ*, and also by the diminution of the charge of the vials between lifting up the wire *bt* and lifting up the wires *Rr* and *Mm*; and indeed it seemed, from some trials I made with a heavier electrometer fastened to *Pp*, as if the experiments were not more exact, if so much so, when the vials were charged stronger, as when they were charged in the usual degree.

I now proceed to relate the experiments I have made.

265] Exp. III. This experiment was made with a view to discover whether the quantity of redundant fluid communicated to

* [Difference of potentials about 11·8. See Art. 329 and Note 10.]

the body *B* was different according to the different situations in which it was placed in respect of the vial *A*, or according to the different shape of the wire *sSRr* by which it was touched, or according to the different parts in which it was touched by that wire. The body which I used for this purpose was a square tin plate, 12 inches each way, and the different ways in which it was tried are drawn in figure 17, which represents a plan of the disposition of the whole apparatus, in which the letters *B*, *d*, *D*, δ, *t*, *m*, *M*, *N*, *a*, *A*, *S*, *R* and *r* represent the same things as in fig. 14.

Fig. 17. [Scale $\frac{1}{48}$.]

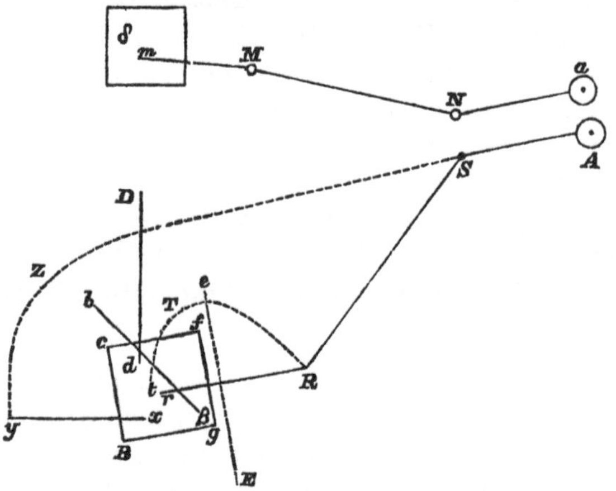

266] 1st Way. The tin plate was placed in a vertical plane so as to be represented in the plan by the line *bβ*, the wires *Rr* and *Dd* when let down resting on the edge of the plate as in the figure.

2nd. The tin plate was placed horizontal, as represented by the square *Bcfg*, the plate being placed so that the wire *Rr* touched it near the middle. N.B. The wire *Rr* was bent at right angles about $\frac{3}{4}$ of an inch from the end *r*, so that $\frac{3}{4}$ of an inch was in a vertical situation, and the rest horizontal. Consequently the wire touched the plate only by its extremity.

3rd. The same as the last, except that the wire *Rr* touched the plate not far from the side *fg*, and pretty near the middle of that side.

4th. The same as the last, except that a cross wire *eε* was fastened horizontally across the wire *Rr*, so as to be parallel to the side *fg*, and about one inch distance from it.

5th. The plate in the same situation as before, but the wire *Rr* was bent into an arch, like *tTR*, only the plane of that arch was vertical. The wire touched the plate near the middle.

6th. The plate in the same situation as before, but the wire *Rr* was removed into the situation *yx*, the communication between *y* and *S* being made by the wire *yzS* bent into an arch, as in the figure, the plane of which was vertical. The wire *yx* touched the plate near the middle.

N.B. In all these ways the tin plate was supported on silk lines.

267] The charges of the plate in the different situations were found to be to each other in the following proportions*:

1st Way	11·7,
2nd „	11·7,
3rd „	12·0,
4th „	10·8,
5th „	11·5,
6th „	10·8.

The plate was tried in some of these situations another night, when the charges came out in the following proportions†:

2nd Way	11·9,
3rd „	12·0,
5th „	11·8,
6th „	11·0.

268] It should seem from these experiments that the charge of the tin plate is not exactly the same in all the ways of trying it, as the extremes seem to differ from each other by above $\frac{1}{12}$ part, which is more than could arise from the error of the experiments; but, excepting the 4th and 6th ways, the others seem to differ by less than $\frac{1}{24}$. This I think we may be well assured of, that no sensible error can arise in the following experiments from any small difference in the manner in which the bodies are touched by the wire.

269] Exp. IV. These experiments were made with intent to see whether the charge of a body of a given shape and size was

* [Art. 470, Dec. 17, 1771. The numbers there found are here multiplied by a constant, so as to make the result by the 3rd way equal to 12.]

† [Art. 468.]

the same whatever materials it consisted of, as it ought to be according to Prop. XVIII.*, and also to see how far the charge of a flat plate depended on its thickness†. The substances used for this purpose were all flat plates about one foot square. The results of the experiments are given in the following Table‡:

Names of substances used.	Mean side of square.	Thickness.	Charge.	Side increased by 1† of thickness.	Reduced charge.
A tin plate	12·00	·02	11·92	12·03	11·89
A hollow plate composed of tin plates soldered together	11·03	1·01	12·30	12·38	11·92
Another of the same kind, but thinner	11·62	·37	12·08	12·11	11·97
A piece of pasteboard such as used for the covering of books	12·02	·087	11·95	12·14	11·81
A piece of Portland stone	12·00	·40	12·44	12·53	11·91
A sandstone known in London by the name of Bremen stone	12·04	·42	12·44	12·60	11·84
A slate such as used for the covering of houses	12·00	·16	12·20	12·21	11·99

N.B. The three pieces of stone were all ground flat, and of an uniform thickness.

270] As it would have been difficult to try the following substances by themselves, I coated panes of crown-glass with them on one side and tried them in that manner, which, as glass does not conduct electricity, seems as unexceptionable as it would have been to have tried them by themselves, supposing it had been possible to have done so.

Names of substances with which the glass was coated.	Mean side of square.	Thickness of glass.	Thickness of coating.	Charge.	Reduced charge.
Gold leaf	11·98	·056		11·87	11·89
Thin tin-foil	11·96	·058	·00113	11·55	11·59
Several folds of thick tin-foil stuck together with gum-water	11·98	·056	·017	11·93	11·95
Gum Arabic laid on in the form of gum-water and suffered to dry	12·05	·064		12·17	12·12
The same mixed up with a good deal of salt	11·96	·061		11·95	11·99
Charcoal powder mixed with a little gum-water	12·04			11·95	11·91
Water thickened with a little gum	11·96	·061		11·80	11·84

* [Art. 68.] † [Prop. XXI. Art. 73.]

‡ [Arts. 293, 471, 480, 481.]

The last mentioned substance was quite fluid, but had sufficient tenacity to prevent its flowing immediately to the lowest part of the plate. In those substances in which the thickness of the coating is not set down it was not measured, but the thickness was small.

271] All these things were supported on the pillars of baked wood and waxed glass described at [Art. 255]. The panes of glass were laid on these pillars with their coated sides uppermost, so that the wires *Rr* and *Dd* fell on their coated sides. As many of the substances used were but imperfect conductors of electricity, I fastened bits of tin-foil about an inch square on the places on which the wires *Rr* and *Dd* touched the plate in order to make the electric fluid spread more readily over it, and I satisfied myself beforehand that with this precaution they conducted readily enough for my purpose, as I found by discharging a Leyden vial, and making these substances part of the circuit.

272] It appears from these experiments that the charge of a thick plate is greater than that of a thin one of the same base, as might be guessed from the theory*, and it seems to be equal to that of a very thin one whose side exceeds that of the thick one by about $1\frac{1}{3}$ of its thickness. Let us therefore increase the mean side of each of these plates by $1\frac{1}{3}$ of its thickness, where that quantity is worth regarding, and alter the charge found by experiment in the ratio of 12 inches to the side thus increased, which will give us the charge of a plate of the same materials and shape whose increased side is 12 inches, when the charge of each substance will stand as in the last column of the preceding Table. These numbers do not differ from each other by more than what may fairly be supposed owing to the error of the experiment, and therefore I think we may conclude—firstly, that the charge of a body of a given shape and size is the same whatever materials it consists of, and, though the experiment was tried only with square plates, yet I think there can be no doubt but the case will be the same with bodies of any other shape; secondly, that the charge of any thin plate is very nearly the same whatever its thickness may be, provided its thickness is very small in respect of its breadth or smallest diameter; and there can be no doubt also but what this will hold good in thin plates of any shape, though it was tried only

* [Note 20.]

with square ones; and thirdly, if the plate is square and its thickness is several times less than its side, though not small enough to be disregarded, its charge is equal to that of a very thin square plate whose side exceeds that of the former by about $1\frac{1}{3}$ of its thickness.

This last circumstance seems far from being repugnant to the theory; but as I do not know how to calculate the charge of such a plate within tolerably near limits, I shall not trouble the reader any further about it.

273] Exp. V. This experiment was made with a view to find what proportion the charges of similar bodies of different sizes bear to each other, and whether it is the same that it ought to be by the theory on a supposition that the electric attraction and repulsion is inversely as the square of the distance, and that the bodies are connected to the jar by which they are electrified by canals of incompressible fluid. It was tried by taking two circular tin plates of 9 inches diameter, and comparing the charge of these two circles together with that of one of twice the diameter. The circles were placed in a vertical situation, and were disposed as in fig. 18, where the letters *D*, *δ*, *m*, *T*, *M*, *N*, *a*, *A*, *S* and *R* stand for the same things as in fig. 17. *Bb* and *Cc* are the two small circles placed parallel to each other, which, as they are in a vertical

Fig. 18.

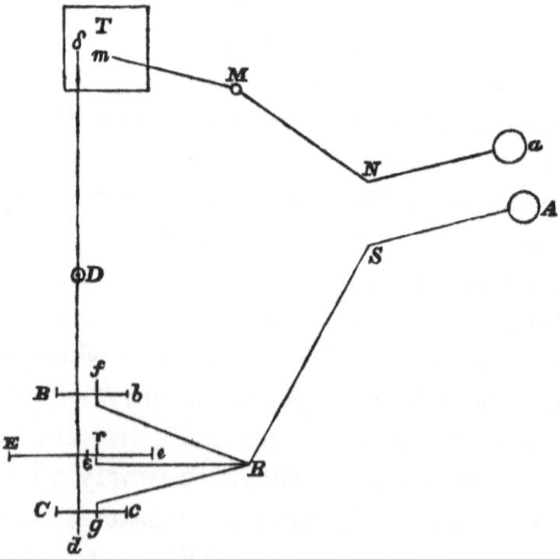

situation, appear in the plan as straight lines. *Rf* and *Rg* are the wires by which they are electrified, which are bent near *f* and *g* so as to enable them to rest on the edges of the circles. *Dd* is the wire for making a communication between them and the trial plate; *Ee* is the large circle placed half way between the two small ones, and *Rr* the wire by which it is electrified. But it must be observed, that in trying the large circle the two small circles and the wires *Rf* and *Rg* are taken away and the wire *Rr* put in their room; and in like manner, when the small circles are tried, the circle *Ee* and the wire *Rr* are removed.

274] It must be observed that the charge of the two small circles together will not be as much as double the charge of one circle, unless the distance of the two circles from each other is extremely great. In order, therefore, to know better what allowance to make on this account, I tried the experiment with the two small circles placed at three different distances, namely, at 18, 24, and 36 inches from each other, the circles being always placed so that the middle point between them was at the same distance from *D*. Their charges came out in the following proportion*:

The large circle	1·000,
The two small ones at 36 inches distance	·899,
" " " 24 " "	·859,
" " " 18 " "	·811.

275] I repeated the experiment in the same manner, except 1st, that the distances of the vials from the circles and trial plate were different from what they were before, namely, in the foregoing experiment the distance *Ta* from the middle of the trial plate to the vial *a* was 87 inches and *εA*, or the distance from the center of *Ee* to the vial *A*, was 106 inches, whereas in this experiment *Ta* was 98 inches and *εA* 63 inches; the distance *Tε* was 83 inches in both experiments; and 2ndly, that I placed a frame of wood about 5 feet square under the circles 14 inches from the ground. The reason of these alterations will be shewn by and by†. Their charges came out as follows:

The large circle	1·000,
The two small ones at 36 inches distance	·894,
" " " 24 " "	·840,
" " " 18 " "	·798.

276] Let us now endeavour to find out what proportion the charges ought to bear to each other by the theory on the above-mentioned supposition of their being connected by canals of in-

* [Arts. 452, 454, 472—475.]

† [Arts. 277, 339, 474 and Note 17.]

compressible fluid, and of the electrical attraction and repulsion being inversely as the square of the distance. This cannot be done exactly without knowing the manner in which the redundant fluid is disposed in the circles, which I am not acquainted with, but if we suppose the fluid to be spread uniformly over the plates, it will appear, by calculating according to Prop. XXX. [Art. 141], that their charges should be in the following proportion:

The large circle			....	1·000,
The two small ones at	36	inches	distance	·933,
„ „ „	24	„	„	·911,
„ „ „	18	„	„	·890.

If we suppose that the whole redundant fluid is collected in the circumference, they should be as follows:

The large circle			....	1·000,
The two small ones at	36	inches	distance	·890,
„ „ „	24	„	„	·844,
„ „ „	18	„	„	·805;

and if we suppose that $\frac{11}{24}$ of the whole redundant fluid is collected in the circumference, and the remainder, or $\frac{13}{24}$, spread uniformly, they should be as follows:

The large circle			....	1·000,
The two small ones at	36	inches	distance	·920,
„ „ „	24	„	„	·890,
„ „ „	18	„	„	·863.

277] I think this latter proportion of the charges much the most likely to agree with the truth*, as it appears from an experiment which will be mentioned hereafter, that the charge of a circular plate bears the same proportion to that of a globe that it would do if the fluid was disposed in that manner. But it must be observed that in these calculations the circles are supposed to be placed at an infinite distance from the vial by which they are electrified, and also from any other over- or under-charged body, whereas in these experiments the circles were at such a distance from the vial that their repulsion on the canal by which they communicated with it was sensibly less than if it was infinite, and moreover the attraction of the under-charged trial plate on the

* I would not be understood by this to suppose that the fluid is actually disposed in this manner in a circular plate, but only that the charges will bear the same proportion to each other that they ought to do on this supposition.

wire *mMNn* has some tendency to increase the quantity of fluid in the circles, and the repulsion of the circles tends to diminish the quantity of fluid in the trial plate, and moreover the floor and walls of the room will be made under-charged near the circles and over-charged near the trial plate, which will also have some tendency to alter the quantity of fluid in the circles and trial plate.

It was with a view to find out what error could proceed from these causes that I tried the experiment in the two different ways above mentioned. It will be shewn, however, in the appendix*, that the first two of these causes cannot produce any sensible alteration in the experiment, and that it is not likely that the last should. This is also confirmed by the near agreement of the results in both ways of trying the experiment, as the difference in the proportion of the charges in these two ways of trying the experiment was not greater than what might well be owing to the error of the experiment.

278] It seems reasonable to conclude, therefore, that the proportion which the charges ought to bear to each other in the theory on the supposition of their being connected by canals of incompressible fluid, and of the electrical attraction and repulsion being inversely as the squares of the distances, must be nearly as in the last Table, and therefore it should seem that the observed charges of the two small plates were rather less in proportion to that of the large one than they ought to have been by theory on the above-mentioned supposition; but the difference is not great, and perhaps not more than what may be owing to our not being able to compute the true proportion with sufficient accuracy, and to the error of the experiment, though I am more inclined to think that the difference is real. This, however, can by no means be looked upon as a sign of any error in the theory, but, on the contrary, I think that the difference being so small is a strong sign that the theory is true. For it cannot be expected that the charges of bodies connected together by wires should bear exactly the same proportion to each other that they should do if they were connected by canals of incompressible fluid; and, indeed, the third experiment shews that they do not, as the charge of the tin plate was found to be a little different according to the situation in which it was placed and the disposition of the wire by which it

* [Art. 188, and Notes 17 and 21.]

was touched, which should not be the case if it was connected to the vial by a canal of incompressible fluid.

279] Exp. VI. This experiment was made with the same view as the last, and consisted in comparing the charge of two brass wires together, with that of a single one of twice the length and thickness. The small wires were 3 feet long and $\frac{1}{10}$th of an inch thick; they were placed horizontal and parallel to each other, as represented by the lines *Bb* and *Cc* in fig. 18, and were tried at three different distances from each other, viz.:—18, 24, and 36 inches. The long wire was 6 feet long and $\frac{1}{5}$th of an inch in thickness, and was placed in the same direction as the small ones, as represented by *Ee*. They were electrified by the same wires and in the same manner as the circles, only they were placed so as to be touched by the wires *fR*, *rR*, and *gR*, very near their extremities *b*, *e*, and *c*. Their charges were as follows:—

The long wire	1·000,
The two short ones at 36 inches distance	·903,
„ „ „ 24 „ „	·860,
„ „ „ 18 „ „	·850.

280] The charges of the two small wires at the several distances of 36, 24, and 18 inches ought by theory to have been to that of the long wire in a proportion between that of ·923, ·905, and ·883 to 1 and that of ·893, ·860, and ·835 to 1, supposing them to be connected to the vial by canals of incompressible fluid, but, as it should seem from the next experiment, ought in all probability to approach much nearer to the former proportion than the latter. The observed charges were actually between these two proportions, but approached much nearer to the latter, so that they agreed as nearly with the computation as could be expected*.

281] Exp. VII. Being a comparison of the proportional charges of several bodies of different shapes: the result is as follows:—

A globe 12·1 inch in diameter	1·000
A tin circle 18·5 „ „	·992
A tin plate 15·5 inches square	·957
An oblong tin plate 17·9 inches by 13·4 inches	·965
A brass wire 72 inches long and ·185 thick	·937
A tin cylinder 54·2 inches long and ·73 in diameter.	·951
A tin cylinder 35·9 inches long and 2·53 in diameter	·999

* [Arts. 453, 476, 477, 683, and Note 13.]

The globe was the same that was used in the first experiment. The wire and cylinders were placed in the same manner as the large wire in the preceding experiment, and were touched in the same manner *.

282] Remarks on this experiment.

First, the proportion which the charge of the circular plate bears to that of the globe agrees very well with the theory, for by Prop. XXIX. [Art. 140] the proportion should be between that of ·76 to 1 and that of 1·53 to 1, and the observed proportion is that of ·992 to 1. We may conclude also from this experiment that the charge of a circular plate is to that of a globe of the same diameter as 12 to $18\frac{1}{2}$, which by the above-mentioned proposition is the proportion which ought to obtain if $\frac{13}{24}$ of the whole quantity of redundant fluid in the plate was spread uniformly [over the surface], and the remainder, or $\frac{11}{24}$, was spread uniformly [round the circumference], that is, if the value of p in that proposition equals $\frac{13}{11}$ †.

283] 2ndly. The charge of a square plate is to that of a circle whose diameter equals the side of the square, as 1·53 to 1, or its charge is to that of a circle whose area equals that of the square as 1·02 to 1 ‡.

284] 3rdly. The charge of the oblong plate is very nearly equal to that of a square of the same area, and consequently as the length of the trial plates used in these experiments never differed from their breadth (whether the trial plate was more or less drawn out) in a greater proportion than those of this oblong plate do, and as the charges of similar bodies of different sizes are as their corresponding diameters, or sides, I think we may safely conclude that the charges of these trial plates were as the sides of a square of the same area, agreeable to what was said in [Art. 247].

285] 4thly. By Prop. XXXI. [Art. 150] the charge of a cylinder whose length $= L$ and diameter $= D$ is to that of a globe whose diameter $= L$ in a ratio between that of 1 to $\log_e \frac{2L}{D}$ and that of 2 to $\log_e \frac{4L}{D}$, and therefore the charges of the brass wire, long

* [Arts. 478, 682.]

† [Arts. 654, 681, and Note 2.]

‡ [Arts. 479, 682, and Note 22.]

cylinder and short cylinder, should be to that of the globe, supposing them to be connected with the vial by which they were electrified by canals of incompressible fluid, in a ratio between that of ·894, ·896 and ·887 to 1 and that of 1·619, 1·573 and 1·469 to 1. The observed charges are as ·966, ·980 and 1·028 to 1, which are between the two above-mentioned proportions, but approach much nearer to the former than the latter, as might have been expected; so that the observed charges agree very well with the theory*.

286] 5thly. If we suppose that the redundant fluid is disposed in the same manner in a cylinder, whether the length is very great in respect of the diameter or not, it is reasonable to suppose that the charges of the brass wire, long cylinder and short cylinder, should be to each other in a proportion not much different from that of ·894, ·896 and ·887, or that of ·966, ·968 and ·959. The observed charges do not differ a great deal from that ratio, only the charges of the two cylinders, especially the shorter, are rather greater in proportion to that of the brass wire than they ought [to be], so that according to this supposition the observed charges do not agree exactly with computation. But if we suppose that the redundant fluid is spread less uniformly in a cylinder whose length is not very great in proportion to its diameter than in another, that is, that there is a greater proportion of the redundant fluid lodged near the extremities, which seems by no means an improbable supposition, the observed charges may perhaps agree very well with what they should be by theory, if they were connected by canals of incompressible fluid.

287] With regard to the small disturbing causes mentioned in [Art. 277], as the length of the brass wire bears so great a proportion to its distance from the trial plate and to its distance from the ground, it is possible that its effect in increasing the deficiency of fluid in the trial plate may be sensibly less, and also that the increase of charge, which it receives itself from the ground near it being under-charged, may be sensibly different from what it would be if it had been of a more compact shape, so that perhaps some alterations may have been made in the experiments by these two causes. I should imagine, however, that they could be but small. It must be observed that the first of these two causes tends to make the charge of the wire appear

* [Note 12.]

greater than it really was, and consequently to make the observed charges appear to agree nearer with the theory than they really did. Which way the second cause should operate I cannot say.

On the whole it should seem as if the true charge of a cylinder whose length is L and diameter D is to that of a globe whose diameter is L nearly as $\frac{9}{8}$ to natural logarithm $\frac{2L}{D}$, or as ·489 to Tabular log. $\frac{2L}{D}$.

288] Exp. VIII. Let AB, ab and eg (Fig. 19) be three equal thin parallel plates equidistant and very near to each other, and let Cf, the line joining their centers, be perpendicular to their planes, and let all three plates communicate with each other and be positively electrified: it may easily be shewn that according to the theory the quantity of redundant fluid in the middle plate will be many times less than that in either of the outer plates, or than that which it would receive by the same degree of electrification if placed by itself. I therefore took three tin plates, each 12 inches square, and placed them as above described, and electrified them by means of a wire fixed to a Leyden jar, the end of the wire being formed in such manner as to touch all three plates at once. As soon as the electrifying wire was taken away I drew away the outer plates, and at the same time approached a pair of cork balls to the middle plate in the same manner as I did to the globe in the first experiment and observed how much they separated, care being taken to take away the electricity of the outer plates as soon as drawn away. I then removed the outer plates and, by the same means that I used in the first experiment, made the quantity of redundant fluid in the jar less than before in a given ratio, and by means of this jar electrified the middle plate by itself and approached the cork balls as before. In this manner I proceeded

Fig. 19.

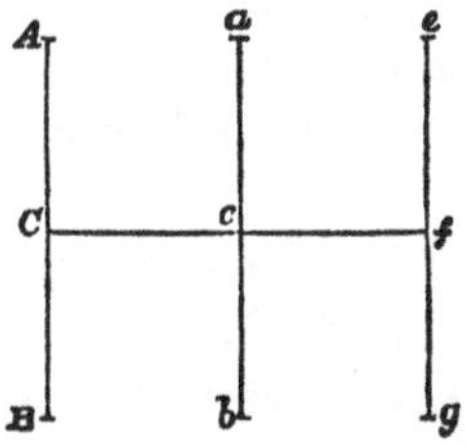

till I found how much it was necessary to diminish the quantity of redundant fluid in the jar in order that the corks might separate as much as before, and consequently how much less the quantity of redundant fluid in the middle plate when placed between the two other plates was than that which it would have received by the same degree of electrification if placed by itself*.

The result was that when the distance of the outer plates was $\begin{cases} 1 \cdot 15 \\ 1 \cdot 65 \end{cases}$ inches, the quantity of redundant fluid in the middle plate was about $\begin{cases} 8 \\ 7 \end{cases}$ times less than it would be if electrified in the same degree when placed by itself.

289] It is plain that according to the theory the quantity of redundant fluid in each of the outer plates should be the same, and that the quantity in the middle plate should be such that the repulsion of AB and ab together on the column cf shall be equal to that of the plate ϵg thereon in the contrary direction, and the redundant fluid in each of the outer plates is not much more than one-half of that which it would receive by the same degree of electrification if placed by itself. Now it will appear by computing, according to the principles delivered in Prop. XXX. [Art. 141], that the quantity of redundant fluid in the middle plate will be so excessively different according to the different manner in which the fluid is disposed in the plates that there is no forming any tolerable guess how much it ought to be; but if we suppose that part of the redundant fluid in each plate is spread uniformly and the rest collected in the circumference, and that in the outer plates the part that is spread uniformly is $\frac{11}{24}$ of the whole, as we supposed in Experiment V., the quantity of redundant fluid in the middle plate when the distance of the outer plates is 1·15 inches will not agree with observation, unless we suppose that not more than the 21st part of it is spread uniformly; but if we suppose that $\frac{4}{7}$ of the redundant fluid in the outer plates is spread uniformly the quantity in the middle plate will agree with observation, if we suppose that about $\frac{1}{9}$ of it is spread uniformly and the rest collected in the circumference.

When the distance of the outer plates is 1·65 inches there is no need of supposing so great a proportion of the fluid in the middle

* [Art. 542 and Note 23.]

plate to be disposed in the circumference in order to reconcile the theory with observation.

N.B. The more uniformly we suppose the fluid to be spread in the outer plates and the less so in the middle, the greater should be the quantity in the middle plate.

The above computations were made on the supposition that the plates were circles of 14 inches diameter, that is, nearly of the same area that they actually were of.

290] It will appear by just the same method of reasoning that was used in the remarks on the 22nd Proposition [Art. 74], that a vastly greater proportion of the redundant fluid in the middle plate will be collected near its circumference than would be if the outer plates were taken away, and perhaps this circumstance may make the fluid in the outer plates be spread more uniformly than it would otherwise be, so that it seems not improbable that the fluid in the plates may be disposed in such manner as to make the experiment agree with the theory.

The circumstance of its being necessary to suppose a greater proportion of fluid in the middle plate to be lodged in the circumference when the plates are at the smaller distance from each other than when they are at the greater agrees very well with the theory, for it is plain that the nearer the outer plates are to each other the greater proportion of the fluid in the middle plate should be lodged in the circumference.

On the whole I see no reason to think that the experiment disagrees with the theory, though the middle plate was certainly more overcharged than I should have expected.

GENERAL CONCLUSION.

291] The 1st experiment shews that when a globe is electrified the whole redundant fluid therein is lodged in or near its surface, and that the interior parts are intirely, or at least extremely nearly, saturated, and consequently that the electric attraction and repulsion is inversely as the square of the distance, or to speak more properly, that the theory will not agree with experiment on the supposition that it varies according to any other law.

292] The 2nd experiment shews that this circumstance of the whole redundant fluid being lodged in or near the surface obtains also in other shaped bodies, as well as in the globe, conformably to the supposition made in the remarks at the end of Prop. IX. [Art. 41]. These two experiments, at the same time that they determine the law of electric attraction and repulsion, serve in some measure to confirm the truth of the theory, as it is a circumstance which, if it had not been for the theory, one would by no means have expected.

293] From the 4th experiment it appears, first, that the charge of different bodies of the same shape and size, all ready conductors of electricity, is the same, whatever kind of matter they are composed of; and secondly, that the charge of thin plates is very nearly the same whatever thickness they may be of, provided it is very small in respect of their breadth or smallest diameter; but if their thickness bears any considerable proportion to their breadth, then their charge is considerably greater than if their thickness were very small. These two circumstances are perfectly conformable to the theory, and are a great confirmation of the truth of it.

294] The remaining experiments contain an examination whether the charges of several different sized and different shaped bodies bear the same proportion to each other, which they ought to do according to the attempts made in different parts of these papers to compute their charges by theory, supposing, as we have shewn to be the case, that the electric attraction and repulsion is inversely as the square of the distance: with regard to this it must be observed that, as in computing their charges I was obliged to make use of a supposition, which certainly does not take place in nature, it would be no sign of any error in the theory if their actual charges differed very much from their computed ones; but, on the other hand, if the observed charges agree very nearly with the computed ones, it not only shews that the actual charges of different bodies bear nearly the same proportion to each other that they would do if they were connected by canals of incompressible fluid, but is also a strong confirmation of the truth of the theory. Now this appears to be the case, for, first the charge of a tin plate was found to be nearly, though not quite, the same in whatever part it was touched by the electrifying wire, or in whatever direction it was placed in respect of the jar by which it was electrified.

Secondly, the charge of a single plate or wire was found to bear nearly, though, in the first case, I believe, not quite the same proportion to two similar plates or wires of half the diameter or length which it ought to do according to computation. Thirdly, the proportion which the charges of a thin circular plate and of three cylindrical bodies of different lengths and diameters bear to that of a globe agree with computation; but it must be observed that, as the proportion of the charges of the bodies to that of the globe is determined by the theory within only very wide limits, their agreement cannot be looked upon as so great a confirmation of the theory as it would otherwise be, yet as their shapes are so very different I think that their agreement, even within those limits, may be considered as a considerable confirmation of it.

PART *

[EXPERIMENTS ON COATED PLATES.]

295] This part consists chiefly of experiments made to determine the charges of plates of glass and other electric substances coated in the manner of Leyden vials. The method I used in doing this was nearly of the same nature as that by which I determined the charges of the other sort of bodies in the preceding part, but the apparatus was more compact and portable and is represented in Fig. 20, where *Hh* is a horizontal board lying on the ground, *Ll* and *Ll* are two upright pillars supporting the two horizontal bars *Nn* and *Pp*, both at the same height above the ground, and parallel to each other.

To these two bars are fastened four upright sticks of glass covered with sealing wax; they are represented in the figure and shaded black, but are not distinguished by letters to avoid confusion. To these sticks of glass are fastened four horizontal pieces of wire *Aa*, *Bb*, *Dd*, and *Ee*, and to *Bb* is fastened another wire *mM* supported at the further end by a stick of waxed glass.

Rr is a wooden bar reaching from the wire *Ee* to the pillar *Ll*, and along the upper edge of this bar runs a wire, one end of which is wound round the wire *Ee* and the other reaches to the ground and serves to make a communication between *Ee* and the ground. *Cc* and *Kk* are two wires fastened firmly together at *k* serving to electrify the plate. They are moveable upon *K* as a center where they communicate with the inside coating of one or more large glass jars, and the same electrometer that was used in the former experiments is fastened to the prime conductor by which the jars

* [Not numbered by Cavendish.]

Fig. 20.

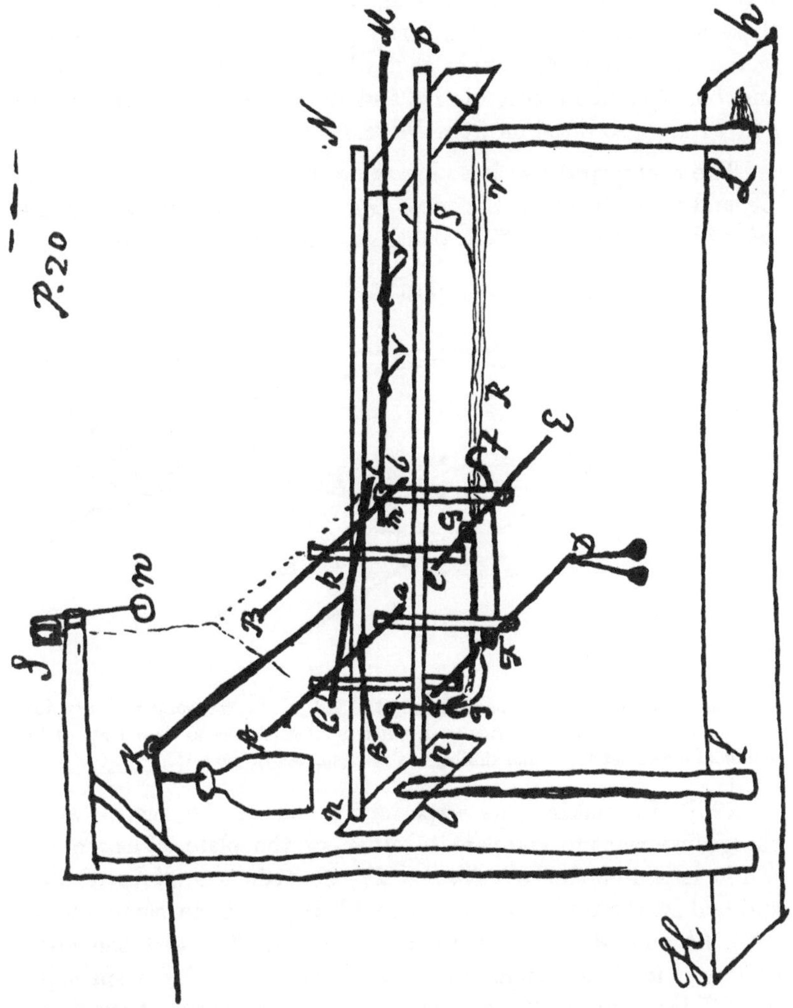

are electrified, in order that they may be charged to the same degree each time.

To the ends *C* and *c* of the wire *Cc* is fastened a silk string, as represented in the figure, passing over the pulley *S*, with a counterpoise *w* at the other end which serves to lift *Cc* from off the wires *Aa* and *Bb*, or to let it down upon them at pleasure. *Gg* is a wire the end *G* of which is bent into a ring, through which passes the wire *Ee*, so that *Gg* turns upon *Ee* as a center. *Ff* is a wire turning in the same manner as *Dd*. The ends *g* and *f* of these wires are fastened by silk strings to *C* and *c* as represented in the figure, in such manner that when *Cc* rests on the wires *Aa* and *Bb*, *Gg* and *Ff* rest on *Dd* and *Ee*, but on lifting up *Cc*, *Gg* and *Ff* are also lifted off from *Dd* and *Ee*.

The counterpoise *w* is so heavy as to overcome the weight of *Cc*, and to lift it up till the wires *Gg* and *Ff* bear against *Aa* and *Bb*, which prevents *Cc* from rising any higher.

Fig. 20 *a*.

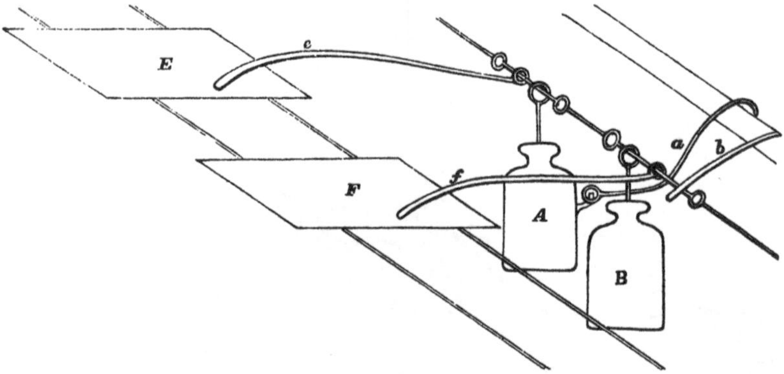

[Note. This Figure was found among the MS. It is not numbered, nor does any part of the MS. seem to refer to it, but it is inserted here to show some of the details of a piece of apparatus similar to that described in the text.]

296] In making the experiment one of the plates whose charges we want to compare together, or the plate *B* as we will call it, is laid on the bars *Nn* and *Pp*, between the sticks of glass and end *N*, the upper coating thereof being made to communicate with *Bb* and *Mm* by a wire *V* resting on *Mm*, and the lower coating is made to communicate with the ground by a springing wire *S* fastened to *Rr*, and by its elasticity bearing against the lower coating of the plate.

Another coated plate is laid on the same bars between the sticks of glass and *n* by way of trial plate, the upper coating of which communicates with *Aa* by the wire *β*, and the lower coating communicates with *Dd* by the springing wire *δ*. A pair of pith balls also, such as were used in the former experiments, were suspended from *D* as represented in the figure.

In trying the experiments, the jars, and consequently the wire *Cc*, are charged, the wire *Cc* being all that time lifted up as high as it will go by means of the counterpoise. When the jars are charged to the proper degree as shown by the electrometer, the wire *Cc* is let down on the wires *Aa* and *Bb* by lifting up the counterpoise. This instantly charges both the coated plates, for when *Cc* rests on *Aa* and *Bb*, and consequently *Ff* and *Gg* rest on *Ee* and *Dd*, the lower coatings of both plates communicate with the ground, and their upper coatings with *Cc*.

Immediately after this the counterpoise is let go, by which means *Cc* is lifted up, and *Gg* and *Ff* along with it, till the two last mentioned wires bear against *Aa* and *Bb*, so that immediately after the coated plates are charged, the communication between them and the wire *Cc*, by which they were electrified, is taken away, and at the same time the communication between the lower coating of the trial plate and the ground is taken away, and immediately after that a communication is made between the upper coating of the plate *B* and the lower coating of the trial plate, and also a communication is made between the upper surface of the trial plate and the ground, so that the upper coating of the trial plate and the lower coating of the plate *B* both communicate with the ground, and the upper coating of *B* and the lower coating of the trial plate communicate with each other and the wire *Dd*.

Consequently, if the quantity of redundant fluid communicated to the wires *Bb* and *Mm* and the upper side of the plate *B* together is equal to the deficient fluid on the under side of the trial plate, they and the wire *Dd* will be neither over nor undercharged after the operation is completed; but if the redundant fluid in them exceeds the deficient fluid on the lower side of the trial plate, *Dd* will be overcharged, and the pith balls will separate positively. On the other hand, if it is less than the deficient fluid, the pith balls will separate negatively.

297] The trial plate consisted of a flat plate of glass, or other electric substance, the lower surface of which was coated all over with tinfoil, but on the upper side there was only a small coating of tinfoil. I had also flat plates of brass of different sizes which I could lay on the upper surface, and slip backwards and forwards, and thereby increase or diminish the size of the upper coating at pleasure, for the area of the upper coating is equal to the area of the plate of brass added to that of so much of the tinfoil as is left uncovered by the brass*.

By this means I could increase or diminish the quantity of deficient fluid on the lower side of the trial plate at pleasure, for I could alter the size of the upper coating at pleasure, and the quantity of deficient fluid on the under side of the plate is not much greater than it would be if the lower coating was no greater than the upper, and consequently depends on the size of that upper coating.

As it is necessary that the trial plate should be insulated, it was not laid immediately on the bars *Nn* and *Pp*, but was supported by sticks of waxed glass fastened to those bars.

Having by these means found what size it was necessary to give to the upper coating of the trial plate in order that the pith balls should separate positively just sensibly, and what size it was necessary to give to it that they might separate as much negatively, I removed the plate *B* and placed the plate or plates which I intended to compare with it (or the plate *b* as I shall call it) in its room and repeated the experiment in just the same manner as before. Then, if I found that the size which it was necessary to give to the upper coating of the trial plate in order to exhibit the same phenomena was the same as before, I concluded that the charge of the plate *b* was the same as that of *B*. If, on the other hand, I found that it was necessary to make the area of the upper coating of the trial plate greater or less than before in any ratio, I concluded that the charge of *b* was greater or less than that of *B* in the same ratio, for the quantity of deficient fluid on the lower side of the trial plate will be pretty nearly in proportion to the area of the upper coating.

N.B. In the following experiments it was always contrived so that the charges of the plates to be compared together should be pretty nearly alike, so that if the quantity of deficient fluid on the lower surface of the trial plate was not exactly in this proportion, it would make very little error in the proportion of the charges.

298] The method above described is that which I made use of in my first experiment, but I afterwards made use of another

* N.B. In order to estimate how much of the tinfoil was left uncovered, I drew parallel lines upon it at small equal intervals from each other, and took notice which of these lines the edge of the brass plate stood at. [Arts. 442, 488.]

method a little different from this, and which I found more exact, though rather more complicated, namely, for each set of plates that I wanted to compare together I prepared two trial plates, which I shall call L and l, not coated as that above described, but in the usual way, namely, with the coatings of the same size on both sides*.

The first of these plates, or L, was of such a size that when used as a trial plate with the plate B or b on the other side, the quantity of deficient fluid in it was rather more than ought to be in order that the pith balls should just separate negatively, and the second plate l was rather greater than it ought to be in order that they should just separate positively.

I also prepared a sliding plate of the same kind as the trial plate used in the former method, but whose charge was many times less than that of the plate B or b. This sliding plate I placed along with the plate B or b on the side N, and on the other side I placed the trial plate L and found what size it was necessary to give to the coating of the sliding plate in order that the balls should just separate negatively. I then removed the plate B and put b in its room, and found what sized coating it was necessary to give to the sliding plate in order that the balls should separate the same as before. Having done this, I removed the trial plate L and put l in its room, and tried each of the plates B and b as before, finding what coating it was necessary to give to the sliding plate that the balls might just separate positively.

Having done this, if I found that it required the coating of the sliding plate to be of the same size in order to exhibit the same phenomena in trying the plate B as in trying b, it is plain that the charges of B and b must be both alike, but if I found that it was necessary to give less surface, one square inch for instance, to the coating of the sliding plate in trying B than in trying b, then it is plain that the charge of B exceeds that of b by a quantity equal to that of the charge of the sliding plate when its surface is one square inch, supposing, as is very nearly the case, that the charge of the sliding plate is in proportion to the surface of its upper coating.

In this way of trying the experiment, it is plain that, in order to determine the proportion which the charges of B and b bear to each other, we must first know what proportion the charge of the

* [Art. 457.]

sliding plate, when its coating is of a given size, bears to that of *B*. This I found by finding what sized coating must be given to the sliding plate that its charge should be equal to that of another plate, the proportion of whose charge to that of *B* I was acquainted with.

It is plain that, if it is necessary to give one inch less surface to the coating of the sliding plate in trying *B* than in trying *b* when the trial plate *L* is made use of, it will be necessary to make the same difference in the surface of the sliding plate when the trial plate *l* is made use of, so that I might have saved the trouble of making two trial plates. However, for the sake of more accuracy, I always chose to make two trial plates and to take the mean of the results obtained by means of each trial plate for the true result.

299] One reason why this method of trying the experiment is more exact than the former, or that by means of a sliding plate only, is that in the former method I was liable to some error from inaccuracy in judging how much of the tinfoil coating of the trial plate was left uncovered by the sliding brass plate, whereas in this method, as the charge of the sliding plate is but small in respect of that of *B*, it was not necessary to be accurate in estimating its surface. But I believe the principal reason is that an error which will be taken notice of by and by, and which proceeds from the spreading of the electricity on the surface of the glass, is greater in a sliding plate than in one coated in the usual manner.

In general I think it required scarcely so great an increase of the charge of the trial plate to make a sensible alteration in the degree of separation of the pith balls in the following experiments as in the preceding, and therefore it should seem as if these experiments were capable of rather more exactness than the former, but this was not the case, as the different trials were found not to agree together with quite so much exactness in these experiments as the preceding. For this reason, and also because they were attended with less trouble, I repeated the experiments oftener, as I not only compared each plate with the trial plate for more times together as I did in the preceding experiments, but in general I repeated the experiment on several different days.

300] The circumstance which gave me the most trouble in these experiments was the spreading of the electricity on the surface of the glass. To understand this, let *ABab*, Fig. 21, be a flat plate of coated glass, *cd* and *CD* being the two coatings, and let *CD* be positively electrified, and let *cd* communicate with the ground.

Fig. 21.

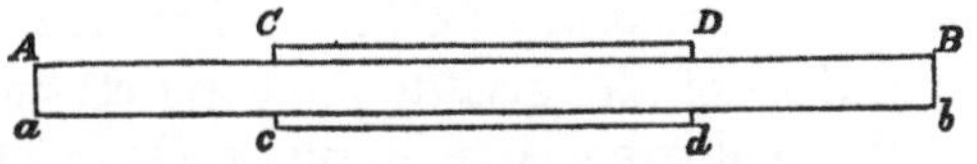

It is plain that the electric fluid will flow gradually from *CD* and spread itself all round on the surface of the glass, and nearly the same quantity of fluid will flow from the opposite side of the glass into *cd*, so that those parts of the glass which are not coated gradually become charged, those parts becoming so soonest which are nearest the edge of the glass.

On discharging the plate the uncoated part of the glass gradually discharges itself, as on the side *AB* the fluid will flow gradually from the uncoated part of the glass into *CD*, and on the opposite side it will flow into the uncoated part of the glass from *cd*.

301] There is a great deal of difference in this respect between different kinds of glass, as on some kinds it spreads many times faster than on others. The glass on which it spreads the fastest of any I have tried is a thin kind of plate-glass, of a greenish colour, much like that of crown-glass, and which I have been told is brought from Nuremberg*. On the English plate-glass it does not spread near so fast, but there is a great deal of difference in that respect between different pieces. On the crown-glass it spreads not so fast as on the Nuremberg, but I think faster than on the generality of English plate-glass. On white glass I think it spreads as slowly as any.

302] The way in which I compared the velocity with which it spread on different plates was as follows†. I took away the wire *Ff* (Fig. 20) and placed the plate which I wanted to try where the plates *L* or *l* used to be placed, the lower coating communicating as usual with *Dd* by the wire δ, but the wire β being drawn up by a silk string so as not to touch the upper coating. The wire *Cc* is suffered to rest on *Aa* and the jars electrified.

* [Art. 497.]

† [See Arts. 485, 486, 487. Also 494 to 499.]

When they are sufficiently charged β is let down on the upper coating, which instantly charges the plate to be tried, and immediately the wire *Gg* is lifted up from *Dd*, but not high enough to touch *Aa*. Consequently, immediately after the plate is charged, the communication between *Dd* and the ground is taken away, and consequently as fast as any fluid flows from the uncoated part of the under surface of the glass to the lower coating, some fluid will flow into *Dd* and overcharge it, and consequently make the pith balls separate.

303] In order to prevent, if possible, the ill effects proceeding from this spreading of the electricity, I took some coated plates of glass, and covered all the uncoated part with cement to the thickness of $\frac{1}{4}$ or $\frac{1}{2}$ an inch, as in Fig. 22, which represents a section

Fig. 22.

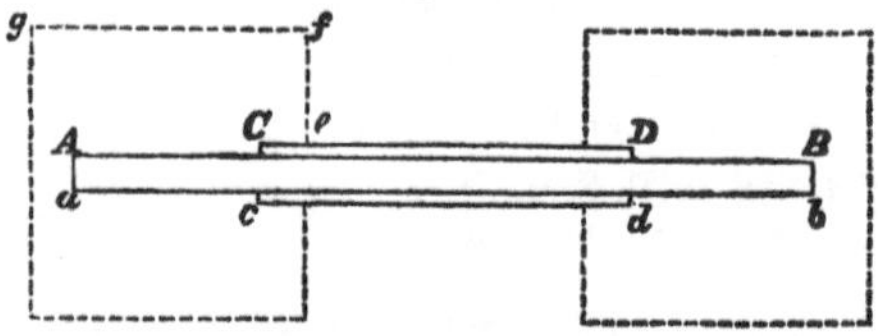

passing through the middle of the plate perpendicular to its plane, and in which the glass plate and coatings are represented by the same letters as before, and the dotted lines represent the cement*. Thinking that it would be impossible for the electricity to spread between the cement and the glass, in which case this method must have been perfectly effectual, as it would be necessary for the electricity to spread itself not only on the perpendicular surface *ef*, but also to some distance on the horizontal surface *fg*, before the quantity of redundant fluid lodged on the surface of the cement could bear any sensible proportion to that in the coating *CD*.

304] The result was that in dry weather the electricity seemed to spread as fast on those plates which were covered with cement as on the others, but in damp weather not so fast, the difference between dry and damp weather being less in those plates which were covered with cement than the others; and besides that there seemed as much difference between the swiftness with which it spread on the surface of the Nuremberg and English plates

* [Art. 184.]

after they were covered with cement as before, which shows plainly that the electricity spread between the cement and the glass, and not on the surface or through the substance of the cement. It could not be owing, I think, to its passing through the substance of the glass, for if it was, there would hardly be much difference in the uncoated plates between damp and dry weather, whereas, in reality, there was a very great one.

I also tried what effect varnishing the glass plates would have, but I did not find that it did better, if as well, as covering them with cement.

305] As there seemed, therefore, to be very little advantage in covering the plates with cement or varnishing them, and as it was attended with a good deal of trouble, I did not make use of those methods, but trusted only to letting the wires down and up pretty quick, so as to allow very little time for the electricity to spread on the surface of the plates, and this I have reason to think was sufficiently effectual, as I never found much difference in the divergence of the pith balls, whether the wires were let down and up almost as quick as I could, or whether they were suffered to rest a second or two at bottom.

306] As the wire *Cc* is suffered to rest so short a time on *Aa* and *Bb*, it is plain that the lower coatings of the trial plate and plate to be tried must have a very free communication with the ground and the outside coating of the jars, or else there would not be time for them to receive their full charge. I accordingly took care that the wires which made the communication should be clean and should touch each other in as broad a surface as I could conveniently. As for the method I took to have a ready communication with the ground, it is described in [Art. 258].

307] Besides this gradual spreading of the electricity on the surface of the glass, there is another sort which is of much worse consequence, as I know no method of guarding against it, namely, the electricity always spreads instantaneously on the surface of the glass to a small distance from the edge of the coating, on the same principle as it flies through the air in the form of a spark. This is visible in a dark room, as one may see a faint light on the surface of the glass all round the edges of the coating, especially if the glass is thin, for if it is thick it is not so visible*.

* [See Art. 532, Feb. 1, 1773.]

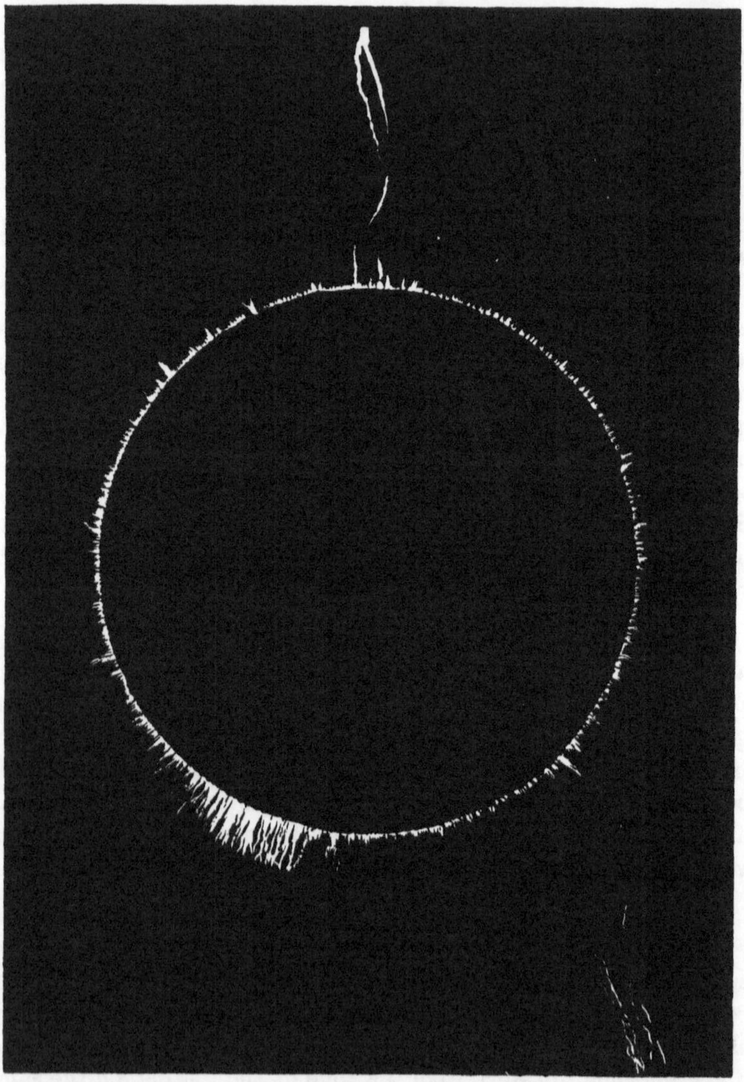

[From a photograph taken in the Cavendish Laboratory of a plate of glass with a circular tinfoil coating on one side, a larger coating being applied to the other side of the glass. The electrification of the coatings was produced by an induction coil.]

308] There is another circumstance which shows this instantaneous spreading of the electricity, namely, after having charged and discharged a coated plate of glass a great many times together without cleaning it, I have frequently seen a narrow fringed ring of dirt on the glass all round the coating, the space between the ring and the coating being clean, and in general about $\frac{1}{10}$ inch broad*. This must in all probability have proceeded from some dirt being driven off from the tinfoil by the explosions, and deposited on the glass about the extremity of that space over which the electricity spreads instantaneously, and therefore seems to show that the distance to which the electricity spreads instantaneously is not very different from $\frac{1}{10}$ of an inch.

309] From some experiments which will be mentioned by and by†, I am inclined to think that the distance to which the electricity spreads instantaneously is about $\frac{7}{100}$ of an inch when the thickness of the glass is about $\frac{1}{8}$ of an inch and about $\frac{9}{100}$ of an inch when its thickness is about $\frac{1}{15}$ of an inch; or more properly the quantity of redundant fluid which spreads itself on the surface of the glass is the same that it would be if the distance to which it spread was so much, and that the glass in all parts of that space was as much charged as it is in the coated part.

310] If I charged and discharged a coated plate several times running, in the dark, with intervals of not many seconds between each time, I commonly observed that the flash of light round the edges of the coating was stronger the first or second time than the succeeding ones, which seems to shew that the electricity spreads further the first or second time than the succeeding ones. Accordingly I frequently found in trying the following experiments that the pith balls would separate rather differently the first or second time of trying any coated plate than the succeeding one. Observing that I now speak of the half dozen trials which, as I said in [Art. 299], I commonly took with the same plate immediately after one another.

311] Before I proceed to the experiments it may be proper to remind the reader‡ that if a plate of glass or other non-conducting substance, either flat or concave on one side and convex on the

* [See Art. 538, Feb. 13, 1773.] † [See below, Arts. 314 to 323.]

‡ [See Art. 166, Prop. xxxiv., Cor. vi.]

other, provided its thickness is very small in respect of its least radius of curvature, is coated on each side with plates of metal of any shape, of the same size and placed opposite to each other, its charge ought by the theory to be equal to that of a globe whose diameter is equal to the square of the semidiameter of a circle whose area equals that of the coating divided by twice the thickness of the glass, supposing the coated plate and globe to be placed at an infinite distance from any over or undercharged body, and to be connected to the jar by which they are electrified by canals of incompressible fluid; provided also that the electricity does not penetrate to any sensible depth into the substance of the glass, and that the thickness of the glass bears so small a proportion to the size of the coating that the electricity may be considered as spread uniformly thereon.

312] It was before said that the electricity spreads instantaneously to a certain distance on the surface of the glass, so that the surface of the glass charged with electricity is in reality somewhat greater than the area of the coating. Therefore, if the plate is flat, let the area of the coating be increased by a quantity which bears the same proportion to the real coating as the quantity of redundant fluid spread on the surface of the glass beyond the extent of the coating does to that spread on the coated part of the glass. That is, let the area of the coating be so much increased as to allow for the instantaneous spreading of the electricity, and let a circle be taken whose area equals that of the coating thus increased. I call the square of the semidiameter of this circle, divided by twice the thickness of the glass expressed in inches, the computed charge of the plate, because, according to the abovementioned suppositions, its charge ought to be equal to that of a globe whose diameter equals that number of inches.

313] In like manner, in what may more properly be called a Leyden vial, that is, where the glass is not flat, but convex or concave, let a circle be taken whose area is a mean between that of the inside and outside coatings, allowance being made for the spreading of the electricity. I call the square of the semidiameter of this circle, divided by twice the thickness of the glass, the computed charge of the vial. In like manner, if the real charge of any plate is found to be equal to that of a globe of x inches in diameter, I shall call its real charge x.

I now proceed to the experiments.

314] I procured ten square pieces of plate-glass all ground out of the same piece of glass, three of them 8 inches each way and about $\frac{21}{100}$ inch thick; three more of about the same thickness 4 inches each way, the rest were as near to $\frac{1}{3}$ of that thickness as the workman could grind them, one being 8 inches long and broad, and the other 4 inches. They were not exactly of the same thickness in all parts of the same piece, but the difference was not very great, being no where greater than $\frac{1}{8}$ of the whole. The mean thickness was found both by actually measuring their thickness in different parts by a very exact instrument and finding the mean, and also by computing it from their weight and specific gravity and the length and breadth of the piece*. The mean thickness, as found by these two different ways, did not differ in any of them by more than 2 thousandths of an inch.

315] All these plates were coated on each side with circular pieces of tinfoil, the opposite coatings being on the same size and placed exactly opposite to each other. The mean thickness of the plates, which for more convenience I have distinguished by letters of the alphabet, together with the diameters of the coatings, and their computed charge, supposing the electricity not to spread on the surface of the glass, are set down in the following table†.

Plate.	Mean thickness.	Diameter of coating.	Computed charge.
A	·2112	6·57	25·5
B	·2132	6·6	25·5
C	·2065	6·5	25·6
D	·2057	2·155	2·82
E	·2065	2·16	2·82
F	·2115	2·175	2·80
H	·07556	6·8	76·5
K	·07712	2·265	8·31
L	·08205	2·335	8·31
M	·07187	2·195	8·38

316] The sizes of the coatings were so adjusted that the computed charges of *D*, *E*, and *F* are all very nearly alike. Those of *K*, *L* and *M* were intended to be three times as great as those of

* A cubic inch of water was supposed in this calculation to weigh 253⅓ grains Troy. [See Arts. 592, 593.]

† [Art. 482.]

the former, and consequently the diameters of their coatings nearly the same. The computed charges of *A*, *B* and *C* were intended to be three times as great as those of *K*, *L* and *M*, and consequently the diameter of their coatings about three times as great, and the computed charge of *H* was intended to be three times as great as that of *A*. By some mistake, however, the coatings of *K*, *L* and *M* were made rather too small, but the error is very trifling.

317] My first trials with these plates were to examine whether the charge of the three plates *D*, *E*, and *F* together was sensibly less when they were placed close together than when they were placed at 6 inches distance from each other, that is at as great a distance as my machine would allow of. I could not perceive any difference. This is conformable to the theory, as is shown in [Art. 185]. I chose to make the experiment with these three plates, as the difference should be more sensible with them than with any of the others.

318] Secondly. I compared together each of the plates *D*, *E* and *F*. I could not perceive any sensible difference in their charges*.

Thirdly. The charge of the plate *K* was found to exceed that of the three plates *D*, *E* and *F* together in the proportion of 1·016 to 1. The charge of *L* was not sensibly different from that of *K*, and that of *M* very little different.

Fourth. The charge of each of the plates *A*, *B* and *C* was to that of the three plates *L*, *K* and *M* together, as 0·905 to 1.

Fifth. The charge of *H* was equal to that of the three plates *A*, *B* and *C* together.

Therefore the charges of *D*, *K*, *A* and *H* were to each other as 1, 3·05, 8·28 and 24·9†.

319] It appears, therefore, that the proportion which the charge of *K* bears to that of *D*, and which *H* bears to that of *A*, is very nearly the same as that of their computed charges, but the proportion which the charge of *A* bears to that of *K* is near $\frac{1}{10}$ part less than it ought to be.

This in all probability proceeds from the effect of the instantaneous spreading of the electricity bearing a greater proportion to

* [See Art. 489, Feb. 4, 1772.]

† [See also Arts. 656 to 658.]

the whole in the plate K than it does in A, the diameter of whose coating is near three times as great.

320] In order to form some judgment, if possible, how great the effect of this instantaneous spreading of the electricity was, I took off the coatings from the plates A and B*, and put on others of just the same area in the form of a rectangular parallelogram (that of A was 6·414 long and 5·310 broad, and that of C 6·398 long and 5·201 broad), and compared their charges with that of the plate B, whose charge, as was before said, was just the same as those of these two plates before their coatings were altered.

321] I then took off these coatings†, and on A I put a square coating 6·388 each way with slits cut in it, as in Fig. 23, each $\frac{4}{10}$ broad, so as to divide it into 9 smaller squares, each 1·863 inches

$\frac{8}{64}$ $\frac{1}{8}$

Fig. 23

6388
.8
3)5.588
1.863

Fig. 24

* [C, See Art. 536.]

† [Art. 537.]

each way. The narrow communications marked in the figure between these squares were each $\frac{1}{10}$ of an inch broad.

On *C* I put an oblong coating 6·377 long and 6·343 broad, with four parallel slits cut in it, as in Fig. 24, each $\frac{4}{10}$ broad, the narrow space left between these slits and the outside being $\frac{1}{10}$ broad. Having done this, I compared their charges with that of the plate *B* as before.

It must be observed that the area of these slit coatings was somewhat less than that of the circular or oblong ones, but their whole circumference, including the circumference of the slits, is more than three times as great as that of the circular or oblong ones, so that the surface of glass charged by means of the instantaneous spreading of the electricity was more than three times as great in these coatings as the former, and consequently the quantity of that surface may be determined thereby, supposing that, if it was not for the spreading of the electricity on the surface, the charge of a coated plate would be the same whatever shape its coating is of, provided the area of the coated surface is given.

322. In order to find whether the electricity spread to the same distance upon thin glass as thick, I also took off the coatings from the plate *H*, and in its room put on first a square coating 6·03 inches each way, and then an oblong one 6·708 long and 6·514 broad, with four slits in it, as in Fig. 24, each $\frac{4}{10}$ broad, and ascertained the proportion which its charge with each of these coatings bore to that with the circular coating by comparing it with another plate, the proportion of whose charge to that of the circular coating I had before ascertained*.

323] It appeared from these experiments that if we suppose the electricity to spread instantaneously about ·07 of an inch on the thick glass plates such as *A* and *C*, and about ·09 on the thin ones, not only the charges of *A*, *C* and *H* with the three different coatings, but also the charges of all the plates will agree very well with the theory, as will appear by the following table; whereas, if we suppose that the electricity does not spread sensibly on the surface of the glass, the charge of the plate *H* with the slit coating would be greater in proportion to its charge with the circular or oblong coating than it ought to be in the ratio of 7 to 6, and the error in the plates *A* and *C* would not be much less.

* [Arts. 659—663.]

324] Plates with circular coatings.

Plates.	Diameter.	Increased diameter.	Thickness.	Computed charge.	Observed charge.
D	2·155	2·295	·2057	3·20	3·21
E	2·16	2·3	·2065	3·20	3·21
F	2·175	2·315	·2115	3·17	3·21
K	2·265	2·445	·07712	9·69	9·74
L	2·335	2·515	·08205	9·63	9·74
M	2·195	2·375	·07187	9·81	9·84
A	6·57	6·71	·2112	26·6	26·6
B	6·6	6·74	·2132	26·6	26·6
C	6·5	6·64	·2065	26·7	26·6
H	6·8	6·98	·07556	80·6	79·8

325] The same plates with other coatings.

Plates.	Area of coating.	Circumference.	Area which electricity spreads over.	Increased area.	Computed charge.	Observed charge.
A with oblong	34·1	23·4	1·64	35·74	26·9	26·8
A with slits	31·8	73·5	5·15	36·95	27·8	27·8
C with oblong	33·3	23·2	1·62	34·92	26·9	27·0
C with slits	30·4	76·5	5·35	35·75	27·5	27·7
H with oblong	36·4	24·1	2·17	38·57	81·2	80·7
H with slits	33·3	80·1	7·21	40·51	85·3	85·5

By the observed charge in the foregoing table, I mean only the proportion which the observed charges bore to each other, not the real observed charges. [See Art. 671.]

326] From the circumstance of the light mentioned in [Art. 307], it appears plainly that the electricity does actually spread instantaneously to a small distance on the surface, and from the rings of dirt taken notice of in Art. 308 it seems likely that the distance to which it spreads is not very different from what we have here supposed; moreover, if the distance to which the electricity spreads is such as we have supposed, the charges of all these plates bear very exactly the same proportion to each other that they ought to do by theory, whereas if the distance to which the electricity spreads is different from that here assigned, and consequently the proportion of the charges of different plates to each other different from that furnished by theory, it seems very strange that their charges should all have happened to agree with computation, notwithstanding that their thickness and the size and shape of their coatings are so very different. I think therefore that we may fairly infer both that the distance here assigned to the spreading of the electricity is right, and that, if it was not for this spreading of the electricity, the charge of any plate of glass would be as the square of the radius of the circle equal in area to the coated surface divided by twice the thickness of the glass, that is, that the actual charges are in proportion to the computed ones.

327] Though it seems likely from these experiments that the electricity spreads further on the surface of thin glass than it does on thick, yet I can not be sure that it does, as the difference observed is not greater than what might proceed from the error of the experiment. However, as there seems nothing improbable in the supposition, I shall suppose in the following pages that it does really do so.

328] When I say that the electricity spreads $\frac{7}{100}$ of an inch on the surface of the glass, I mean that the quantity of electricity thereby spread on the uncoated part of the glass is the same that it would be if it actually spread to that distance, and if all that part of the glass which it spread over was charged in the same degree as the coated part, and consequently that the charge of the plate is the same as if the size of the coating was increased by a ring drawn round it ·07 of an inch broad, and that the electricity was prevented from spreading any further. But I would by no means be understood to mean that no part of the electricity spreads to a greater distance than that, as it seems very likely that it does so, but that the part furthest from the coating is less charged with electricity than that nearest to it.

329] What is said above must be understood of the distance to which the electricity spreads with that degree of strength which I commonly made use of in my experiments, but I also made some trials with the plates *A* and *C* to determine to what distance it would spread with two other degrees of electricity.

If a jar with Lane's electrometer fixed to it* was charged to the higher degree, it would discharge itself when the knobs of the electrometer were at ·053 inches distance; when it was charged to the lower degree, it discharged itself when they were at about half that distance, or at ·027 of an inch; and when it was charged to

* [Art. 540, Feb. 16, 1773.]

[Lane's electrical machine, with discharging electrometer. From his paper in the *Phil. Trans.* 1767, p. 451. For Cavendish's form of discharging electrometer, see Art. 405.]

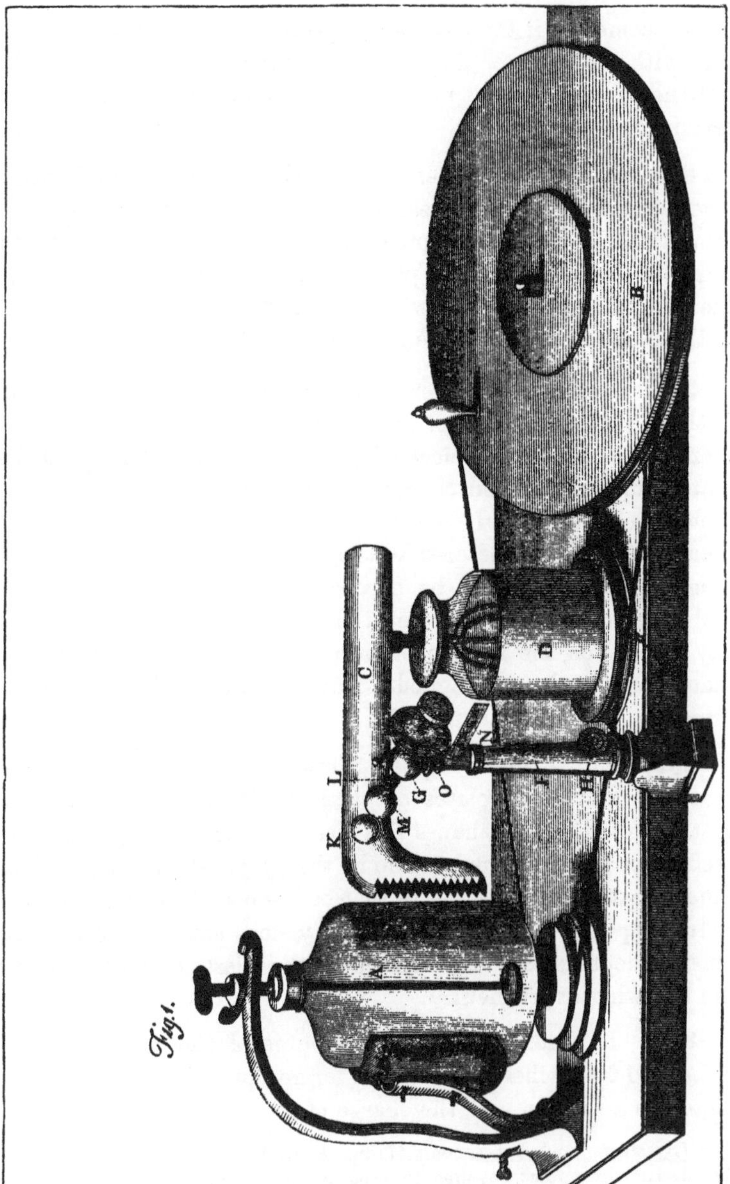

the usual degree, it discharged itself, as was before said, at ·04 of an inch, so that the usual degree of electricity was about a mean between these two*.

It seemed as if the electricity spread about $\frac{1}{30}$ of an inch further with the stronger degree of electricity than with the weaker, but the experiment was not accurate enough to determine it with certainty.

330] I made an experiment of the same kind to determine whether the electricity spread to the same distance on crown-glass as on this. It seemed to spread about $\frac{8}{100}$ of an inch on it, that is, rather less than on the plate *H*, though its thickness was, of the two, rather less. But whether this difference is real, or owing to the error of the experiment, I cannot tell.

331] There seems no reason, from the foregoing experiments, to think that the charge of any of these plates is sensibly greater than it would be if the electricity was disposed uniformly on their coated surfaces, as their charges agree very well together without such a supposition. If we suppose that the charges of any of them are sensibly greater than they would be if the fluid was disposed uniformly, it will be necessary to suppose that there is a still greater difference between the distance to which the electricity spreads on the surface of thin plates and that of thick ones than what we have assigned. But I shall speak more on this subject at the end of Art. [365].

332] But though it appears from the foregoing experiments that the charges of plates of glass of different thicknesses with coatings of different shapes and sizes bear the same proportion to each other that they ought to do by theory, yet their charge is many times greater in proportion to that of a globe than it ought to be on a supposition that the electricity does not penetrate to any sensible depth into the substance of the glass, as will appear by the following experiment.

333] In order to compare the charge of the plate *D* with the globe of $12\frac{1}{10}$ inches used in the former part, I made two plates coated as a Leyden vial, the charge of each of which was about $\frac{1}{2}$

* [By Macfarlane's experiments (*Trans. R. S. Edin.* Vol. XXVIII. Part II. 1878) the electromotive force required to produce sparks between flat disks at those distances would be 14, 11·8, and 9 units respectively.]

that of D, each consisting of two plates of glass cemented together and coated on their outside surfaces with circular pieces of tinfoil about $1\frac{4}{5}$ inch in diameter*.

I then compared the charge of each of these double plates with that of the globe in the same manner that I compared together the charges of different bodies in the former part, the only difference being that, in trying either of these double plates, I made a communication between the lower coating of the plate and the ground, the wires Mm and Dd (Fig. 14) being contrived so that they were sure to fall on the upper coating†.

By this means the charge of each of these double plates was found to be just equal to that of the globe. The charge of the plate D was then compared with that of the two double plates together, and was found to be less than that in the proportion of 263 to 272, and consequently the charge of the plate D is to that of the globe as 26·3 to 13·6.

334] Before we go further it will be proper to consider what effect the three circumstances taken notice of in Art. 277 will have in altering the proportion of the charge of the double plate to that of the globe. With regard to the two first, it appears that the charge of the globe and double plate will neither of them be sensibly different from what they would be if they were placed at an infinite distance from the jar by which they are electrified, and moreover, in trying the globe, the repulsion of the redundant fluid in the globe increased the deficience of fluid in the trial plate as much as the attraction of the trial plate increased the quantity of redundant fluid in the globe‡, so that it required the same size to be given to the trial plate as it would have done if the globe and trial plate had exerted no attraction or repulsion on each other; and in trying the coated plate, the coated plate could not sensibly increase the deficience in the trial plate, nor could the attraction of the trial plate sensibly increase the redundance in the coated plate, so that neither of these two causes had any tendency to alter the proportion of the charges of the globe and coated plate to each other.

* If they had been made of a single piece of glass, the coatings must have been so small as would have been inconvenient unless the glass had been of a greater thickness than could have been easily procured. [Arts. 446, 451, 649, 653, 654.]

† [Arts. 455, 456, 478.]

‡ [Note 17.]

335] But the third cause will have a sensible effect, for in trying the globe the floor and sides of the room near it would be made undercharged, which would increase the charge of the globe, whereas in trying the coated plate the floor would not be made sensibly undercharged, nor, if it was, would it have any sensible effect in increasing the charge of the plate.

So that the charge of the globe bore a sensibly greater proportion to that of the coated plate than it would have done if it had been placed at an infinite distance from any other bodies.

How much the charge of the globe should be increased hereby I can not tell, but I should imagine it should be at least by $\frac{1}{15}$th part, for if the room had been spherical and 16 feet in diameter (about its real size) and the globe placed in its center, it should have been increased as much as that*, and as the globe was really placed three times as near to the floor as to the ceiling†, I suppose the effect to have been still greater.

* Let the globe $Bb\beta$, whose centre is C, be insulated in the hollow globe $Dd\delta$ concentric with [it]. Let the inner globe be pos. electrified by the canal BE not

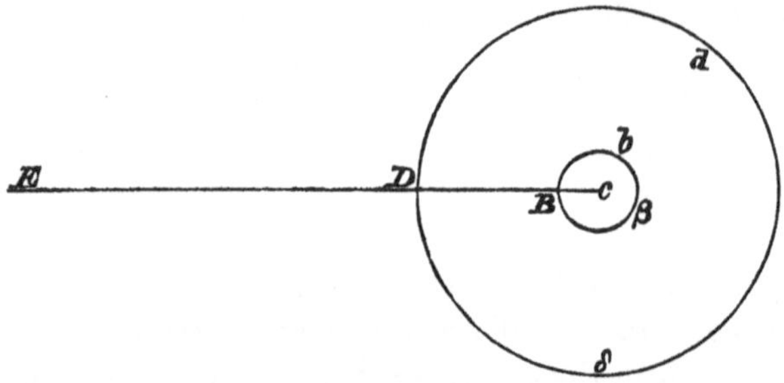

communicating with the outer globe, and let the outer globe communicate with the ground. The quant. defic. fluid in the outer globe must be equal to the redundant in the inner globe, and the attraction of the outer globe on the canal BE is to the repulsion of the inner one thereon as

$$\frac{1}{CD} : \frac{1}{BC},$$

and therefore the quantity of redun. fluid in the inner globe is to that which it would contain if the outer globe were away as

$$\frac{1}{BC} : \frac{1}{BC} : \frac{1}{DC} :: DC : DC - BC.$$

If room was spherical, 16 feet in diameter, globe in middle of it, its charge should be increased in ratio of 16 to 15 by reason of undercharged floor, &c.

† [This is the only indication of the height of the room. The circles were suspended by silk strings from a horizontal bar (Art. 466) 87·5 inches from the floor. By Art. 474 the platform 14 inches high diminished the height of the bodies in the

336] In order to find out, if possible, how much the charge of the globe was increased hereby, I made four flat plates of a mixture of rosin and bees wax*, about 4 inches square and ·22 thick, and coated each of them with circles of about 1·8 inches in diameter, and compared the charge of each of them separately with that of a circular plate of tin, 9·3 inches in diameter. I then compared the charge of two of these plates together with that of a tin circular plate 18½ inches in diameter, and lastly I compared the charge of all together with that of a circle of 36 inches diameter†.

337] By a mean of the different experiments it appears that the charge of each of the rosin plates was alike, and that the charge of any one of them was to that of the circle of 9·3 inches as 10·34 to 9·3, that the charge of the circle of 18½ inches was to that of two of the rosin plates together as 20·19 to 21·96, and that the charge of the circle of 36 inches was to that of all four plates as 43·75 to 42·06.

But the charge of the four plates together will not be exactly four times the charge of one plate singly, as some allowance must be made for the charge of the wire connecting their upper surfaces, and, besides that, the charge of the plates when placed close together will not be quite so great as if placed at a distance from each other‡.

By trying the charge of all four rosin plates together by the machine, Fig. 20, both when placed close together and at as great a distance from each other as I could, I found their charge when close together to be to their charge when placed at a distance nearly as 41 to 41½, and, from some other experiments I made, I am inclined to think that the charge of each of the wires which connected the upper coatings of the plates was to that of one plate alone as 28 to 930§.

ratio of 2 to 3. Hence the height of the center of the bodies from the floor was 42 inches, and the height of the room 4×42 inches, or 14 feet. This would agree with the height of the top of the circle of 18 inches being 51 inches from the floor (Art. 472).]

* These plates are non-conductors of electricity, and may be charged as Leyden vials. The manner in which I made them will be described in the following pages [Arts. 373, 514]. My reason for making them of these materials is that the charge of such a plate is much less than that of a plate of glass of the same dimensions.

† It must be observed that, in the two last mentioned comparisons, the rosin plates were placed close together and their upper surfaces connected by a piece of wire.

‡ [Art. 557.]

§ [Arts. 555, 558, also 443.]

From these circumstances, I am inclined to think that the charge of two plates together is to that of one plate alone as 21·96 to 10·34, and that the charge of the four plates together is to that of one alone as 42·06 to 10·34, and consequently that the charges of the tin circles of 9·3 inches, $18\frac{1}{2}$ inches and 36 inches are to each other as 9·3, 20·19 and 43·75*.

338] Though I do not know how to calculate how much the charge of the circles ought to be increased by the attraction of the undercharged ground, yet I think there can be little doubt but that if the charge of the plate of $18\frac{1}{2}$ inches is increased in any ratio whatever as that of x to $x - 18\frac{1}{2}$, the charge of the plate of 36 inches will be increased in the ratio of x to $x - 36$, and that of the plate of 9·3 inches in the ratio of x to $x - 9{\cdot}3$; therefore if we suppose that the charge of the $18\frac{1}{2}$ inch plate is increased in the ratio of 9 to 8, or of $166\frac{1}{2}$ to $166\frac{1}{2} - 18\frac{1}{2}$, the charges of the three plates should be to each other as

$$\frac{36 \times 166\frac{1}{2}}{130\frac{1}{2}}, \quad \frac{18\frac{1}{2} \times 166\frac{1}{2}}{148} \text{ and } \frac{9{\cdot}3 \times 166\frac{1}{2}}{157{\cdot}2},$$

that is, as 43·37, 19·65 and 9·3,

which agrees very nearly with experiment, and nearer so than it would have done if we had supposed the charge of the $18\frac{1}{2}$ inch plate to have been increased in any other proportion which can be expressed in small numbers†.

339] I think we may conclude therefore that the charge of the 12·1 inch globe was increased by the attraction of the undercharged ground nearly in the proportion of 9 to 8, for I think there can be little doubt but that the charge of the globe must be increased thereby in nearly the same ratio as that of the $18\frac{1}{2}$ inch plate, and therefore we may conclude that the charge of the plate D is to the charge which the 12·1 inch globe would receive, if it was placed at a great distance from any over or under-charged matter, nearly in the proportion of 26·3 to 12·1, or, in other words, the charge of the plate D is 26·3, which is rather more than eight times greater than it ought to be if the electric fluid did not penetrate into the glass. I shall speak further as to the cause of this in [Art. 349].

340] In order to try the charge of what Æpinus‡ calls a plate of air, I took two flat circular plates of brass, 8 inches in diameter

* [Art. 649.] † [Art. 652, and Note 24.] ‡ [*Mém. Berl.* 1756, p. 119.]

and $\frac{1}{4}$ thick, and placed them on the bars *Nn* and *Pp* of the machine (Fig. 20), the two plates being placed one over the other, and kept at a proper distance from each other by three small supports of sealing-wax placed between them, the supports being all of the same height, so that the plates were exactly parallel to each other. Care was also taken to place the plates perpendicularly over each other, or so that the line joining their centers should be perpendicular to their planes.

The lowermost plate communicated with the ground by the wire *RS*, and the uppermost communicated with *Mm* by the wire *V*, just as was done in trying the Leyden vials.

I then found its charge, or the quantity of redundant fluid in the uppermost plate, in the usual manner, by comparing it with the plate *D*, and found it to be to that of *D* as*...

341] As I was desirous of trying larger plates than these, and was unwilling to be at the trouble of getting brass plates made, I took two pieces of plate-glass† $11\frac{1}{2}$ inches square, and coated each of them on one side with a circular plate of tinfoil 11·5 inches in diameter, and placed them on the machine as I did the brass plates in the former experiment, with the tinfoil coatings turned towards each other, and kept at the proper distance by supports of sealing-wax as before, care being taken that the tinfoil coatings should be perpendicularly over each other.

For the more easy making a communication between the circular coating of the lower plate and the ground, and between that of the upper plate and the wire *Mm*, I stuck a piece of tinfoil on the back of each plate, communicating by a narrow slip of the same metal with the circular coatings on the other side.

I then tried the charge as before, the lower plate communicating with the ground and the upper with the wire *Mm*.

As glass does not conduct electricity, it is plain that the quantity of electric fluid in the pieces of tinfoil will be just the same that it would be if the glass was taken away, and the pieces of tinfoil kept at the same distance as before.

* The memoranda I took of that experiment are lost, but to the best of my remembrance the result agreed very well with the following experiment.

† [Art. 517.]

The distance of the two circular coatings of tinfoil was measured by the same instrument with which I measured the thickness of the plates of glass, and may be depended on to the 1000th or at least to the 500th part of an inch*.

342] In this manner I made the experiment with the plates at four different distances, namely ·910, ·420, ·288 and ·256, and when I had made a sufficient number of trials with the plates at each distance, I took off these circular coatings and put on smaller, namely of 6·35 inches diameter, and tried the experiment as before with the plates at ·259 inches distance. The result of the experiments is given in the following table:

343†]

No. of Experiment.	Distance of the tinfoil coatings.	Diameter of the coatings corrected for the spreading of electricity.	Computed charge.	Observed charge.	Observed charge by computed charge.	Diameter of coatings by distance of ditto.
1	·910	11·5	18·2	27	1·49	12·6
2	·420	......	39·4	52	1·32	27·4
3	·288	......	57·4	72·1	1·26	40
4	·256	......	64·6	78·3	1·21	45
5	·259	6·35	19·5	26·5	1·36	24·5

It is plain that some allowance ought to be made in these trials for the spreading of the electricity on the surface of the glass. In the above table I have supposed it to spread ·05 of an inch, but the effect is so small that it is of very little signification whether that allowance is made or not.

344] In my former paper [Art. 134] I expressed a doubt whether the air contained between the two plates in this experiment is overcharged on one side and undercharged on the other, as is the case with the plate of glass in the Leyden vial, or whether the redundant and deficient fluid is lodged only in the plates, and that the air between them serves only to prevent the electricity from running from one plate to the other, but the following experiment shows that the latter opinion is true.

I placed the two brass plates on the machine (Fig. 20), and tried their charge as before, except that, after having charged the plates‡, I immediately lifted up the upper plate by a silk string so as to separate it two or three inches from the lower one, and let it

* [See Art. 459, "Bird's instrument," and "dividing machine," Art. 517. Also 594, 595.]

† [See Arts. 669, 519.]

‡ [Arts. 511, 516, Dec. 18, 26, 1772.]

down again in its place before I found its charge by making the communication between *Bb* and *Dd* and between *Aa* and *Ee*.

The way I did this was that as soon as I had let down the wire *Cc* on *Aa* and *Bb*, and thereby charged the plates, I lifted it up again half way so as to take away the communication between *Cc* and the upper plate &c., but did not lift it quite up, so as to make the communication between *Bb* and *Dd*, and between *Aa* and *Ee*, till after I had separated the upper plate from the lower, and put it back in its place.

I could not perceive any sensible difference in the charge, whether I lifted up the upper plate in the above-mentioned manner, or whether I tried its charge without lifting it up.

345] It is plain that in lifting up the upper plate from the lower and letting it down again, the greatest part of the air contained between the two plates must be dissipated and mixed with the other air of the room, so that if the air contained between the two plates was overcharged on one side and undercharged on the other, the charge must have been very much diminished by lifting up the upper plate and letting it down again, whereas, as I said before, it was not sensibly diminished.

I think we may conclude, therefore, that redundant and deficient fluid is lodged only in the plates, and that the air between them serves only to prevent the electricity from running from one plate to the other.

346] As this is the case, the charge of these plates ought, according to the theory, to be equal to that of a globe whose diameter equals the square of the radius of the plate or circular coating divided by twice their distance, that is, to their computed charge, provided the electricity is spread uniformly on the surface of the plates, and therefore in reality the numbers in the last column but one ought to be rather greater than in the last but two, and moreover the less the distance of the plates is in proportion to the diameter of the coating, the less should be the proportion in which those numbers differed, and if the distance is infinitely small in proportion to the diameter, the proportion in which those numbers differ, should also be infinitely small.

347] This will appear by inspecting the table to be the case, only it seems from the manner in which the numbers decrease, that they would never become equal to unity though the distance of the plates was ever so small in respect of their diameter, and I should think, or rather I imagine, would never be less than 1·1, so that it seems as if the charge of a plate of air was rather greater in proportion to that of the globe than it ought to be, and I believe nearly in the proportion of 11 to 10*.

348] The reason of this, I imagine, is as follows. It seems reasonable to conclude from the theory that when a globe or any other shaped body is connected by a wire to a charged Leyden vial, and thereby electrified, the quantity of redundant fluid in the globe will bear a less proportion to that on the positive side of the jar than it would do if they could be connected by a canal of incompressible fluid†, but in all probability when a plate of air is connected in like manner to the Leyden vial, the quantity of redundant fluid on its positive side will bear nearly the same proportion to that in the vial that it would do if they were connected by a canal of incompressible fluid, and consequently the charge of the plate of air in these experiments ought to bear a greater proportion to that of the globe than if they had been connected to the vial by which they were electrified by canals of incompressible fluid.

349] It was said in Art. 339 that the charges of the glass plates were rather more than eight times greater than they ought to be by the theory, if the electric fluid did not penetrate to any sensible depth into the glass. Though this is what I did not expect before I made the experiment, yet it will agree very well with the theory if we suppose that the electricity, instead of entering into the glass to an extremely small depth, as I thought most likely when I wrote the second part of this work‡, is in reality able to enter into the glass to the depth of $\frac{7}{16}$ of the whole thickness of the glass, that is, to such a depth that the space into which it can not penetrate is only $\frac{1}{8}$ of the thickness of the glass, as in that case it is evident that the charge should be as great as it would be if the thickness of the glass was only $\frac{1}{8}$ of its real thickness, and the electricity was unable to penetrate into it at all.

350] There is also a way of accounting for it without suppos-

* [Art. 670.]

† This seems likely from Appendix, Coroll. 5 [Art. 184].

‡ [Refers to Art. 132.]

ing the electricity to enter to any sensible depth into the glass, by supposing that the electricity at a certain depth within the glass is moveable, or can move freely from one side of the glass to the other.

Thus, in Fig. 25, let *ABDE* be a section of the glass plate perpendicular to its plane, suppose that the electricity from with-

Fig. 25.

out can penetrate freely into the glass as far as the line *ab* or *ed* but not further, suppose too that within the spaces $ab\beta\alpha$ and $ed\delta\epsilon$ the electric fluid is immoveable, but that within the space $\alpha\beta\delta\epsilon$ it is moveable, or is able to move freely from the line $\alpha\beta$ to $\delta\epsilon$. Then will the charge of the plate be just the same as on the former supposition, provided the distances $a\alpha$ and $e\epsilon$ are each $\frac{1}{18}$ of the thickness of the plate*.

351] But I think the most probable supposition is that there are a great number of spaces within the thickness of the glass in which the fluid is alternately moveable and immoveable.

Fig. 26.

Thus let *ABDE* (Fig. 26) represent a section of the plate of glass as before, and let the glass be divided into a great number of spaces by the parallel lines *ab*, $\alpha\beta$, *ed*, $\epsilon\delta$, &c., and suppose that in the two outermost spaces *ABba* and *EDde* the fluid is moveable, that in the two next spaces $ab\beta\alpha$ and $ed\delta\epsilon$ it is immoveable, and

* The only reason why I suppose the electric fluid to be able to enter into the glass from without as far as the lines *ab* and *ed* is that Dr Franklin has shewn that the charge resides chiefly in the plate of glass and not in the coating, and consequently that the electricity is able to penetrate into the glass to a certain depth. Otherwise it would have done as well if we had supposed the fluid to be immoveable in the whole spaces $AB\beta\alpha$ and $ED\delta\epsilon$, and that the distance $A\alpha$ and $E\epsilon$ are each $\frac{1}{18}$ of *AE*.

that in the two next spaces it is moveable, and so on. The charge will be the same as before, supposing the sum of the thickness of the spaces in which the electricity is immoveable to be $\frac{1}{8}$ of the whole thickness of the glass, as it is shewn that the charge of such a plate will be the same as that of a plate in which the electricity is entirely immoveable, whose thickness is equal to the sum of the thicknesses of those spaces in which we supposed the fluid immoveable*.

352] It must be observed that in those spaces in which we supposed the fluid to be moveable, as in the space *ABba* for example, though the fluid is able to move freely from the plane *Ab* to *ab*, that is, though it moves freely in the direction *Aa* or *aA*, or in a direction perpendicular to the plane of the plate, yet it must not [be] able to move lengthways, or from *A* to *B*, for if it could, and one end of the plate *AE* was electrified, some fluid would instantly flow from *AE* to *BD*, and make that end overcharged, which is well known not to be the case. The same thing must be observed also with regard to the two former ways of explaining this phenomenon.

353] The chief reason which induces me to prefer the latter way of accounting for it is that in the two former ways the thickness of the spaces in which the fluid is moveable must necessarily be very considerable. In thick glass, for example, in a plate of the same thickness as *D*, it must be not less than $\frac{8}{100}$ of an inch in the first way of explaining it, and in the second way it must be still greater. Now if the electric fluid is able to move through so great a space in the direction *AE*, it seems extraordinary that it should not be able to move in the direction *AB*, whereas in the latter way of accounting for it the thickness of the spaces in which the electricity is moveable may be supposed infinitely small, and consequently the distance through which the electricity moves in the direction *AE* also infinitely small.

354] Another thing which inclines me to this way of accounting for it is that there seems some analogy between this and the power by which a particle of light is alternately attracted and repelled many times in its approach towards the surface of any refracting or reflecting medium. See Mr Michell's explanation

* [Prop. xxxv. Art. 169, and Note 15.]

of the fits of easy reflection and transmission in Priestley's *Optics*, page 309.

355] To whichever of these causes it is owing that the charges of these plates are so much greater than they should be if the electric fluid was unable to enter into the plate, it was reasonable to expect that the greater the force with which the plate was electrified, the greater should be the depth to which the electric fluid penetrates into the glass, or the greater should be the thickness of the spaces in which we supposed the fluid to be moveable, and consequently in comparing the charge of the plate D with the circle of 36 inches diameter, or with any other body, the greater the force with which they are electrified the greater proportion should the charge of the glass plate bear to that of the circle.

356] I therefore compared the charge of the plate D with that of the circle of 36 inches with electricity of two different degrees of strength, namely the same which I made use of in [Art. 329], in trying whether the distance to which the electricity spread on the surface of glass was different according to the strength of the electricity.

The way in which I compared their charges was just the same that I made use of in comparing the rosin plate with the tin circles in [Art. 337]. The event was that I could not perceive that the proportion which their charges bore to each other with the stronger degree of electricity was sensibly different from what they did with the weaker*.

357] But it must be remembered that it seemed from the experiment related in [Art. 329], that the electricity spread $\frac{1}{30}$ of an inch further on the surface of the glass with the stronger degree of electricity than with the weaker. The difference of charge owing to this difference in the spreading of the electricity is $\frac{1}{16}$ part of the whole, so that it seems that if the electricity had been prevented from spreading on the surface of the glass, the proportion of the charge of the glass plate to that of the tin circle would have been less with the stronger degree of electricity than with the weaker, and that nearly in the proportion of 16 to 17.

* [Arts. 547, 551, 553, also Arts. 451, 463, 526, 535, 538, 664.]

358] I also made an experiment to determine whether the charge of a coated plate of glass bore the same proportion to that of another body when the electricity was very weak as when it was of the usual strength*.

For this purpose I first found what proportion the charge of a tin cylinder 15 feet long and 17 inches in circumference bore to that of the two plates *D* and *E* together when the electricity was very weak. This I did in the manner represented in Fig. 27,

Fig. 27.

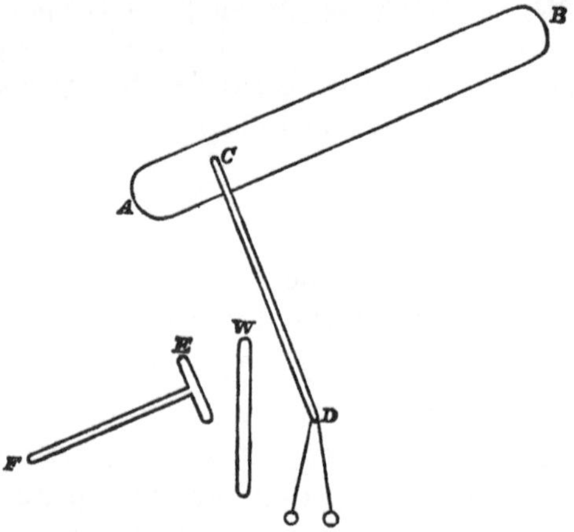

where *AB* is the tin cylinder supported horizontally by non-conductors. *DC* is a brass wire 37 inches long and about $\frac{1}{6}$ inch in diameter supported also horizontally by non-conductors, the end *C* being in contact with the cylinder, and a pair of fine pith balls being suspended from the other end *D*. *FE* is a piece of wire communicating with the prime conductor, and between it and *DC* is suspended by a silk string the wire *W* in a vertical situation.

359] The cylinder *AB*, and consequently the wire *DC*, were first electrified negatively to such a degree as to make the pith balls separate to the distance of one diameter of the balls. The prime conductor and wire *FE* being then charged to the usual degree, as shewn by the usual electrometer hung down from it, one end of the wire *W* was brought in contact with *E* so as to be electrified by it, and was then immediately removed and

* [Arts. 539, 666.]

brought in contact with *DC* so as to communicate its electricity to the cylinder*.

Now I found that if the wire *W* was 29 inches long, and $\frac{1}{6}$ in diameter, and its electricity was twice communicated in this manner to the cylinder, the pith balls would separate as much positively as they before did negatively, consequently the cylinder *AB* and the wires *DC* and *W* together, when electrified to such a degree as to make the pith balls separate one diameter, contain as much electricity as the wire *W* alone does when electrified in the usual degree.

The cylinder *AB* was then removed, and the two glass plates *D* and *E* placed under the wire *DC* in its room, their upper coatings communicating with *DC*, and their lower coatings with the ground, and the operation performed as before. I found that I was obliged to change the wire *W* for one of the same thickness, and only 22 inches long, in order that the pith balls should separate the same as before.

360] Therefore the charge of the two plates *D* and *E* and the wire *DC* together is to that of the tin cylinder and wire *DC* together as the charge of a wire $\frac{1}{6}$ inch thick and 22 inches long to that of a wire of the same thickness and 29 inches long, that is, as 1 to 1·26, and consequently, as the charge of the wire *DC* is but small in comparison of that of the two plates, the charge of the two plates will be to that of the tin cylinder pretty nearly in the same proportion of 1 to 1·26†.

Having thus found what proportion the charges of the plates and cylinder bear to each other when electrified in a very weak degree, I tried what proportion they bore with the usual degree of electrification.

361] To this purpose I placed the two plates on the machine represented in fig 20 between *M* and *m* in the usual manner, and on the other side I placed a sliding coated plate, and found as usual what size must be given to the coating of this plate that

* [See plan at Art. 539.]

† I believe the true proportion is between that of 1 to 1·28 and that of 1 to 1·37, but as the experiment is not capable of much accuracy, I think it needless to trouble the reader with the computation. [See Art. 666 and Note 25.]

the pith balls should just separate positively, and what size must be given to it that they should just separate negatively.

I then removed the two plates and suspended the tin cylinder so as to touch the wire *Mm*, but without touching any other part of the machine, and found what size it was necessary to give to the coating of the sliding plate that the pith balls should separate as before.

By this means the charge of the tin cylinder was found to be to that of the two plates as 1·33 to 1. Therefore the charge of the two plates seems to bear pretty nearly the same proportion to that of the cylinder whether the electricity is of the usual strength or very weak. But if we suppose that the electricity spreads ·07 inches on [the] surface of glass with the usual degree of electrification, and that it does not spread sensibly with the weak degree of electrification, then the proportion which the charge of the glass plates bears to that of the cylinder should be less with the usual degree of electrification than with the weak one, and that by about $\frac{1}{8}$ part.

This difference, however, is not more than what might very well proceed from the error of the experiment.

362] On the whole, I am uncertain whether the charge of a glass plate would really bear a rather less proportion to that of a globe or other body when the electricity is strong than when it is weak, provided the electricity was prevented from spreading on the surfaces as it should seem by these experiments, or whether it was not rather owing partly to the error of the experiment, and partly to there not being so much difference in the distance to which the electricity spreads on the surface of the glass according to the different degree in which it is electrified, as I imagined.

If the first of these suppositions is true, I do not know how to reconcile it with the theory, except by supposing that the greater the force with which the plate is electrified the less is the depth to which the electricity penetrates into the glass, or the less is the thickness of the spaces in which we supposed the fluid to be moveable.

Though it seemed natural to expect that the electric fluid should penetrate further into the glass, or that the fluid within the glass should move through a greater space when the glass was strongly electrified than when weakly, that is, when the force with which the fluid was impelled was great than when it was small, yet it is not strange that it should be otherwise, as it is very possible that the electric fluid may penetrate with great freedom to a certain depth within the glass, and that no ordinary force shall be able to impel it sensibly further, and in like manner it is very possible that the fluid may be able to move with perfect ease in the space *aε* (Fig. 27) and yet that no ordinary force shall be able to move the fluid at all beyond that space.

But it would be very strange that the fluid should penetrate to a less depth within the glass, or that the fluid within the glass should move through a less space when the glass is strongly electrified than when weakly.

363] The reader perhaps may be tempted from this circumstance to think that the reason of the actual charge of the glass plates so much exceeding their computed charge is not owing to the electric fluid penetrating into the glass, or to any motion of the fluid within the glass, but to some error in the theory. But I think the experiments on the plate of air [Art. 344] form a strong argument in favour of its being owing to the penetration of the electric fluid into, or its motion within the glass, for it appears plainly from these experiments that the electric fluid does not penetrate into the air, and on account of the fluidity of the air it seems very improbable that the electric fluid within the air should be able to move in the manner we supposed it to do within the glass; whereas it appears plainly from Dr Franklin's analysis of the Leyden vial, that the electric fluid does actually penetrate into the glass.

Therefore as this excess of the observed charge above the computed does not take place in the plate of air, where it could not do it consistently with the theory, but does in the glass plate, where it may do so consistently with the theory, I think there seems great reason to think that it is not owing to any defect in the theory, but to some such motion of the electricity as we have supposed.

364] I could not find that there was any difference in the proportion which the charge of a glass plate bore to that of another body whether they were electrified positively or negatively*.

365] It was said in Art. [331], that there seemed no reason to think that the charge of the plate D, or of any other of those glass plates was sensibly greater than it would be if the electricity was spread uniformly on their surfaces, whereas the charge of most of the plates of air was found very considerably greater than it would be on that supposition. But this is by no means inconsistent, for according to the first way of accounting for the great excess of the real charge of those plates above the computed, namely supposing that the electricity penetrates into the glass to the depth of $\frac{7}{16}$ of its thickness, the increase of its charge on account of the electricity being not spread uniformly, should be not greater than it would be if the glass was only $\frac{1}{8}$ of its real thickness, and the electricity was unable to penetrate into it at all, and therefore should not be greater than it is in a plate of air in which the thickness is $\frac{1}{84}$ of the diameter, and should therefore in all probability be quite imperceptible.

And by Prop. XXXVI. [Art. 170], the increase of charge should hardly be much, if at all, greater according to the second or third way of accounting for this phenomenon.

366] In order to try† whether the charge of coated glass is the same when hot as when cold, I made use of the apparatus in Fig. 28, where *ABCba* represents a short thermometer tube with a ball *BCb* blown at the end and another smaller ball near the top. This is filled with mercury as high as the bottom of the upper ball, and placed in an iron vessel *FGMN* filled with mercury as high as *FN*. Consequently the ball *BCb* was coated as a Leyden vial, the mercury within it forming the inside coating, and that in the vessel *FGMN* the outer one.

In trying it, I set the vessel *FGMN* on the wooden bars of the machine represented in Fig. 20, near the end *NP*, and dipt a small iron wire bound round the wire *Mm* into the mercury within the tube, so as to make a communication between the wire *Mm* and the inside coating, the outside coating, or the mercury in *FGMN*, being made to communicate with the ground.

* [Art. 463.]

† [Art. 556, March 21, 1773. See also Arts. 548, 549, 680.]

~~would not find that there was any difference in the electric whether the charge of a glass plate bore the same proportion to that of another body whether they were electrified positively or negatively~~ (interlined: or proportion which)

~~To P.~~ P. 116

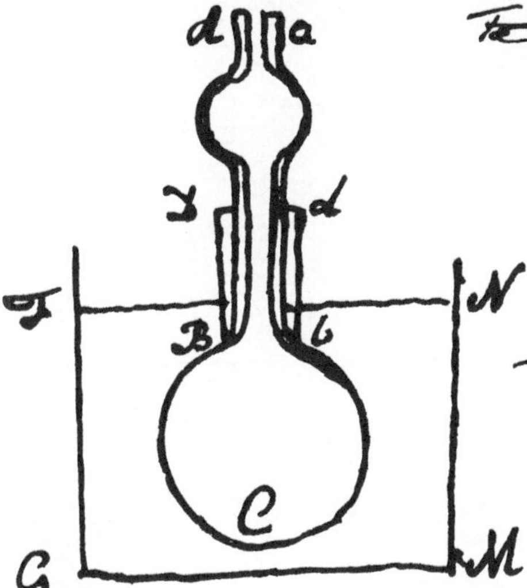

Fig. 28

I order to try whether the charge of coated glass is the same when hot as when cold I made use of the appar. in Fig. 28 where ABCba represents a short thermom tube with a ball BCb blown at the end & another smaller ball near the top. This is filled with ☿ as high as the bottom of the upper ball & placed in an iron vessel FGMN filled with ☿ as high as FN consequently the ball BCb was ~~charged~~ coated as a Leyden vial the ☿ within

~~P. 119~~

To face p. 180.

It was heated by a lamp placed under *FGMN*, and its charge was frequently tried while heating by comparing with a sliding coated plate placed on the other end of the wooden bars.

When it was sufficiently heated, the lamp was taken away, and the charge frequently tried in the same manner while cooling, a thermometer being dipt every now and then into the mercury in *FGMN* to find its heat.

367] As it was apprehended that the electricity might spread further on the surface of the glass while hot than while cold, a paper coating *DBbd* was fastened on the tube, so that as the outside coating was made to extend as far as *Dd*, that is three or four inches above the mercury in *FGMN*, where the tube was very little heated, and as the inside coating reached still higher, that is to the bottom of the upper ball, no sensible error could proceed from thence.

The use of the upper ball was to prevent the mercury within the tube from overflowing when hot.

368] By a mean between the experiments made while the ball was heating and while cooling, its charge answering to the different degrees of heat was as follows.

Heat.	Charge.	Difference of heat.	Difference of charge.
55	100		
		102	4
157	104		
		65	12
222	116		
		73	20
295	136		
		10	5
305	141		

369] At 295° the electricity passed through the glass pretty freely, but at 305° much faster. It appears, therefore, that the charge of glass is considerably greater when heated to such a degree as to suffer the electricity to pass through than when cold, but that its charge does not begin to be sensibly increased till it is heated to a considerable degree*.

370] *On the charges of plates of several different sorts of glass, and also of plates of some other substances which do not conduct electricity, charged in the manner of Leyden vials.*

* [Note 26.]

The result of the experiments I made on this subject is contained in the two following tables:—

TABLE OF GLASS PLATES*.

	Thickness.	Diameter.	Ditto corrected.	Computed charge.	Observed charge.	Observed charge by computed charge.	Specific gravity.
Flint glass ground flat	·2115	2·23	2·37	3·32	26·3	7·93	3·279
Ditto a thinner piece	·104	2·215	2·385	6·84	52·3	7·65	3·284
Plate glass P	·127	2·85	3·02	8·98	71·9	8·01	2·752
W	·172	3·435	3·585	9·34	74·8	8·01	2·787
G	·1848	3·575	3·725	9·38	75·5	8·05	2·973
N	·106	2·12	2·29	6·18	51·4	8·31	2·682
O	·106	2·505	2·675	8·44	75	8·89	2·514
Q	·076	2·065	2·245	8·29	76·5	9·23	2·504
Crown glass	·0682	3·495	3·675	24·76	211·3	8·54	2·537
Ditto another piece	·0659	3·43	3·61	24·72	208·7	8·44	2·532
Crown glass ground	·07	2·035	2·215	8·76	76·5	8·73	} 2·535
Part of same piece	·0693	3·54	3·72	24·96	215·1	8·62	
Mean of the 10 pieces used in former experiments						8·22	2·678

371] Plates of other substances†.

	Thickness.	Diameter.	Computed charge.	Observed charge.	Observed charge by computed.
Gum Lac	·125	4·23	17·89	80	4·47
Mixture of rosin and bees wax. Plate 1	·4845	3·75	3·63	13·5	3·72
2	·192	3·355	7·22	25·2	3·49
3	·103	4·247	21·89	69	3·15
4	·103	4·525	24·85	78·9	3·18
5	·103	1·79	3·89	13	3·34
Dephlegmated bees wax. Plate 1	·303	3·78	5·90	24·5	4·16
2	·120	3·525	12·95	46·1	3·56
3	·063	2·74	14·90	50·5	3·39
Plain bees wax	·119	3·475	12·69	51·3	4·04

372] The coatings of all these plates were circular.

In computing the charge of the glass plates, the diameter of the coating was corrected on account of the spreading of the electricity as in the fourth column, the electricity being supposed to spread ·07 of an inch if the thickness is ·21 and ·09 if the thickness is ·08, and so on in proportion in other thicknesses. But no correction is made in computing the charges of the other plates, as I was uncertain how much to allow.

* [See Art. 673.]

† [See Art. 674.]

373] The method I used in making all the plates of the second table was this. I first cast a round plate of the substance, three or four times as thick as I intended it should be, and rather thinner near the edges than in the middle, taking care to cast it as free from air bubbles as I could.

I then heated it between two thick flat plates of brass, till it was become soft, and then pressed it out to the proper thickness by squeezing the plates together with screws*. In order to prevent its sticking to the brass plates, I put a piece of thin tinfoil between it and each plate, and I found the tinfoil did not stick to it so fast but what I could get it off without any danger of damaging them.

374] The heat necessary to melt shell lac is so great as to make it froth and boil; which makes it impossible to cast a plate of it free from air bubbles. The plate mentioned in the preceding table was as free from them as I could make it. It contained, however, a great quantity of minute bubbles, but no large ones.

375] Bees wax melts with a heat of about 145°. If it is then heated to a degree rather greater than that of boiling water, it froths very much, and seems to lose a good deal of watery matter, and if it is kept at this heat till it has ceased frothing, it will then bear being heated to a much higher degree without frothing or boiling. Bees wax thus prepared I call dephlegmated.

In order that the plates of dephlegmated bees wax should all be equally so, I dephlegmated some bees wax with a pretty considerable heat, and suffered it to cool and harden, and out of this lump I made all three plates, taking care in casting them not to heat them more than necessary.

I used the same precautions also in casting the plates of a mixture of rosin and bees wax, the proportion of the rosin to the bees wax was forgot to be set down.

What are called in the table the 4th and 5th plate of rosin and bees wax are in reality the same plate as the 3rd, only with a smaller coating.

376] It appears from these experiments, first, that there is a very sensible difference in the charge of plates of the same

* [Art. 514.]

dimensions according to the different sort of glass they consist of, the charge of the plates *O* and *Q*, which consisted of the greenish foreign plate glass mentioned in [Art. 301] being the greatest in proportion to their computed charge of any, next to them the crown glass, and the flint glass being the least of all.

Secondly. The charge of the Lac plate is much less in proportion to its computed charge than that of any glass plate, and that of a plate of bees wax, or of the mixture of rosin and bees wax still less.

But it must be observed that there is a very considerable difference between the three different plates of dephlegmated bees wax in that respect. The same thing, too, obtains in the mixture of rosin and bees wax*.

377] As the proportion of the real charge to the computed is greater in the thick plates than the thin ones, one might be inclined to think that this was owing to the electricity being not spread uniformly. But as the difference seems to be greater than could well proceed from that cause, I am inclined to think that it must have been partly owing to some difference in the nature of the plates. Perhaps it may have been owing to some of the plates having been less heated, and consequently having suffered a greater degree of compression in pressing out than the others.

378] The piece of ground crown glass mentioned in the first of the foregoing tables was made out of a piece of crown glass about $\frac{1}{4}$† of an inch thick, and ground down to the thickness mentioned in the table, care being taken by the workman to take away as much from one side as the other, so that the plate consisted only of the middle part of the glass.

My reason for making it was that as there appears to be a considerable difference in the charge of different sorts of glass, it was suspected that there might possibly be a difference between the inside of the piece and the outside, and if there had, it would have affected the justness of the experiments with the ten pieces of glass ground out of the same piece.

But by comparing the charges of the plates of crown glass with those of the two other pieces of crown glass in the table,

* [Note 27.]

† There are pieces of that thickness sometimes blown for the use of the Opticians.

there does not seem to be any difference which can be depended on with certainty.

The experiment indeed would have been more satisfactory if the piece of ground glass and the pieces with which it was compared had been all made out of the same pot. But as it would have been difficult procuring such pieces, and as I have found very little difference in the specific gravity of different pieces of crown glass, and as I am informed it is all made at the same glass house, I did not take that precaution.

379] Let two or more flat plates of different non-conducting substances, as *AabB*, *BbcC* and *CcdD*, (Fig. 29) be placed close together and coated in the manner of a single plate with the coatings *Ee* and *Ff*. Let the charge of the plate *AabB*, supposing it placed by itself and coated in the usual manner, be equal to that of a plate of glass whose thickness is *A* and whose coatings are of the same size as those of *AabB*.

Fig. 29.

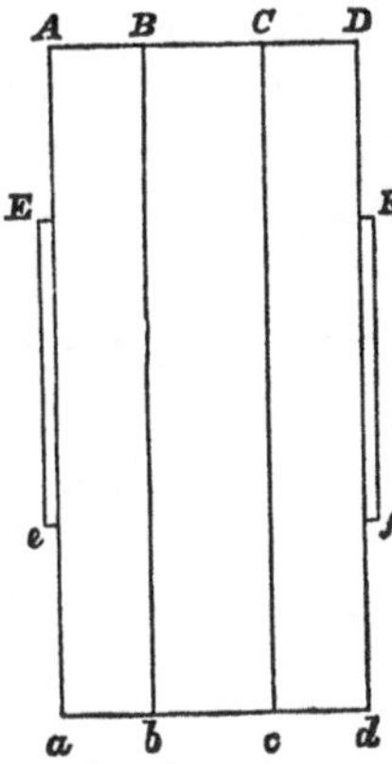

In like manner let the charge of *BbcC* be equal to that of a plate of the same glass whose thickness is equal to *B*, and let that of *CcdD* equal that of one whose thickness is *C*.

Then whichever of the three ways of accounting for the excess of the real charge of glass plates above the computed we prefer, it is a necessary consequence of our theory that the charge of this compound plate *AadD* should be equal to that of a single plate of glass whose thickness equals $A + B + C$, and whose coatings are of the same size as *Ee* and *Ff*.

380] In like manner if two or more plates of the same kind of glass are placed together and coated as above, the charge of this compound plate should be equal to that of a single plate of the same glass whose thickness is equal to that of all the plates together. This appears from the following experiments to be the case, for

1st. I took the three plates of glass *A*, *B* and *C**, and laid them on one another, having first taken off their old coatings and coated the outside surfaces as in fig. 29 with circles of tinfoil 6·6 inches in diameter. The charge of this compound plate was found to be to that of the three plates *D*, *E* and *F* together as ·944 to 1. The sum of the thicknesses of *A*, *B* and *C* together is ·6309, and the computed charge of a plate of that thickness with coatings 6·6 in. diameter is to that of *D*, *E* and *F* together, allowing in the same manner as in [Art. 328] for the instantaneous spreading of the electricity, as ·94 to one. So that the charge of this compound plate is exactly the same that it ought to be according to the foregoing rule.

381] 2ndly, I made a plate of a mixture of rosin and bees wax†, about 8 inches square and somewhat more than ·12 thick, and coated it with circles 6·61 in. diameter. Its charge was found to be to that of the plates *K*, *D* and *E* together as 56 to 55, and therefore should be equal to that of a plate of glass of the same kind as *K* whose thickness is ·345 and the diameter of whose coatings is the same as those of the rosin plate, namely 6·61 inches.

This plate was then inclosed between the glass plates *B* and *H*‡, the coatings being first taken off, and the outside surfaces of *B* and *H* coated with circles 6·6 inches in diameter. Its charge was found to be to that of *K* as 7·56 to 8.

According to the foregoing rule, its charge should be the same as that of a plate of glass of the same kind as *B* ·634 of an inch thick with coatings 6·6 inches in diameter, and should therefore be to that of *K* as 7·34 to 8, which is very nearly the same that it was actually found to be.

382] On the charges of such Leyden vials as do not consist of flat plates of glass.

These experiments were made with hollow cylindrical pieces of glass, open at both ends, and coated both within and without with pieces of tinfoil surrounding the cylinder in the form of a ring, the breadth of the ring being everywhere the same, and the inside and outside coatings being of the same breadth, and placed exactly opposite to each other. Only as the inside diameter of the two

* [Arts. 534, 544, 546, 677.]

† [Arts. 548, 678.]

‡ [Arts. 552, 679.]

thermometer tubes was too small to admit of being coated in this manner, they were filled with mercury by way of inside coating.

The thickness of the glass was found by suspending the cylinder by one end from a pair of scales with its axis in a vertical position, and the lower part immersed in a vessel of water, and finding the alteration of the weight of the cylinder according as a greater or less portion of it was under water*.

383] The result of the experiments is contained in the following table†.

	Mean thickness.	Mean outside semi-diameter.	Length of coating.	Computed charge.	Observed charge.	Observed charge by computed.	Specific gravity.	Outside diameter by thickness.
Part of a jar of flint glass	·084	1·62	4·4	85·9	717	8·35	3·254	19·3
A cylinder of ditto	·0704	·645	9·86	87·1	650	7·46	3·281	9·2
Thermometer tube I.	·094	·14	11	11·0	80·2	7·31	3·098	1·5
,, ,, II.	·130	·16	15·5	11·1	80·7	7·26	3·243	1·24
Cylinders of green bottle glass 1	·045	·50	7·16	77·2	754	9·77	2·665	11·3
Cylinders of green bottle glass 2	·060	·53	8·55	76·6	690	9	2·664	8·8
Cylinders of green bottle glass 3	·078	·48	7	40·8	353	8·65	2·665	6·2

The lengths of the coating here set down are the real lengths. But in computing the charges of the white jar and cylinder and the three green cylinders, these lengths were increased on account of the spreading of the electricity according to the same supposition as was used in computing the charges of the flat plates.

But in computing the charges of the thermometer tubes no correction was made, as I was uncertain how much to allow, but as the length of their coatings is so great, this can hardly make any sensible error.

384] It should seem from these experiments as if the proportion of the real to the computed charge was rather less in a cylinder in which the thickness of the glass is $\frac{1}{6}$ of the semidiameter than in one in which it is only $\frac{1}{11}$, and most likely rather less in that than in a flat plate, but then it seems to be not much less in a cylinder in which the inside diameter is many times less than

* [Art. 594.]

† [See Art. 676, and Note 28.]

the outside, that is, in which the thickness of the glass is almost equal to the outside semidiameter, than it is in the first mentioned cylinder.

Nothing certain, however, can be inferred as to this point, as in all probability the four pieces of flint glass used in these experiments and the two flat pieces used in [Art. 370] did not consist exactly of the same kind of glass, as indeed appears from their specific gravities.

385] The three green cylinders, indeed, were all made at the same time and out of the same pot, so that it seems difficult to suppose that there should be any difference of that kind between them*. But then I had no flat plates to compare them with.

On the whole, I think we may with tolerable certainty infer that the ratio of the real to the computed charge is not very different from what it is in flat plates, whatever is the proportion which the thickness of the glass bears to the diameter of the cylinder, though it seems to be not exactly the same.

* Though it seems not likely that there should be any difference in the nature of the glass of which the three green cylinders consisted, yet I am not sure that there was not, for the inside of the glass, that is, that part which was nearest to the inside surface, was manifestly more opaque and of a different colour from the outside, and the separation between these two sorts of glass appeared well defined, so that the cylinder seemed to consist of two different coats of glass lying one over the other. The distinction was the most visible in those cylinders which consisted of the thickest glass and in the thickest part of those cylinders. The specific gravities, however, do not indicate any difference in the nature of the glass. What was the reason of the above-mentioned appearance I cannot tell.

WHETHER THE FORCE WITH WHICH TWO BODIES REPEL IS AS THE SQUARE OF THE REDUNDANT FLUID, TRIED BY STRAW ELECTROMETERS *.

386] If two bodies, *A* and *B*, placed near to each other, are both connected to the same overcharged Leyden jar, and the force with which this jar is electrified is varied, everything else remaining unaltered, the force with which *A* and *B* repel each other ought by the theory to be as the square of the quantity of redundant fluid in the jar, supposing the distance of the bodies *A* and *B* to remain unaltered. For the quantity of redundant fluid in *A* is directly as the quantity of redundant fluid in the jar, and therefore the force with which each particle of redundant fluid in *B* is repelled by *A* is also directly as the quantity of redundant fluid in the jar, and therefore as the number of particles of redundant fluid in *B* is also as the quantity of redundant fluid in the jar, the force with which *B* is repelled by *A* is as the square of the quantity of redundant fluid in the jar.

387] In order to try whether this was the case, I made use of the following apparatus †.

CD (Fig. 31) is a wooden rod 43 inches long, covered with tinfoil and supported horizontally by non-conductors. At the end

Fig. 31.

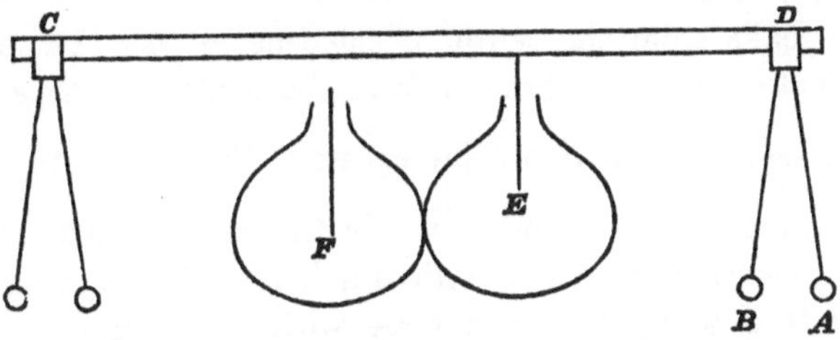

* [Title supplied from Cavendishes Index to his experiments, Art. 563.]
† [Arts. 563, 567, also Art. 525.]

C is suspended, as in the figure, the electrometer described in Art. 249, and at the other end D is suspended a similar electrometer, only the straws reached to the bottom of the cork balls A and B, but not beyond them, and were left open so as to put in pieces of wire, and thereby increase their weight and the force with which they endeavoured to close. The lower ends of these wires when used were just even with the bottom of the cork balls, and were kept in that situation by wax, the wax being cut off even with the bottom of the corks, so as to leave no roughnesses to carry off the electricity. In like manner, when the wires were not used, the ends of the straws were closed up with wax.

388] The proportion which the force with which the balls of this electrometer endeavoured to close when the wires were inserted bore to that with which they endeavoured to close without the wires was thus found. The weight of the straw $\left\{\begin{matrix} A \\ B \end{matrix}\right.$ with its ball and centre pin but without its wire was found to be $\left\{\begin{matrix} 7{\cdot}6 \\ 6{\cdot}65 \end{matrix}\right.$ grains, and the distance of its center of gravity from the center of suspension was $\left\{\begin{matrix} 5{\cdot}36 \\ 5{\cdot}285 \end{matrix}\right.$ inches, as was found by balancing it on the edge of a knife. Consequently the force with which this straw when put in its place, endeavours to descend towards the perpendicular, supposing it to be removed to a given distance from it, was as $\left\{\begin{matrix} 7{\cdot}6 \times 5{\cdot}36 \\ 6{\cdot}65 \times 5{\cdot}285 \end{matrix}\right.$.

The weight of the wire inserted was $\left\{\begin{matrix} 12{\cdot}05 \\ 10 \end{matrix}\right.$ grains, and half its length was $\left\{\begin{matrix} 1{\cdot}23 \\ 1{\cdot}00 \end{matrix}\right.$ inches, so that as the distance of the bottom of the cork balls from the center of suspension was 11·1 inches, the distance of its center of gravity from the center of suspension was $\left\{\begin{matrix} 9{\cdot}87 \\ 10{\cdot}1 \end{matrix}\right.$ inches, and therefore the excess of the force with which the ball endeavours to descend towards the perpendicular when the wire is inserted above that with which it endeavours to descend without [the wire] is to the force with which it endeavours to descend without the wire as $\left\{\begin{matrix} 12{\cdot}03 \times 9{\cdot}87 \\ 10 \times 10{\cdot}1 \end{matrix}\right.$ to $\left\{\begin{matrix} 7{\cdot}6 \times 5{\cdot}36 \\ 6{\cdot}65 \times 5{\cdot}285 \end{matrix}\right.$, or as

$\left\{\begin{matrix}2{\cdot}92\\2{\cdot}88\end{matrix}\right.$ to one. Therefore the force with which the electrometer endeavours to close when the wires are inserted is to that with which it endeavours to close without the wires as 3·9 to 1.

389] *E* and *F* are two coated Leyden vials, nearly of the same size. The outside coatings of both communicate with the ground, and the inside coating of *E* communicates with *CD*, but not that of *F*.

390] The way in which I tried the experiment was as follows. I first compared the electrometer *C* with the electrometer *D* without the wires, and found that when the jar *E* was electrified to such a degree as to make *D* separate $\left\{\begin{matrix}13\\12\end{matrix}\right.$ divisions, *C* separated $\left\{\begin{matrix}14\frac{1}{4}\\13\frac{1}{4}\end{matrix}\right.$ divisions, so that the same degree of electrification which made *C* separate $\left\{\begin{matrix}13\\12\end{matrix}\right.$ divisions made *D* separate $\left\{\begin{matrix}14\frac{1}{4}\\13\frac{1}{4}\end{matrix}\right.$ divisions.

I then put the wires into the electrometer *D*, and put the larger of the two vials in the place of *E*, and electrified *E* and consequently the rod *CD* and the two electrometers till *D* separated $\left\{\begin{matrix}13\\12\end{matrix}\right.$ divisions.

The wire by which *E* was electrified was then immediately taken away and a communication made between *E* and *F*, so that the redundant fluid in *E* and *CD* and the electrometers was communicated to *F*.

It was found that the electrometer *C* then separated $\left\{\begin{matrix}15\frac{1}{4}\\14\end{matrix}\right.$ divisions.

The experiment was then repeated in the same manner, except that the smaller vial was placed at *E*. It was found that if *E* was electrified till *D* separated $\left\{\begin{matrix}13\\12\end{matrix}\right.$ divisions, then on making a communication between *E* and *F*, *C* separated $\left\{\begin{matrix}13\frac{1}{4}\\12\frac{1}{2}\end{matrix}\right.$ divisions.

391] From hence we may conclude that if the vials had been exactly equal and *E* had been electrified till *D* separated $\left\{\begin{matrix}13\\12\end{matrix}\right.$

divisions, then on making a communication between E and F, C would have separated $\left\{\begin{matrix}14\frac{1}{4}\\ 13\frac{1}{8}\end{matrix}\right.$ divisions.

But it appears from the first mentioned part of the experiment, that the same degree of electrification which makes C separate $\left\{\begin{matrix}14\frac{1}{4}\\ 13\frac{1}{8}\end{matrix}\right.$ divisions is sufficient to make D without the wires separate $\left\{\begin{matrix}13\\ 11\frac{7}{8}\end{matrix}\right.$ divisions. From whence it appears that if the jars are exactly equal, and one of them is electrified till the electrometer D with the wires separates $\left\{\begin{matrix}13\\ 12\end{matrix}\right.$ divisions, and its electricity is then communicated to the other vial, the electricity will be of that degree of strength which is necessary to make the same electrometer without the wires separate $\left\{\begin{matrix}13\\ 11\frac{7}{8}\end{matrix}\right.$ divisions, that is, very nearly the same as before, or as it did with the wire before the communication of the electricity.

But if the vials are equal, the quantity of redundant fluid in the first vial, after its electricity is communicated to the second, will be very little more than half of what it was before the communication, for the quantity of redundant fluid in the rod DC and the electrometers is trifling in comparison of that in the vial*, and consequently it appears that the distance to which the electrometer with the wires in it separates with a given quantity of redundant fluid in the vial is very nearly the same as that to which it separates without the wires when there is only half that quantity of redundant fluid in the vial.

Therefore as the force with which the electrometer endeavours to close by its weight when the wires are in is to that with which it endeavours to close without the wires as 3·9 to 1, it appears that the force with which the balls of the electrometer are repelled with a given quantity of redundant fluid in the vial, is to that with which they are repelled when there is only half that quantity of redundant fluid in the vial as 3·9 to 1 (supposing the distance

* [In a sentence which Cavendish has scored out in his MS. we read—]

The charge of the two vials together was found to be 2168 inches. The diameter of the rod CD was at a medium about $\frac{3}{4}$ of an inch. [This would make the computed charge of the rod 9·7 inches.—Ed.]

of the balls to be the same in both cases), that is, very nearly as the square of the quantity of redundant fluid in the vial, the difference being not more than what might very easily be owing to the error of the experiment. So that the experiment agrees very well with the theory.

392] It was found that if the communication was made between the two vials by a piece of metal, the electricity was diminished so suddenly as to set the straws a vibrating, and it was some time before they stopt, for which reason the communication was made by a piece of moist wood, which, though it communicates the electricity of one vial to the other very quickly, did not do it so instantaneously as to make the straws vibrate much.

393] The electricity of the vial was found to waste very slowly, so that it could not be sensibly diminished during the small time spent in communicating the electricity from one vial to the other and reading off the divisions, so that no sensible error could proceed from that cause.

394] I tried the experiment before in the same manner, and with the same electrometers, except that the straws were not gilt, but only moistened with salt. It then seemed as if the force with which the balls of the electrometer were repelled with a given quantity of redundant fluid in the vial was to that with which they were repelled with only half that quantity in the vial as 4 to $\frac{9}{8}$.

As I suspected that this small difference from the theory was owing to the straws not conducting sufficiently readily, I gilt the straws, when, as was before shewn, the experiment agreed very well with theory.

It must be observed that if the straws do not conduct sufficiently readily, the balls of the electrometer will not be so strongly electrified and will not separate so much as they ought to do, and in all probability the difference will be greater in the stronger degree of electricity, in which the electricity wastes much faster, than it is in the weaker, and will therefore diminish the degree of separation more in the stronger degree of electricity than in the weaker, and will therefore make the force with which the balls repel with the stronger degree of electricity appear to be less in proportion to that with which they repel with the weaker degree than it ought to be.

AN ACCOUNT OF SOME ATTEMPTS TO IMITATE THE EFFECTS OF THE TORPEDO BY ELECTRICITY. BY THE HON. HENRY CAVENDISH, F.R.S.*

395] Although the proofs brought by Mr Walsh†, that the phenomena of the torpedo are produced by electricity, are such as leave little room for doubt; yet it must be confessed, that there are some circumstances, which at first sight seem scarcely to be reconciled with this supposition. I propose, therefore, to examine whether these circumstances are really incompatible with such an opinion; and to give an account of some attempts to imitate the effects of this animal by electricity.

396] It appears from Mr Walsh's experiments, that the torpedo is not constantly electrical, but hath a power of throwing at pleasure a great quantity of electric fluid from one surface of those parts which he calls the electrical organs to the other; that is, from the upper surface to the lower, or from the lower to the upper, the experiments do not determine which; by which means a shock is produced in the body of a person who makes any part of the circuit which the fluid takes in its motion to restore the equilibrium.

397] One of the principal difficulties attending the supposition, that these phenomena are produced by electricity, is, that a shock may be perceived when the fish is held under water; and

* From the *Philosophical Transactions* for 1776, Vol. LXVI. Part I. pp. 196—225. *Read* Jan. 18, 1775.

† [*Philosophical Transactions*, 1773, pp. 461—477. Of the Electric Property of the Torpedo. In a letter from John Walsh, Esq., F.R.S., to Benjamin Franklin, Esq., LL.D., F.R.S., &c. *Read* July 1, 1773.]

in other circumstances, where the electric fluid hath a much readier passage than through the person's body. To explain this, it must be considered, that when a jar is electrified, and any number of different circuits are made between its positive and negative side, some electricity will necessarily pass along each; but a greater quantity will pass through those in which it meets with less resistance, than those in which it meets with more. For instance, let a person take some yards of very fine wire, holding one end in each hand, and let him discharge the jar by touching the outside with one end of the wire, and the inside with the other; he will feel a shock, provided the jar is charged high enough; but less than if he had discharged it without holding the wire in his hands; which shews, that part of the electricity passes through his body, and part through the wire. Some electricians indeed seem to have supposed that the electric fluid passes only along the shortest and readiest circuit; but besides that such a supposition would be quite contrary to what is observed in all other fluids, it does not agree with experience. What seems to have led to this mistake is, that in discharging a jar by a wire held in both hands, as in the above-mentioned experiment, the person will feel no shock, unless either the wire is very long and slender, or the jar is very large and highly charged. The reason of which is, that metals conduct surprisingly better than the human body, or any other substance I am acquainted with; and consequently, unless the wire is very long and slender, the quantity of electricity which will pass through the person's body will bear so small a proportion to the whole, as not to give any sensible shock, unless the jar is very large and highly charged.

398] It appears from some experiments*, of which I propose shortly to lay an account before this Society, that iron wire conducts about 400 million times better than rain or distilled water; that is, the electricity meets with no more resistance in passing through a piece of iron wire 400,000,000 inches long, than through a column of water of the same diameter only one inch long. Sea water, or a solution of one part of salt in 30 of water, conducts 100 times, and a saturated solution of sea salt about 720 times better than rain water.

* [Arts. 576, 577, 684, 687.]

399] To apply what hath been here said to the torpedo; suppose the fish by any means to convey in an instant a quantity of electricity through its electric organs, from the lower surface to the upper, so as to make the upper surface contain more than its natural quantity, and the lower less; this fluid will immediately flow back in all directions, part over the moist surface, and part through the substance of its body, supposing it to conduct electricity, as in all probability it does, till the equilibrium is restored: and if any person hath at the time one hand on the lower surface of the electric organs, and the other on the upper, part of the fluid will pass through his body. Moreover, if he hath one hand on one surface of an electric organ, and another on any other part of its body, for instance the tail, still some part of the fluid will pass through him, though much less than in the former case; for as part of the fluid, in its way from the upper surface of the organ to the lower, will go through the tail, some of that part will pass through the person's body. Some fluid also will pass through him, even though he does not touch either electric organ, but hath his hands on any two parts of the fishes body whatever, provided one of those parts is nearer to the upper surface of the electric organs than the other.

400] On the same principle, if the torpedo is immersed in water, the fluid will pass through the water in all directions, and that even to great distances from its body, as is represented in Fig. 1, where the full lines represent the section of its body, and

Fig. 1.

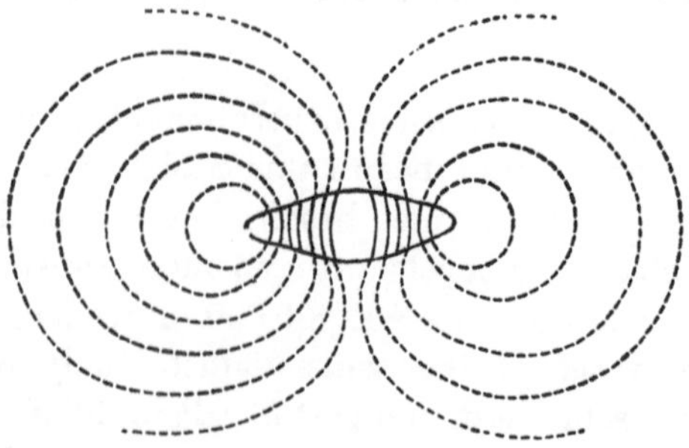

the dotted lines the direction of the electric fluid; but it must be observed, that the nearer any part of the water is to the fishes body, the greater quantity of fluid will pass through it. Moreover, if any person touches the fish in this situation, either with one hand on the upper surface of an electric organ, and the other on the lower, or in any other of those manners in which I supposed it to be touched when out of the water, some fluid will pass through his body; but evidently less than when the animal is held in the air, as a great proportion of the fluid will pass through the water: and even some fluid will pass through him, though he does not touch the fish at all; but only holds his hands in the water, provided one hand is nearer to the upper surface of the electric organs than the other.

401] The second difficulty is, that no one hath ever perceived the shock to be accompanied with any spark or light, or with the least degree of attraction or repulsion. With regard to this, it must be observed, that when a person receives a shock from the torpedo, he must have formed the circuit between its upper and lower surface before it begins to throw the electricity from one side to the other; for otherwise the fluid would be discharged over the surface of the fishes body before the circuit was completed, and consequently the person would receive no shock. The only way, therefore, by which any light or spark could be perceived, must be by making some interruption in the circuit. Now Mr Walsh found, that the shock would never pass through the least sensible space of air, or even through a small brass chain. This circumstance, therefore, does not seem inconsistent with the supposition that the phenomena of the torpedo are owing to electricity; for a large battery will give a considerable shock, though so weakly charged that the electricity will hardly pass through any sensible space of air; and the larger the battery is, the less will this space be. The principle on which this depends will appear from the following experiments.

402] I took several jars of different sizes, and connected them to the same prime conductor, and electrified them in a given degree, as shewn by a very exact electrometer; and then found how near the knobs of an instrument in the nature of Mr Lane's electrometer must be approached, before the jars would discharge themselves. I then electrified the same jars again in the same degree as before, and separated all of them from the conductor except one. It was found, that the distance to which the knobs must be approached to discharge this single jar was not sensibly less than the former. It was also found, that the divergence of the electrometer was the same after the removal of the jars as before, provided it was placed at a considerable distance from them: from which last circumstance, I think we may conclude, that the force with which the fluid endeavours to escape from the single jar is the same as from all the jars together*.

403] It appears, therefore, that the distance to which the spark will fly is not sensibly affected by the number or size of the jars, but depends only on the force with which they are electrified; that is, on the force with which the fluid endeavours to escape from them: consequently, a large jar, or a great number of jars, will give a greater shock than a small one, or a small number, electrified to such a degree, that the spark shall fly to the same distance; for it is well known, that a large jar, or a great number of jars, will give a greater shock than a small one, or a small number, electrified with the same force.

404] In trying this experiment, the jars were charged very weakly, insomuch that the distance to which the spark would fly was not more than the 20th of an inch. The electrometer† I used consisted of two straws, 10 inches long, hanging parallel to each other, and turning at one end on steel pins as centers, with cork balls about $\frac{1}{4}$ of an inch in diameter fixed on the other end. The way by which I estimated the divergence of these balls, was by seeing whether they appeared to coincide with parallel lines placed behind them at about 10 inches distance; taking care to hold my eye always at the same distance from the balls, and not less than thirty inches off. To make the straws conduct the better, they were gilded, which causes them to be much more regular in their effect. This electrometer is very accurate; but can be used only when the electricity is very weak. It would be easy, however, to make one on the same principle, which should be fit for measuring pretty strong electricity.

405] The instrument by which I found to what distance the spark would fly is represented in Fig. 2; it differs from Mr Lane's electrometer ‡ no otherwise than in not being fixed to a jar, but

* [Art. 604.] † [Art. 249.] ‡ [Art. 329.]

made so as to be held in the hand. The part *ABCDEFGKLM* is of baked wood, the rest of brass; the part *GKL* being covered with tinfoil communicating with the brass work at *FG*; and the

Fig. 2.

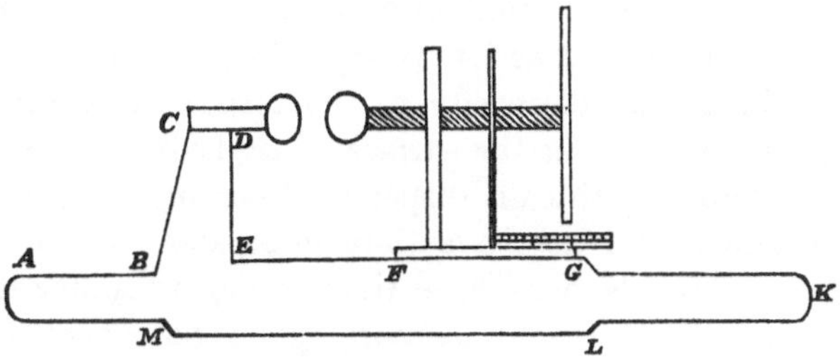

part *ABM* being also covered with a piece of tinfoil, communicating with the brass work at *CD*.

406] I next took four jars, all of the same size; electrified one of them to a given degree, as shewn by the electrometer; and tried the strength of the shock which it gave; and found also to what distance the spark would fly. I then took two of the jars, electrified them in the same degree as before, and communicated their electricity to the two remaining. The shock of these four jars united, was rather greater than that of the single jar; but the distance to which the spark would fly was only half as great*.

407] Hence it appears, that the spark from four jars, all of the same size, will not dart to quite half so great a distance as that from one of those jars electrified in such a degree as to give a shock of equal violence; and consequently the distance to which the spark will fly is inversely in a rather greater proportion than the square root of the number of jars, supposing them to be electrified in such a degree that the shock shall be of a given strength. It must be observed, that in the last mentioned experiment, the quantity of electric fluid which passed through my body was twice as great in taking the shock of the four jars, as in taking that of the single one; but the force with which it was impelled was evidently less, and I think we may conclude, was only half as great. If so, it appears that a given quantity of electricity, impelled through our body with a given force, produces a

* [Arts. 573, 610, 613.]

rather less shock than twice that quantity, impelled with half that force; and consequently, the strength of the shock depends rather more on the quantity of fluid which passes through our body, than on the force with which it is impelled.

408] That no one could ever perceive the shock to be accompanied with any attraction or repulsion, does not seem extraordinary; for as the electricity of the torpedo is dissipated by escaping through or over the surface of its body, the instant it is produced, a pair of pith balls suspended from any thing in contact with the animal will not have time to separate, nor will a fine thread hung near its body have time to move towards it, before the electricity is dissipated. Accordingly I have been informed by Dr Priestley, that in discharging a battery he never could find a pair of pith balls suspended from the discharging rod to separate. But, besides, there are scarce any pith balls so fine, as to separate when suspended from a battery so weakly electrified that its shock will not pass through a chain, as is the case with that of the torpedo.

409] In order to examine more accurately, how far the phenomena of the torpedo would agree with electricity, I endeavoured to imitate them by means of the following apparatus. *ABCFGDE*, Fig. 3, is a piece of wood, the part *ABCDE* of which

Fig. 3.

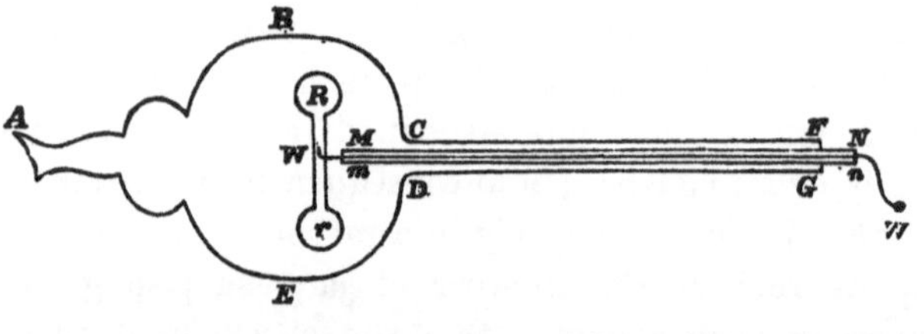

is cut into the shape of the torpedo, and is $16\frac{3}{4}$ inches long from *A* to *D*, and $10\frac{3}{4}$ broad from *B* to *E*; the part *CFGD* is 40 inches long, and serves by way of handle. *MNmn* is a glass tube let into a groove cut in the wood. *Ww* is a piece of wire passing through the glass tube, and soldered at *W* to a thin piece of pewter *Rr* lying flat on the wood, and intended to represent the upper surface of the electric organs. On the other side of the wood there is placed such another glass tube, not represented in the figure, with a wire passing through it, and soldered to another piece of pewter of the same size and shape as *Rr* intended to represent the lower surface of those organs. The whole part *ABCDE* is covered with a piece of sheep's skin leather.

410] In making experiments with this instrument, or artificial torpedo as I shall call it, after having kept it in water of about the same saltness as that of the sea, till thoroughly soaked, I fastened the end of one of the wires, that not represented in the drawing for example, to the negative side of a large battery, and when it was sufficiently charged, touched the positive side with the end of the wire *Ww*; by which means the battery was discharged through the torpedo: for as the wires were inclosed in glass tubes, which extended about an inch beyond the end of the wood *FG* no electricity could pass from the positive side of the battery to the negative, except by flowing along the wire *Ww* to the pewter *Rr*, and thence either through the substance of the wood, or along the wet leather, to the opposite piece of pewter, and thence along the other wire to the negative side. When I would receive a shock myself, I employed an assistant to charge the battery, and when my hands were in the proper position, to discharge it in the above mentioned manner by means of the wire *Ww*. In experiments with this torpedo under water, I made use of a wooden trough; and as the strength of the shock may, perhaps, depend in some measure on the size of the trough, and on the manner in which the torpedo lies in it, I have, in Fig. 4,

Fig. 4.

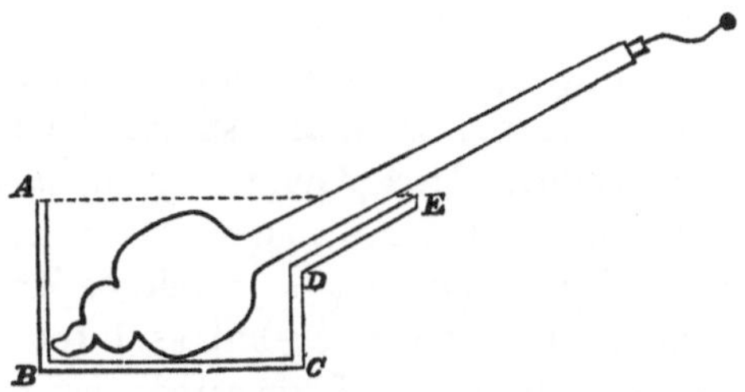

given a vertical section of it; the torpedo being placed in the same situation as in the figure. *ABCDE* is the trough; the length *BC* is 19 inches; the depth *AB* is 14; and the breadth is 13; consequently, as the torpedo is two inches thick in the thickest part, there is about $5\frac{1}{2}$ inches distance between its sides and those of the trough.

411] The battery was composed of 49 jars, of extremely thin glass, disposed in 7 rows, and so contrived that I could use any number of rows I chose. The outsides of the jars were coated with tinfoil; but as it would have been very difficult to have coated the insides in that manner, they were filled with salt water. In a battery to answer the purpose for which this was intended, it is evidently necessary that the metals serving to make the communications between the different jars should be joined quite close: accordingly care was taken that the contacts should be made as perfect as possible. I find, by trial, that each row of the battery contains about $15\frac{3}{4}$ times as much electricity, when both are connected to the same prime conductor, as a plate of crown glass, the area of whose coating is 100 square inches, and whose thickness is $\frac{55}{1000}$ of an inch; that is, such that one square foot of it shall weigh 10 oz. 12 dwts.; and consequently, the whole battery contains about 110 times as much electricity as this plate*.

412] The way by which this was determined, and which, I think, is one of the easiest methods of comparing the quantity of electricity which different batteries will receive with the same degree of electrification, was this: First of all, supposing a jar or battery to be electrified till the balls of the above-mentioned electrometer separated to a given distance, I found how much they would separate when the quantity of electricity in that jar or battery was reduced to one-half. To do this, I took two jars, as nearly equal as possible, and electrified one of them till the balls separated to a given degree, and then communicated its electricity to the other; and observed to what distance the balls separated after this communication. It is plain, that if the jars were exactly equal, this would be the distance sought for; as in that case the quantity of electricity in the first jar would be just half as much after the communication as before; but as I could not be sure that they were exactly equal, I repeated the experiment by

* I find, by experiment, that the quantity of electricity which coated glass of different shapes and sizes will receive with the same degree of electrification, is directly as the area of the coating, and inversely as the thickness of the glass; whence the proportion which the quantity of electricity in this battery bears to that in a glass or jar of any other size, may easily be computed. [See Art. 584. The charge of the first row of jars was 64538, and that of the whole battery about 481000 inches of electricity.]

electrifying the second jar, communicating its electricity to the first, and observing how far the balls separated; the mean between these two distances will evidently be the degree of separation sought, though the jars were not of the same size. Having found this, I electrified one row of the battery till the balls separated to the first distance, and repeatedly communicated its electricity to the plate of coated crown glass, taking care to discharge the plate each time before the communication was made, till it appeared by the electrometer, that the quantity of electricity in that row was reduced to one-half. I found it necessary to do this between 11 or 12 times, or $11\frac{1}{4}$ times as I estimate it. Whence the quantity of electric fluid in the row may be thus determined.

413] Let the quantity in the plate be to that in the row as x to 1; it is plain, that the electricity in the row will be diminished each time it is communicated to the plate, in the proportion of 1 to $1+x$, and consequently after being communicated $11\frac{1}{4}$ times will be reduced in the proportion of 1 to $(1+x)^{11\frac{1}{4}}$; therefore, $(1+x)^{11\frac{1}{4}}=2$; and $1+x=\overline{2}^{\frac{1}{11\frac{1}{4}}}$. Whence the value of x may easily be found by logarithms. But the readiest way of computing it, and which is exact enough for the purpose, is this: multiply the number of times which you communicated the electricity of the row to the plate, by 1,444; and from the product subtract the fraction $\frac{1}{2}$; the remainder is equal to $\frac{1}{x}$, or the number of times by which the electricity in the row exceeds that in the plate*.

414] The way by which I estimated the strength of the charge given to the battery, was taking a certain number of jars, and electrifying them till the balls of the electrometer separated to a given distance, and then communicating their electricity to the battery. This method proved very convenient; for by using always the same jars, I was sure to give always the same charge with great exactness; and by varying the number and size of the jars, I could vary the charge at pleasure, and besides could estimate pretty nearly the proportion of the different charges to each other. It was also the only convenient method which occurred to me; for I could not have done it conveniently by charging the

* [Arts. 441, 582.]

whole battery till an electrometer suspended from it separated to a given distance; because in most of the experiments the electricity was so weak, that a pair of fine pith balls suspended from the battery would separate only to a very small distance; and counting the number of revolutions of the electrical machine is a very fallacious method.

415] I found, upon trial*, that though a shock might be procured from this artificial torpedo, while held under water, yet there was too great a disproportion between its strength, when received this way, and in air; for if I placed one hand on the upper, and the other on the lower surface of the electric organs, and gave such a charge to the battery, that the shock, when received in air, was as strong as, I believe, that of the real torpedo commonly is; it was but just perceptible when received under water. By increasing the charge, indeed, it became considerable; but then this charge would have given a much greater shock out of water than the torpedo commonly does. The water used in this experiment was of about the same degree of saltness as that of the sea; that being the natural element of the torpedo, and what Mr Walsh made his experiments with. It was composed of one part of common salt dissolved in 30 of water, which is the proportion of salt usually said to be contained in sea water. It appeared also, on examination, to conduct electricity not sensibly better or worse than some sea water procured from a mineral water warehouse. It is remarkable, that if I used fresh water instead of salt, the shock seemed very little weaker, when received under water than out; which not only confirms what was before said, that salt water conducts much better than fresh; but, I think, shews, that the human body is also a much better conductor than fresh water: for otherwise the shock must have been much weaker when received under fresh water than in air.

416] As there appeared to be too great a disproportion between the strength of the shock in water and in air, I made another torpedo†, exactly like the former, except that the part *ABCDE* instead of wood was made of several pieces of thick leather, such as is used for the soles of shoes, fastened one over the other, and cut into the proper shape; the pieces of pewter

* [Art. 596.]

† [Arts. 599, 600.]

being fixed on the surface of this, as they were on the wood, and the whole covered with sheep skin like the other. As the leather, when thoroughly soaked with salt water, would suffer the electricity to pass through it very freely, I was in hopes that I should find less difference between the strength of the shock in water and out of it, with this than with the other.

417] For suppose that in receiving the shock of the former torpedo under water, the quantity of electricity which passed through the wood and leather of the torpedo, through my body, and through the water, were to each other as T, B, and W*; the quantity of electricity which would pass through my body, when the shock was received under water, would be to that which would pass through it, when the shock was received out of water, as $\frac{B}{B+T+W}$ to $\frac{B}{B+T}$; as in the first case, the quantity which would pass through my body would be the $\frac{B}{B+T+W}$ part of the whole; and in the latter the $\frac{B}{B+T}$ part. Suppose now, that the latter torpedo conducts N times better than the former; and consequently, that in receiving its shock under water, the quantity of electricity which passes through the torpedo, through my body, and through the water, are to each other as NT, B, and W; the quantity of electricity which will now pass through my body, when the shock is received under water, and out of water, will be to each other as $\frac{B}{B+NT+W}$ to $\frac{B}{B+NT}$; which two quantities differ from each other in a less proportion than $\frac{B}{B+T+W}$ and $\frac{B}{B+T}$: consequently, the readier the body of the torpedo conducts, the greater charge will it require to give the same shock, either in water or out of it; but the less will be the difference between the strength of the two shocks. It should be observed, that this alteration, so far from making it less resembling the real torpedo, in all probability makes it more so; for I see no reason to think, that the real torpedo is a worse conductor of electricity than other animal bodies; and the human body is at least as good, if not a much better conductor than this new torpedo.

* [Arts. 597, 598.]

418] The event answered my expectation; for it required about three times as great a charge of the battery, to give the same shock in air, with this new torpedo as with the former; and the difference between its strength when received under water and out of it, was much less than before, and perhaps not greater than in the real torpedo. There is, however, a considerable difference between the feel of it under water and in air. In air it is felt chiefly in the elbows; whereas, under water, it is felt chiefly in the hands, and the sensation is sharper and more disagreeable. The same kind of shock, only weaker, was felt if, instead of touching the sides, I held my hands under water at two or three inches distance from it.

419] It is remarkable, that I felt a shock of the same kind, and nearly of the same strength, if I touched the torpedo under water with only one hand, as with both. Some gentlemen* who repeated the experiment with me thought it was rather stronger. This shews, that the shock under water is produced chiefly by the electricity running through one's hand from one part to the other; and that but a small part passes through one's body from one hand to the other. The truth of this will appear with more certainty from the following circumstance; namely, that if I held a piece of metal, a large spoon for instance, in each hand, and touched the torpedo with them instead of my hands, it gave me not the least shock when immersed in water; though when held in air, it affected me as strongly if I touched it with the spoons as with my hands. On increasing the charge, indeed, its effect became sensible: and as well as I could judge, the battery required to be charged about twelve times as high to give the same shock when the torpedo was touched with the spoons under water as out of it. It must be observed, that in trying this experiment, as my hands were out of water, I could be affected only by that part of the fluid which passed through my body from one hand to the other.

420] The following experiments were made with the torpedo in air. If I stood on an electric stool, and touched either surface of the electric organs with one hand only, I felt a shock in that hand; but scarcely so strong as when touching it in the same

* [See Art. 601, 27 May, 1775, "Mr Ronayne, Mr Hunter, Dr Priestley, Mr Lane, Mr N[airne."]

manner under water. If I laid a hand on one surface of the electric organs, and with the other touched the tail, I felt a shock; but much weaker than when touching it in the usual manner; that is, with one hand on the upper surface of those organs, and the other on the lower. If I laid a thumb on either surface of an electric organ, and a finger of the same hand on any part of the body, except on or very near the same surface of the organs, I felt a small shock.

In all the foregoing experiments, the battery was charged to the same degree, except where the contrary is expressed: they all seem to agree very well with Mr Walsh's experiments.

421] Mr Walsh found, that if he inclosed a torpedo in a flat basket, open at the top, and immersed it in water to the depth of three inches, and while the animal was in that situation, touched its upper surface with an iron bolt held in one hand, while the other hand was dipped into the water at some distance, he felt a shock in both of them. I accordingly tried the same experiment with the artificial torpedo; and if the battery was charged about six times as high as usual, received a small shock in each hand*. No sensible difference could be perceived in the strength, whether the torpedo was inclosed in the basket or not. The trough in which this experiment was tried was 36 inches long, 14½ broad, and 16 deep; and the distance of that hand which was immersed in the water from the electric organs of the torpedo, was about 14 inches. As it was found necessary to charge the battery so much higher than usual, in order to receive a shock, it follows, that unless the fish with which Mr Walsh tried this experiment were remarkably vigorous, there is still too great a disproportion between the strength of the shock of the artificial torpedo when received under water and out of it. If this is the case, the fault might evidently be remedied by making it of some substance which conducts electricity better than leather.

422] When the torpedo happens to be left on shore by the retreat of the tide, it loosens the sands by flapping its fins, till its whole body, except the spiracles, is buried; and it is said to

* As well as I could judge, the battery required to be charged about 16 or 20 times as high, to give a shock of the same strength when received this way as when received in the usual manner with the torpedo out of water. [Art. 615.]

happen sometimes, that a person accidentally treading on it in that situation, with naked feet, is thrown down by it. I therefore filled a box, 32 inches long and 22 broad, with sand, thoroughly soaked with salt water, to the depth of four inches, and placed the torpedo in it, intirely covered with the sand, except the upper part of its convex surface, and laid one hand on its electrical organs, and the other on the wet sand about 16 inches from it. I felt a shock, but rather weak; and as well as I could judge, as strong as if the battery had been charged half as high, and the shock received in the usual way *.

423] I next took two thick pieces of that sort of leather which is used for the soles of shoes, about the size of the palm of my hand; and having previously prepared them by steeping in salt water for a week, and then pressing out as much of the water as would drain off easily, repeated the experiment with these leathers placed under my hands. The shock was weaker than before, and about as strong as if received in the usual way with the battery charged one-third part as high. As it would have been troublesome to have trod on the torpedo and sand, I chose this way of trying the experiment. The pieces of leather were intended to represent shoes, and in all probability the shoes of persons who walk much on the wet sand will conduct electricity as well as these leathers. I think it likely, therefore, that a person treading in this manner on a torpedo, even with shoes on, but more so without, may be thrown down, without any extraordinary exertion of the animal's force, considering how much the effect of the shock would be aided by the surprise.

424] One of the fishermen that Mr Walsh employed assured him, that he always knew when he had a torpedo in his net, by the shocks he received while the fish was at several feet distance; in particular, he said, that in drawing in his nets with one of the largest in them, he received a shock when the fish was at twelve feet distance, and two or three more before he got it into his boat. His boat was afloat in the water, and he drew in the nets with both hands. It is likely, that the fisherman might magnify the distance; but, I think, he may so far be believed, as that he felt the shock before the torpedo was drawn out

* [Art. 608.]

of water. This is the most extraordinary instance I know of the power of the torpedo; but I think seems not incompatible with the supposition of its being owing to electricity; for there can be little doubt, but that some electricity would pass through the net to the man's hands, and from thence through his body and the bottom of the boat, which in all probability was thoroughly soaked with water, and perhaps leaky, to the water under the boat: the quantity of electric fluid, however, taking this circuit, would most likely bear so small a proportion to the whole, that this effect cannot be accounted for, without supposing the fish to exert at that time a surprizingly greater force than what it usually does.

425] Hitherto, I think, the effects of this artificial torpedo agree very well with those of the natural one. I now proceed to consider the circumstance of the shock's not being able to pass through any sensible space of air. In all my experiments on this head, I used the first torpedo, or that made of wood; for as it is not necessary to charge the battery more than one-third part as high to give the same shock with this as with the other, the experiments were more likely to succeed, and the conclusions to be drawn from them would be scarcely less convincing: for I find, that five or six rows of my battery will give as great a shock with the leathern torpedo, as one row electrified to the same degree will with the wooden one; consequently, if with the wooden torpedo and my whole battery, I can give a shock of a sufficient strength, which yet will not pass through a chain of a given number of links, there can be no doubt, but that, if my battery was five or six times as large, I should be able to do the same thing with the leathern torpedo.

426] I covered a piece of sealing wax on one side with a slip of tinfoil, and holding it in one hand, touched an electrical organ of the torpedo with the end of it, while my other hand was applied to the opposite surface of the same organ. The shock passed freely, being conducted by the tinfoil; but if I made, with a penknife, as small a separation in the tinfoil as possible, so as to be sure that it was actually separated, the shock would not pass, conformably to what Mr Walsh observed of the torpedo.

427] I tried the experiment in the same manner with the Lane's electrometer described in Art. 405, and found that the shock would not pass, unless the knobs were brought so near together as to require the assistance of a magnifying glass to be sure that they did not touch.

428] I took a chain of small brass wire, and holding it in one hand, let the lowest link lie on the upper surface of an electric organ, while my other hand was applied to the opposite surface. The event was, that if the link, held in my hand, was the fifth or sixth from the bottom, and consequently, that the electricity had only four or five links to pass through besides that in my hand, I received a shock; so that the electricity was able to force its way through four or five intervals of the links, but not more. One gentleman, indeed, found it not to pass through a single interval; but in all probability the link which lay on the torpedo happened to bear more loosely than usual against that in his hand. If instead of this chain I used one composed of thicker wire, the shock would pass through a great number of links; but I did not count how many. It must be observed, that the principal resistance to the passage of the electrical fluid is formed by the intervals of the lower links of the chain; for as the upper are stretched by a greater weight, and therefore pressed closer together, they make less resistance. Consequently the force required to make the shock pass through any number of intervals, is not twice as great as would be necessary to make it pass through half the number. For the same reason it passes easier through a chain consisting of heavy links than of light ones.

429] Whenever the electricity passed through the chain, a small light was visible, provided the room was quite dark. This, however, affords no argument for supposing that the phenomena of the torpedo are not owing to electricity; for its shock has never been known to pass through a chain or any other interruption in the circuit; and consequently, it is impossible that any light should have been seen.

430] In all these experiments, the battery was charged to the same degree; namely, such that the shock was nearly of the same strength as that of the leathern torpedo, and which I am inclined to think, from my conversation with Mr Walsh, may be considered as about the medium strength of those of a real one of the same size as this. It was nearly equal to that of the plate of crown glass in Art. 411, electrified to such a degree as to discharge itself when the knobs of a Lane's electrometer were at ,0115 inches distance; whence a person, used to electrical experiments, may ascertain its strength *. The way I tried it was by holding the Lane's electrometer in one hand, with the end resting on the upper surface of the plate, and touching the lower surface with the other hand, while an assistant charged the plate by its upper side till it discharged itself through the electrometer and my body. There is, however, a very sensible difference between the sensation excited by a small jar or plate of glass like this, and by a large battery electrified so weakly that the shock shall be of the same strength; the former being sharper and more disagreeable. Mr Walsh took notice of this difference; and said, that the artificial torpedo produced just the same sensation as the real one.

431] As it appeared, that a shock of this strength would pass through a few intervals of the links of the chain, I tried what a smaller would do. If the battery was charged only to a fourth or fifth part of its usual height, the shock would not pass through a single interval; but then it was very weak, even when received through a piece of brass wire, without any link in it. This chain was quite clean and very little tarnished; the lowest link was larger than the rest, and weighed about eight grains. If I used a chain of the same kind, the wire of which, though pretty clean, was grown brown by being exposed to the air, the shock would not pass through a single interval, with the battery charged to about one-third or one-half its usual strength.

432] It appears, that in this respect the artificial torpedo does not completely imitate the effects of the real one, though it approaches near to it; for the shock of the former, when not stronger than that of the latter frequently is, will pass through four or five intervals of the links of a chain; whereas the real torpedo was never known to force his through a single interval. But, I think, this by no means shews, that the phenomena of the torpedo are not produced by electricity; but only that the battery I used is not large enough. For we may safely conclude, from the experiments mentioned in Arts. 402, 406, 407, that the greater the battery is, the less space of air, or the fewer links of a chain, will

* [Charge of plate = 4100 inches of electricity = 5207 centimetres capacity. Electromotive force = 5·5. See note 10.]

a shock of a given strength pass across. For greater certainty, however, I tried, whether if the whole battery and a single row of it were successively charged to such a degree, that the shock of each should be of the same strength when received through the torpedo in the usual manner, that of the whole battery would be unable to pass through so many links of a chain as that of a single row*. In order to which I made the following machine†.

433] *GM*, Fig. 5, is a piece of dry wood; *Ff*, *Ee*, *Dd*, *Cc*, *Bb*, and *Aa*, are pieces of brass wire fastened to it, and turned

Fig. 5.

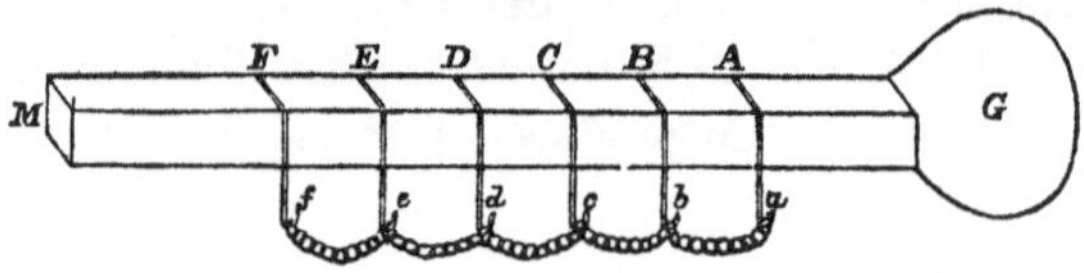

up at bottom into the form of a hook, on which is hung a small brass chain, as in the figure, so as to form five loops, each loop consisting of five links; the part *G* is covered with tinfoil, which is made to communicate with the wire *Aa*. If I held this piece of wood in one hand, with my thumb on either of the wires *Ff*, *Ee*, &c. and applied the part *G* to one surface of an electric organ, while with a spoon, held in the other hand, I touched the opposite surface, I received a shock, provided the battery was charged high enough, the electricity passing through all that part of the chain between *Aa*, and my thumb; so that I could make the shock pass through more or fewer loops, according to which wire my thumb was placed on; but if the charge was too weak to force a passage through the chain, I felt no shock, as the wood was too dry to convey any sensible quantity of electricity. The event of the experiment was, that if I charged the whole battery to such a degree that the shock would but just pass through two loops of the machine, and then charged a single row to such a degree as appeared, on trial, just sufficient to give a shock of the same strength as the former, it passed through all five loops; whether it would have passed through more I cannot tell. If, on the other hand, I gave such a charge to the whole battery, and also to the single row, as was just sufficient to force a passage

* The battery, as was before said, was divided into seven rows, each of which could be used separately.

† [Arts. 605, 607.]

through two loops of the chain, the shock with the whole battery was much stronger than that with the single row.

434] It must be observed, that in the foregoing machine, each loop consisted of the same number of links, and the links of each loop were stretched by the same weight; so that it required no more force to impel the electricity through one loop than another, which was my reason for using this machine rather than a plain chain. Considerable irregularities occurred in trying the above experiments, and indeed all those with a chain; for it frequently happened, that the shock would not pass with the battery charged to a certain degree, when perhaps a minute after, it would pass with not more than three-fourths of the charge. The irregularity, however, was not so great but that, I think, I may be certain of the truth of the foregoing facts; especially as the experiments were repeated several times. The uncertainty was at least as great in the experiments with Lane's electrometer, when the knobs were brought so close together, as is necessary in experiments of this kind.

435] It appears therefore, that if the whole battery, and a single row of it, are both charged in such a degree as to give a shock of the same strength, the shock with the whole battery will pass through fewer loops of the chain than that with the single row; so that, I think, there can be no doubt, but that if the battery had been large enough, I should have been able to give a shock of the usual strength, which yet would not have passed through a single interval of the links of a chain.

436] On the whole, I think, there seems nothing in the phenomena of the torpedo at all incompatible with electricity; but to make a compleat imitation of them, would require a battery much larger than mine. It may be asked, where can such a battery be placed within the torpedo? I answer, perhaps it is not necessary that there should be anything analogous to a battery within it. The case is this; it appears, that the quantity of electric fluid, transferred from one side of the torpedo to the other, must be extremely great; for otherwise it could not give a shock, considering that the force with which it is impelled is so small as not to make it pass through any sensible space of air. Now if such a quantity of fluid was to be transferred at once from one side to the other, the force with which it would endeavour to escape would be extremely great, and sufficient to make it dart through the air to a great distance, unless there was something within it analogous to a very large battery. But if we suppose, that the fluid is gradually transferred through the electrical organs, from one side to the other, at the same time that it is returning back over the surface, and through the substance, of the rest of the body; so that the quantity of fluid on either side is during the whole time very little greater or less than what is naturally contained in it; then it is possible, that a very great quantity of fluid may be transferred from one side to the other, and yet the force with which it is impelled be not sufficient to force it through a single interval of the links of a chain. There seems, however, to be room in the fish for a battery of a sufficient size; for Mr Hunter * has shewn, that each of the prismatical columns of which the electrical organ is composed, is divided into a great number of partitions by fine membranes, the thickness of each partition being about the 150th part of an inch; but the thickness of the membranes which form them is, as he informs me, much less. The bulk of the two organs together in a fish $10\frac{1}{3}$ inches broad, that is, of the same size as the artificial torpedos, seems to be about $24\frac{1}{2}$ cubic inches; and therefore the sum of the areas of all the partitions is about 3700 square inches. Now 3700 square inches of coated glass $\frac{1}{150}$ of an inch thick will receive as much electricity as 30,500 square inches ,055 of an inch thick†; that is, 305 times as much as the plate of crown glass mentioned in Art. 411, or about $2\frac{3}{4}$ times as much as my battery, supposing both to be electrified by the same conductor; and if the glass is five times as thin, which perhaps is not thinner than the membranes which form the partitions, it will contain five times as much electricity, or near fourteen times as [much as] my battery.

437] It was found, both by Dr Williamson ‡ and by a committee appointed by the Philosophical Society of Pensylvania, that the shock of the *Gymnotus* would sometimes pass through a chain, though they never perceived any light. I therefore took

* "Anatomical observations on the Torpedo." By John Hunter, F.R.S. *Phil. Trans.* LXIII. (1773), p. 485. See Art. 614.

† Vide note in p. 202.

‡ "Experiments and Observations on the Gymnotus Electricus, or Electrical Eel." By Hugh Williamson, M.D. Communicated by John Walsh, Esq., F.R.S. *Phil. Trans.* LXV. (1775), p. 94.

the same chain which I used in the foregoing experiments, consisting of 25 links, and suspended it by its extremities from the extreme hooks of the machine described in Art. 433, and applying the end of the machine to the negative side of the battery, touched the positive side with a piece of metal held in the other hand, so as to receive the shock through the chain without its passing through the torpedo; the battery being charged to such a degree that the shock was considerably stronger than what I usually felt in the foregoing experiments. I found that if the chain was not stretched by an additional weight, the shock did not pass at all: If it was stretched by hanging a weight of seven pennyweights to the middle link, it passed, and a light was visible between some of the links; but if fourteen pennyweights were hung on, the shock passed without my being able to perceive the least light, though the room was quite dark; the experiment being tried at night, and the candle removed before the battery was discharged*. It appears, therefore, that if in the experiments made by these gentlemen the shock never passed, except when the chain was somewhat tense, which in all probability was the case, the circumstance of their not having perceived any light is by no means repugnant to the supposition that the shock is produced by electricity†.

* [Art. 613.]

† [See note 29 and preface.]

EXPERIMENTS, 1771.

1st Night*.

438] East plate neg[ative] standing east and west. West plate pos[itive] north and south. East plate touched perpend[icularly] by wire near midd[le]. West wire bearing against north side of west plate.

East.	West.	
10½	12	a small matter positive.
12	10½	nearly same perhaps rather less negative.
12	10	a good deal more.

East plate touched flat near west side.

12	10½	separ[ated] very little, scarce enough to say whether positive or negative.
10½	12	I thought rather more pos.

Position of east and west plates reversed. Plates touched by wires as last time.

10½	12	Seemed to separate rather more than before positive.
12	10½	did but just separate.

439]

2nd Night.

West plate positive.	East plate negative.	Both plates east and west, wires straight.
10¼		Tin, pasteboard, and tin foil, each 12 inches; separated a little and equally negative. With paper of 12 inches, in one or two first trials it seemed to separate much the same. Afterwards it did not separate at all, owing, as was supposed, to its being too dry to conduct well, but after being moistened it seemed to separate like the rest.
13½		The same things being tried, the corks separated more than before and were positive, and I believe pretty equally.

* [Probably the first trials of the apparatus described in Art. 240].

440] 3RD NIGHT.

West.	East.	
pos.	neg.	
10½	12	Separated visibly, I guess about 1 diameter.
–	11¾	Seemed to separate rather less.
–	11½	Scarcely separated.
	paper of 12	Separated much the same as tin of 12.
12	10½	Nearly the same as in first experiment but of [the] 2 separated rather more.
11¾	–	
11½	–	
13½	12	Nearly the same as in first experiment.
	11¾	Seemed as if it was rather more.
	11½	Sensibly more than with 12.

From the two other nights' experiments it seemed as if the positive bottle electrified the plates sensibly stronger than the negative one: why there was not the same difference this night I cannot tell.

Plates east and west. Wires straight.

441] Two pair of large corks were made, each of which was found to separate with the same force. The weight of one pair of them was then made four times as great by the addition of lead to them.

The quantity of electricity in 3rd made vial was then compared by means of these corks with that of a glass plate with circular coating 2·4 inches in diameter and about 06 thick, by touching the glass 8 or 9 times the electricity was reduced from strength requisite to make heavy corks separate to that requisite to make light corks separate, or was reduced to ½, therefore the vial should contain 12 times as much electricity as the glass plate and wire by which communication was made, which was about 12 inches long*.

442] Three coated plates were made

	C	D	F
Thickness	·06031	·05908	·05914 inches.
Diameter of coating ...	1·82	1·79	1·785
Therefore square of diameter of coating by thickness, or computed power of plate	54·92	54·23	53·88

Mean 54·34. (D is cased with cement.)

A circular coating 5·39 inches diameter was made to thick plate in place where its thickness seemed ·178, therefore its computed power is equal to the sum of foregoing three plates. The proportion of thickness to diameter is nearly the same.

* [See Arts. 413, 582.]

Two sliding coated plates were made for trying the foregoing, the trial plates being electrified negatively, the others positively.

	Breadth of trial plate.	
C	12	separated pos.
D	–	rather more.
D	13	scarce at all.
F	12	same as C.
C	17	separate neg.
C	16	do.
C	15	not at all.
D	16	scarce sensibly.
D	17	about as much as C at 16.
F	17	about as much as C.

Therefore F seems to contain about as much electricity as C, and D to contain about $\frac{1}{16}$ more.

443] The three foregoing plates

	Large plate.		No.
placed close together	23	separated pos.	1
placed as far asunder as possible	23	a trifle more	2
The above mentioned large plate	23	rather less than No. 1	3
[Art. 442]	21	same as No. 1.	
	20	same as No. 2.	
	30	separated a very little neg.	

444] Three coated plates were made on thick plate each 1·8 inches diameter, the mean thickness of glass being supposed ·18, therefore the computed power of all three together = 54.

All 3 plates together	4	separated pos.
	5	did not separate.
	11	sep. neg.
	10	did not separate.
With C	12	sep. neg.
	11	did not.
	4	sep. neg.
	5	did not.

N.B. The breadth of the sliding plate is not known.

445] Small sliding plate not drawn out 14 × 9·4.
Large 19 × 13.

Globe hung on silk strings negative.
Sliding plates on waxed glass positive.

		[Equivalent*.]
Globe—plate 19 × 13	did not separate	15·7
14	doubtful	16·3
15	separated pos.	16·9
16	seemed rather more	17·4
Globe—plate 14 × 10·4	separated neg.	12
11·4	doubtful	12·6
12·4	did not separate	13·2

* [This column gives the side of a square equivalent to the trial plate. See Art. 465.]

Pasteboard circle 19·4 inches diameter hung on silk strings.

			[Equivalent.]
Circle—plate 19 ×	14	did not separate	16·3
	15	did rather doubtful	16·9
	16	did very little	17·4
	17	did more	18·
Circle—plate 14 ×	12·7	did	13·2
	13·4	did rather doubtful	13·7
	14·4	did not	14·2*

With circle 1·8 inches diameter on glass ·18 thick it separated a little negatively with plate 19 × 19, and would most likely not separate at 19 × 21 or 19 × 22 = 20 or 20½. Therefore quantity of electricity therein most likely is to that of globe as 20·2 to 12·4 or as 10 : 6.

446] Thickness of double plate of glass at centre of circle = ·285. Diameter of coating = 1·75.

Being tried against small plate not drawn out, separated considerably positive, therefore quantity of electricity therein might perhaps be to that in globe as 11 to 18, and therefore its actual power would be to that of thick plate as 6·6 to 18. The computed power is to that of thick plate as 10·8 to 18.

A coating 1·45 inches diameter was made on thick plate where the thickness is supposed = ·168, therefore computed power = 12·5. This being tried against sliding plates was as follows:

Small sliding plate.	Equivalent.		Large sliding plate.		Equivalent.
3	13·2	separated negative	2	did not sep.	16·9
4	13·7	separated	3	doubtful	17·4
5	14·2	did not	4	separ. pos.	18

therefore quantity of electricity therein seems to be to that of globe as 13·7 to 12·6, or 17·4 to 16·3, *id est* as 14 to 13,

therefore actual power = 11·6

In thick plate 1·8 diam., $\frac{\text{diam. plate}}{\text{thickness}} = 10$

do.	1·45	8·1
double plate		6·14

Trials of Wires.

447] The wires placed horizontally and parallel to each other, one end supported by silk, the other by waxed glass.

The trial wire consisted of iron wires ·14 thick sliding on each other, supported in [the] same manner.

* [The charges of the globe and the circle of 19·4 inches appear from these numbers to be as 28·9 : 30·7. The diameter of the tin circle, 18·5, was probably calculated from these experiments so that its charge might be equal to that of the globe. The correct diameter would have been 19 inches.]

Single wire ·19 inch thick, 96 inches long.

Trial wire drawn out	8 inches	separated neg.
	$10\frac{1}{2}$	did not.
	32	did not.
	34	separated very little pos.

Two wires ·1 inch thick, 48 inches long, placed 36 inches asunder.

Trial wire drawn out	24 inches	separated pos.
	22	did not.
	0	sep. very little neg.
	2	did not.

The same wires at 18 inches distance.

	$17\frac{1}{2}$	sep. pos. rather doubtful.
	18	did.
	16	did.
	$13\frac{1}{2}$	did not.

By these it should [seem] as if trial wire required to be drawn out 9 less with the wires at 36 inches distance than with single wire, and 17 less with two wires at 18 inches, whence I should suppose that [the quantity of] el[ectricity] in these three cases was as 96, 87 and 79.

The trial wire not drawn out was 70 inches, but the straight part of it was only $51\frac{1}{2}$.

448] Wires of half that length tried in the same manner with a shorter trial wire.

Two wires ·1 thick, 24 long, at 18 inches distance.

Trial wire drawn out	1 inch	sep. neg.
	3	very little.
	5	rather doubtful.
	7	did not.
	12	did not.
	14	sep. pos. very little.

The same at 36 inches distance.

[trial wire] at	20	sep. pos.
	18	doubtful.
	16	did not.
	11	did not.
	9	did not.
	7	did a good deal.

Wire 48 inches long, touched by end of touching wire.

[trial wire] at	9	did not.
	7	sep. neg.
	20	did.
	18	did not.

Same wire touched by middle of touching wire.

18	doubtful.
20	did.
9	did not.
7	doubtful.
5	did.

449] From these experiments the quantity of electricity in

long wire touched at end	should	96
... ... middle	seem	94
short wires 36 dist.	to	96
do. 18 —	be as	87

450] Experiments to determine whether the quant. el. in the large circle was the same whether it was supported on waxed glass* or on silk strings, the trial plates, which were of wood covered with tinfoil being supported on waxed glass, the large trial plate drawn out to n inches being expressed by $L-n$, the small ditto by $S-n$.

Large circle supported on silk strings.

L − 5 sep. pos. very sensibly if I staid some time before letting down the wires, but scarce sensibly if I did not.
L − 4 seemed to separate, but rather doubtful if I staid, but not if I did not.
S − 5 sep. neg. if I did not stay, but not if I did.
L − 5 tried again, sep. very little whether I staid or not.

The circle supported on waxed glass.

L − 5 sep. very little whether I staid or not.
S − 5 sep. very little whether I staid or not.

From these experiments there seems no reason to think that there is any sensible difference in the quantity of electricity whether the circle is supported on silk or on waxed glass. I believe the air was moderately but not very dry when these experiments were tried. The next experiment was made the same night.

451] Experiment to determine whether quantity of electricity in coated glass bears the same proportion to that in a non-electric body whether electrification is strong or weak †.

Two pair of corks were made; each separated with rather a less degree of electrification than those used in former experiments. Some lead was then added to those of one pair, so as to double their weight

* [See Art. 255.] † [See Art. 355.]

and consequently to make them require 2ce the force to make them separate.

The plate of glass used was the double plate called *A* in the following experiments, but with coating 1·78 inches diameter.

Tried with light corks.

L – 3	sep. a little pos.
S – 4	as much neg.

Tried with heavy corks.

L – 2	separated pos.
S – 5¾	as much neg.

If these experiments could be depended on as perfectly exact the coated plate should contain $\frac{1}{48}$th part more electricity in proportion when electrified with heavy corks than with light, but this difference is much too small to be depended on.

452] Comparison of two tin circles* 9·3 inches diameter with one of 18·5, the tin plates supported on waxed glass and touched in the same manner as wires, the trial plates supported on silk strings.

The two circles at 36 in. distance.

		Side of square equivalent to trial plate.
S – 1	sep. very little neg.	11·26
S – 2	did not	
L – 1	sep. very little	15·03
L – ¼	doubtful	

Large circle touched by middle of touching wire.

L – 2	sep. very little pos.	15·57
S – 1½	sep. very little neg.	11·83

Do. circle touched by extremity of touching wire.

S – 3½	very little neg.	12·62
L – 4	very little pos.	16·64

Small plates at 36 inches distance tried again.

sep. very little with L – 1, which is the same as before.

Small plates at 24 inches.

	S – 7	very little pos. equivalent to	14·26.
Do. at 18 inches	S – 5½	very little pos.	13·55.

453] A brass wire†, 72 inches long and ·19 thick was then tried, touched by middle of touching wire.

L – 2	sep. pos.	15·57
S – 2½	very little neg.	12·07

* [Art. 273 and notes 11 and 21.] † [Art. 279.]

454] From these experiments it should seem as if el[ectricity] in

Large circle touched at extremity	were as	14·63
... ... at middle		13·55
Two small circles at 36 inches		13·15
do. at 24		12·26
do. at 18		11·55

If the two circles were placed at the same distance from each other in the same manner as in coated plates, and were electrified by wires touching their centers perpendicularly, the quantity of electricity should be

Large circle	14·02
Two at 36	13·15
24	12·72
18	12·28

The quant[ity of] el[ectricity] in the wire 72 inches long and ·19 thick seems to be nearly equal to that in the circle of 18·5 inches. Therefore if we suppose quantity of electricity in a cylinder to be proportional to its length divided by the logarithm

of $\frac{\text{length}}{\text{thickness}}$,	quantity of electricity in cylinder is to that in globe whose diameter = length of cylinder as	·4266	to tab. log.	–	or as	·982	to N log.	–
of $\frac{\text{length}}{\frac{1}{2}\text{ thickness}}$,		·4761		–		1·096		–
of $\frac{\text{length}}{\frac{1}{4}\text{ thickness}}$,		·5259		–		1·211		–

and the quantity of electricity therein is to that in a circle of the same diameter as

·6627	to tab. log. or as	1·526	to N. log.	–*
·74		1·704		–
·8173		1·882		–

455] A trial plate for Leyden vials consisting of two plates with rosin between.

S – 2½	sep. neg.	rather doubtful	3½
L – 1	pos.	rather doubtful	

Double plate A, computed power = 11·04.

L – 3½	sep.	a little pos.	7½
S – 4		a little neg.	

Double plate B, computed power = 11·1.

L – 3	a little pos.	7¼
S – 4¼	a little neg.	

* [See note 12.]

Large circle on silk strings.

L − $3\frac{1}{2}$	a little pos.	8
S − $4\frac{1}{2}$	a little neg.	

Globe on silk strings.

L − $4\frac{1}{2}$	a little pos.	$9\frac{1}{4}$
S − $4\frac{3}{4}$	a little neg.	

456] Therefore the quant. el. in these bodies seems as follows:

Trial plate	$17\frac{1}{2}$
A	18·4
B	18·3
circle	18·5
globe	18·8

Diameter of the globe = 12·1, therefore quantity of electricity in globe is to D° in circle of same diameter as 1·56 to 1*.

457] Two trial plates were made on a piece of the large bit of ground glass, one 2·37 inches diameter on place where the thickness = 1·80, computed power = 31·2; the other 2·57 inches diameter where thickness = 1·90, computed power = 34·8.

The first is called S the other L.

†The plates of ground glass E and F were each coated on one side with a circle 7·95 inches diameter communicating with coating on the other side. These plates were kept from touching by three bits of sealing-wax. When the coatings were kept at distance ·39 from each other this is called plate of air ·39 thick, &c.

A piece of wire of the same thickness as the other was made to slide thereon.

When the plate of air was tried against trial-plate S with wire drawn out 12 inches it is expressed

plate air − S + 12 &c.

Double plates A and B	S + $29\frac{1}{2}$	sep. a little pos.
	L + 17	sep. a little neg.
plate air ·343	S + 0	did not sep.
	[L] + 3	a little pos.
plate air ·39	S + 18	sep. a little pos.
	L + 3	sep. a little neg.
same plate air	L + 38	sep. pos. ¶
	S + 18	did same.

* [See Art. 653 and Preface.]

† [See Art. 341.]

Tried again in afternoon of the same day.

A and B	S + 27	sep. a little pos.
plate air ·39	S + 19	do.
A and B	S + 29	do.
A and B	L + 15	sep. a little neg.
plate air ·39	L + 4	do.

458] The wire not drawn out is about 40 inches, and may therefore contain about 10 cyl.* inc. of electricity, *id est*, as much electricity as is contained in circle of 10 inches diameter. Quantity of electricity in additional wire is supposed to be equal to its length [divided] by 4·4.

Both the trial plates together, whose computed power = 66, is equivalent to 2A + 2B + 80 inches of wire + 45 of additional wire, *id est*, to 73·4 + 20 + 10·1 = 103·5 inches of electricity, therefore 1 inch of computed power in the glass of which trial plates are made should be equivalent to 1·41 inches of electricity.

By the experiment marked ¶ in [457], a difference of computed power in the trial plates = 3·6, which is equivalent to 5·08 inches of electricity, was equivalent to drawing out wire 20 inches, which is supposed = 4·54 inches of air, which is as near an agreement as can be expected.

By a medium of the experiments, the plate of air ·39 thick required wire to be drawn out $11\frac{1}{2}$ inches less than A and B, the different experiments varying from 9 to 14, therefore the plate of air contains 2·6 inches more electricity than A and B, *id est*, it contains 39·3 inches of electricity. The plate of air ·343 seemed by 1 experiment to contain 42·7 of electricity.

Therefore plate of air ·39 contains 4·94 times more electricity than a circle of same diameter, therefore quantity of electricity therein is to that in circle of same diameter as radius to thickness × 2·06 or quantity of electricity = computed power × ·243.

459] Four irregular pieces of glass, N, O, P, Q, were coated with circles. The thickness, specific gravity of glass and diameter of circles are marked in [Art. 370], the thickness of glass being found by taking thickness with calipers at center of proposed circle, and finding a part of outside of same thickness and measuring that part by Bird's instrument†; the computed power of all being just 40. The experiments were tried with sliding wire as former[ly].

Tried with large trial plate.

N	. 2	+ 0	separated constantly neg.
		+ 3	sep. but not certain.
P	. 1	+ 6	sep.
		+ 9	doubtful.
P	. 1	+ 0	did not.
Q	. 1	+ 0	did not.
N	. again	+ 3	sep.
		+ 6	rather doubtful.

* [Probably "circ". See Art. 648.]

† [Arts. 341, 517.]

With small trial plates.

N	+ 9	sep. pos.
	+ 6	did not.
Q	+ 0	sep. considerably.
O	+ 0	sep. considerably, but not so much as Q.
P	+ 6	doubtful.
	+ 9	sep. plainly.

The afternoon when these were tried, hygrometer corks closed in about 20 seconds.

The trial plates being inlarged, tried with large trial plate.

P	+ 42	doubtful.
	+ 39	do.
	+ 24	sep.
	+ 28	doubtful.
O	+ 0	very little, rather doubtful.
Q	+ 0	did not sep.
N	+ 21	sep. a little.
P	+ 24	sep. a little.

With small trial plate.

P	+ 28	did not.
	+ 36	did.
Q	+ 0	separated rather more.
O	+ 9	sep. a little.
	+ 18	sep. about as much as Q at 0.
N	+ 28	sep. a little.

These experiments were tried in the morning. In the afternoon hygrometer corks closed in about 30 seconds.

460] The plate B was coated with a circle 2·79 inches diameter, computed power = 40, and the plate D was coated with a circle 2·73, computed power = 46.

A piece of the white glass was also coated with a circle 2·85 in. diameter where the thickness was ·182, computed power 44·6.

They were tried with the same trial plates.

With large trial plate.

D	+ 0	sep. neg.
	+ 3	very little, rather doubtful.
B	+ 33	very little.
N	+ 21	very little.
D	+ 3	rather doubtful.

With small plate.

B	+ 48	very little.
D	+ 15	do.
N	+ 39	do.
White	+ 32	sep. a little pos.
B	+ 48	did not quite sep.

White	+ 18	nearly same as B.
	+ 24	sep. supposed nearly same as 1st time.
N	+ 27	sep. very little.
	+ 30	nearly same or rather more than W at 24 with large plate.
N	+ 14	sep. a little neg.
W	+ 10	do.
B	+ 32	do.
W	+ 8	do.
N	+ 14	do.

461] The plate A was coated with a circle 2·16 inches diameter, computed power = 22·6; a plate of rosin also, the first which was pressed out after hardening, was coated with a circle 2·51, thickness ·102, computed power = 2·51*; they were tried with the trial plates described in p. 16 [Art. 457].

Tried with small plate.

Rosin	+ 19	sep. a little pos.
A	+ 36	do.
Double plates A & B	+ 36	do.
Rosin	+ 16	do.

With large plate.

Rosin	+ 0	sep. very little, rather uncertain.
A	+ 17	sep. a little, rather uncertain.
	+ 14	sep. a little.
Double plate	+ 15	do.

462] Hence it appears that A contains as much electricity as the two double plates. The rosin plate required the wire to be drawn out 18 inches less than them, therefore rosin plate contains 40·7 inches of electricity, and therefore quantity of electricity therein = comp. power × †.

A contains 36·7 inches of electricity, and therefore as A and B are of the same kind of glass, the quantity of electricity in them = computed power × 1·62 = ·21056, and B contains 64·96 inches of electricity.

The whitish glass plate required the wire to be drawn out 27 inches less than B, D requires 33 less and N requires 14 less, P requires 3 more than N, O 21 less, and Q 37 less than N, therefore W contains 71·2 of electricity, D 72·5, N 68·2, P 67·5, O 73 and Q 76·7.

Therefore

Quant. el. in. ‡ / comp. power		spe. gra.
	D = 1·58	2·973
	W = 1·60	2·787
	B = 1·62	2·674
	P = 1·69	2·752
	N = 1·71	2·682
	O = 1·83	2·514
	Q = 1·92	2·504

* [Should be 61·7.] † [So in MS. See note to Art. 464.]

‡ [The "real charges" here given are in "circular inches," and the computed power is 8 times the true value, so that the numbers here given must be multiplied by 8/1·57 = 5·1 to compare them with those given in Art. 370 The diameters of the coatings in these experiments are not the same as those in Art. 370 which are taken from Arts. 508—515 and 672.]

463] Experiments to determine whether the quantity of electricity in coated plates bore the same proportion to that in other bodies whether el. was weak or strong, or whether it was positive or negative*.

On the side of corks was placed plate A with circle 2 inches in diameter, containing 31 inches of electricity. On the other side there was no coated plate, but the wire was drawn out 23 inches and made to rest at further end on the sliding wooden plates. The heavy corks required more than 2[ce] the force to make them separate than the light ones.

With light corks	S – 0	sep. a very little neg.
heavy	S – 5	sep. a little.
——	L – 5½	sep. a little pos.
light	L – 7	do.

Tried with the usual corks.

with the electricity neg.	L – 2½	sep. a little.
pos.	L – 3½	do.
neg.	L – 2½	do.

According to these experiments the plate should seem to contain $\frac{3\frac{1}{2}}{4\times 31}\times\frac{4}{3}=\frac{7}{186}=\frac{1}{27}$th part more electricity in proportion when electrified by heavy corks than light, and about $\frac{1}{60}$th more when electrified pos. than neg.

464] A plate ·345 inches thick was pressed out of exper. rosin and coated with circle 3·41 inches diameter, therefore computed power = 33·7. This was compared with double plate B by help of the sliding coated plate mentioned in [Art. 442].

	Breadth of coating on sliding plate.	
Rosin	29	sep. a little neg.
—	22	sep. a little pos.
B	20	sep. a little pos.
	26½	sep. a little neg.

Therefore the plate contains $18\cdot 3\times\frac{51}{46\frac{1}{2}}=20$ inches of electricity†.

465] Side of square equivalent to trial plate.

Small plate drawn out to	0 = 10·72 1 = 11·26 2 = 11·80 3 = 12·35	·54
	4 = 12·83 5 = 13·31 6 = 13·78 7 = 14·26 8 = 14·74	·48

* [Art. 355.]

† [This would make the specific capacity of rosin 20 × 5·1/33·7 = 3. The numbers in Art. 462 make it 3·3.]

Large plate drawn out to		
	0 = 14·49	·54
	1 = 15·03	
	2 = 15·57	
	3 = 16·10	
	4 = 16·64	
	5 = 17·12	·48
	6 = 17·60	
	7 = 18·08	
	8 = 18·56	
	9 = 19·04	
	10 = 19·52	
	11 = 20·00	

EXPERIMENTS, 1772*.

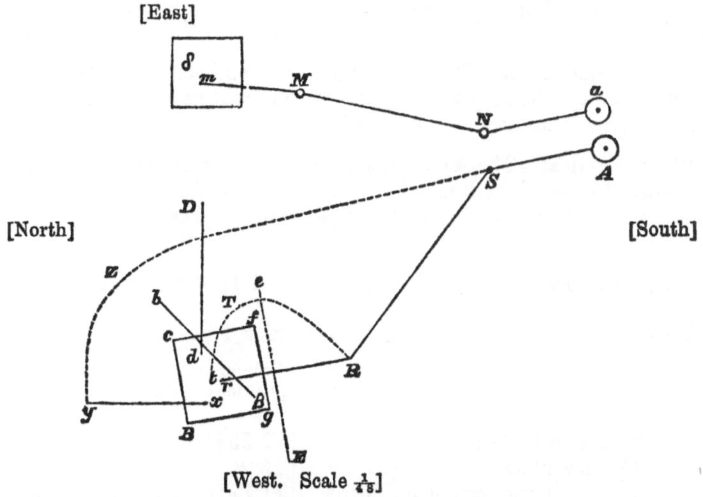

466] Plan of usual disposition of vials and bodies to be tried &c drawn in the true proportion and shape†.

δ is the trial plate, B the body to be tried, A and a the vials, mM and rR the touching wires.

$$rm = 83 \text{ inches}, \quad \left.\begin{matrix} mM \\ rR \end{matrix}\right\} = 27, \quad \begin{matrix} MN = 41, \\ RS = 61, \end{matrix} \quad Na = 24, \quad AC = 10.$$

Height of body and trial plate above ground = 4·2
below horizontal bar = 3·1½.

All the wires were about ·07 thick.

* [This is the heading of this bundle of the Journal, though the dates up to Art. 475 belong to 1771.]

† [See Art. 240.]

467] Comparison of quantity of electricity in a tin plate one foot square, according to the different situations in which it was electrified. The trial plate was suspended on waxed glass, the plate to be tried on silk strings *.

Description of the different ways in which it was tried.

Fore Observation—The plate horizontal and placed as in figure, the touching wire also as in figure but extending to different distances upon the plate [called 2nd and 3rd way in Art. 266].

Bent Wire—the same as former except that the touching wire was bent into the shape *r*TR, the distance *r*R remaining as before, the arch *r*TR being vertical and its greatest distance from the straight line *r*R being about 15 inches [5th way].

Cross Wire—The same except that the touching wire *r*R had a cross wire E*e* placed horizontally fastened on within 3 inches of *r*, the touching wire being of the same length as before, and E*e* 23 inches long. The touching wire was made to extend so much on the plate that E*e* was about 1 inch distant from the edge of the plate [4th way].

Back Observation—The touching wire removed into the situation *xy*, the wire *yz*S being 13 feet 5 inches long and passing nearly perpendicularly over B and at the height of 3′. 7″ above it [6th way].

Plate Vertical—The plate hanging in a vertical plane nearly perpendicular to the right line joining it and the vial. The touching wire touched it about the middle of the upper side [1st way].

468] Sat. Dec. 14 [1771]. Th. 53°. S. H. 19. C. H. + 7†.

Back observation. Touching wire extends 4 inches.

		Side of equivalent square.	diff.	½ sum.
B − 2	sep. a little neg.	12·33		
1½	very little	12·06		
1	very little, rather doubtful	11·78	2·85	10·35
D 3	do.	8·93		
2	sep. pos.	8·45		

Fore observation. Touching wire extends 9 inches over.

D − 5	very little, rather doubtful	9·84		
4	sep.	9·39	2·75	11·21
B − 3	sep. a little	12·85		
2½	very little, rather doubtful	12·59		

* [See Exp. III. Art. 265.]

† [Th.—Fahrenheit's Thermometer: S. H.—Smeaton's Hygrometer. See 'Description of a new Hygrometer by John Smeaton.' *Phil. Trans.* 1771, p. 198.
C.H.—Common Hygrometer.]

Fore observation, wire extends very little over.

	Side of equivalent square.	diff.	½ sum.
B − 3 very little, rather doubtful	12·85	3·01	11·34
B − 3½ sep. a little	13·10		
D − 5 very little, rather doubtful	9·84		
4 sep.	9·39		

Touched by bent wire near middle.

D − 3½ sep. a little	9·17		
4 very little, rather doubtful	9·39	3·46	11·12
B − 3½ extremely little	13·10		
3 rather doubtful	12·85		

Th. 55°. Smeaton's Hygrometer 18½. Common Hygrometer + 5.

469] Monday, Dec. 16 [1771]. Trials of time in which electricity of stone square, &c. was destroyed.

The squares were supported on glass, and a piece of tin foil about 1½ inch square fastened on each corner. On one of these pieces was fastened a wire from which the pith balls were suspended. The square was then electrified by applying a charged vial, and then a wire communicating with the wall was applied to the other piece of tin foil.

With	slate the corks closed in	10″
	Portland	15
	Bremen	8
	gummed glass	5

The stones had been kept in fore room for several days. The gummed glass had been kept in fore room till the gum began to crack. It was then kept in back room for about 5 hours, and then kept in fore room about 1½ hour.

Th. 54. S. H. 22. C. H. + 14.

Hygrometer corks closed in 4′. The glass being then wiped they closed in 7′. Being then suffered to stand uncovered for 2 or 3 hours, the corks closed in 5′.

Th. 54. S. H. 20½. C. H. 11.

470] Tuesday, Dec. 17 [1771]. Th. 53. S. H. 20. C. H. + 9.

Experiments of [Art. 468] continued.

Plate vertical.

			Side sq. equiv.	diff.	½ sum.
C − 2	sep. a little	pos.	10·09		
B − 3½	do.	neg.	13·10	3·14	11·53
D − 5	do.	pos.	9·84		

Fore observation. Wire extends very little over.

		Side square equiv.	diff.	½ sum.
C – 2	sep. a little pos.	10·09	3·25	11·71
D – 5	more	9·84		
B – 4	sep. a little	13·34		

Fore observation. Wire extending 9 inches over.

B – 4	do.	13·34	3·78	11·45
C – 1	do.	9·56		

Bent wire.

C – ½	do.	9·28	3·82	11·24
B – 3½	do.	13·10		

Cross wire.

B – 2	do.	12·33	3·64	10·51
D – 2½	do.	8·69		

Back observation.

D – 2½	do.	8·69	3·64	10·51
B – 2	do.	12·33		

Th. 55. S. H. 18½. C.H. + 6.

	This night	1st night
Plate vertical	11·71 – ·18	
Fore obs. at extremity	– ·0	11·34 – 0
do. wire 9 inches over	– ·26	– ·13
Bent wire	– ·47	– ·22
Cross wire	– 1·20	
Back observation	– 1·20	– ·99

471] Wednesday, Dec. 18 [1771]. Th. 50°. S. H. 17½. C. H. + 3.

Trials of flat plates of different substances about 1 foot square*.

			Side of equiv. square.	diff.	½ sum.
Tin plate	C – 2	sep. about $\frac{1}{10}$ in.	10·09	3·73	11·95
	B – 5	do.	13·82		
Slate	B – 5½	do.	14·06	3·71	12·20
	C – 2½	do.	10·35		
Portland stone	C – 3		10·59	3·70	12·44
	B – 6		14·29		

* [Exp. IV. Art. 269.]

		Side of equiv. square.	diff.	½ sum.
Bremen stone	B − 6 C − 3	14·29 10·59	3·70	12·44
Glass coated with tinfoil	C − 1¾ B − 5	9·96 13·82	3·86	11·89
Pasteboard	B − 5¼ C − 1¾	13·94 9·96	3·98	11·95
Glass coated with salt and gum-water	C − 2 B − 5	10·09 13·82	3·73	11·95
Do. with charcoal powder	B − 5 C − 2	13·82 10·09	3·73	11·95
Hollow tin	C − 2¼ B − 5¼	10·22 13·94	3·72	12·08
Tin plate same as first tried	B − 5 C − 1¾	13·82 9·96	3·86	11·89

Th. 50°. S. H. 17. C. H. + 2½.

The subject is continued in Art. 480.

472] Comparison of two tin circles 9·3 inches in diameter with one of 18·5; the two circles being placed in vertical planes parallel to each other and perpendicular to the vertical plane joining their centers and the trial plate, their centers being both in the above-mentioned plane*.

There was a distinct touching wire to each plate meeting each other at R, the two wires were kept asunder by a slender glass tube, and about 1 inch of the end of the wires bent at right angles horizontally in order to touch the plates by being let fall on their edges. When the large circle was tried, this double touching wire was removed and a single one used in its room, which was sometimes fastened to the middle of the glass tube, and sometimes used without it, as will be expressed.

The height of the top of the circles above floor = 4′. 3″.

The center of the large circle when that was used, or the middle point between the centers of the two small circles when they were used, was $\begin{Bmatrix} 8'.\ 10'' \\ 7'.\ \ 3'' \end{Bmatrix}$ from $\begin{Bmatrix} \text{vial} \\ \text{middle of trial plate} \end{Bmatrix}$.

The circles were suspended by silk strings. The length of the touching wires for the circles was 36 inches.

* [Exp. v., Art. 273.]

473] Monday Dec. 30 [1771]. Th. 50°. S. H. 18.

Two small circles at 18 inches from each other.

		Equivalent.	diff.	$\frac{1}{2}$ sum.	Proportion.
B – 5	sep. about $\frac{1}{10}$	13·82	3·47	12·08	1·000
C – $2\frac{1}{2}$	do.	10·35			

The same at 26 inches distance.

C – 4	do.	11·07	3·44	12·79	1·059
B – $6\frac{1}{2}$	do.	14·51			

at 36 distance.

C – $4\frac{1}{2}$	do.	11·31			
B – $\frac{1}{4}$	do.	11·37	4·08	13·38	1·108
A – $14\frac{1}{2}$	do.	15·42			

Large circle, touching wire being fastened to glass tube.

A – $17\frac{1}{2}$	do.	16·94	4·09	14·89	1·233
B – 3	do.	12·85			

Do. without glass.

B – 3		12·85	4·09	14·89	1·233
A – $17\frac{1}{2}$		16·94			

The Proportion by theory, vide P. 14 of calculations*, are as follows†.

	Calculation.	Experiment.
Small circles at 18	1·000	1·000
at 26	1·044	1·059
at 36	1·074	1·108
single plate	1·160	1·233

474] The same experiments repeated in the same manner except that the distance of the center of the large circle, or of the middle point between the centers of the small ones was 5′. 3″ from the vial, and the middle point of the trial plate 8′. 2″ from vial, and that some boards forming a floor about 4 or 5 feet square was placed under the circles 14 inches from the ground, and that a perpendicular bar of the same breadth as those of the frame was placed 5 inches nearer to the circles than the other, so that the distance of the center of the large circle from the vial and the ground, and also the distance of the nearest small

* ["P. 14 of Calculations" refers to a rough calculation in parcel No. 6, which is an early form of Props. XXIX. and XXX. See Arts. 140—143. "P. 14" contains the following remark, which fixes its date after Art. 456, "By exp. P. 15 [Art. 456] quant. el. in circle is to that in globe of same diam. as 1 : 1·56 :: $\frac{1}{2}$: ·78, therefore $\frac{2+n}{2+2n}=\cdot 78$." Here n is the reciprocal of p in Art. 140.]

† [See Art. 681 and Notes 11 and 21.]

circle from the perpendicular bar when they were placed at 36 inches distance, were diminished in about the ratio of 2 to 3*.

475] Tu. Dec. 31 [1771]. Th. 51°. S. H. 18.

Small circles at 18 inches distance.

					Proportion.
B – 6 sep. C – 4 do.		14·29 11·07	3·22	12·68	1·000
Do. at 26.					
B – ½ sep. B – 8 do.		11·51 15·17	3·66	13·34	1·052
Do. at 36.					
B – 2 do. A – 15¾		12·33 16·07	3·74	14·20	1·120
Large circle with glass.					
A – 18¾ B – 5		17·54 13·82	3·72	15·68	1·237
Do. without glass.					
B – 5½ A – 20		14·06 18·11	4·05	16·08	1·268

Th. 53. S. H. 15⅓.

476] Comparison of 2 wires 3 feet long and $\frac{1}{10}$ inch in diameter with 1 of 6 feet long and ·185 in diameter†.

The wires were placed parallel to each other, horizontal and perpendicular to the horizontal bar. They were touched almost close to one extremity by the same wires and in the same manner as the circles in the former experiment.

That end of the wires near the part which was touched was suspended by silk, the other end was supported on waxed glass. The distances were the same as in [Art. 472].

477] Fr. Jan. 3 (1772). Th. 50°. S. H. 19½. C. H. + 2.

Short wires at 18 inches distance.

				Proportion.
C – 4 sep. B – 8 do.	11·07 15·17	4·10	13·12	·847
at 24.				
B – ½ A – 14	11·51 15·15	3·64	13·33	·860

* [Art. 275.] † [Exp. VI., Arts. 279 and 683.]

Short wires at 36 inches distance.

				Proportion.
A – 15½ B – 1½	15·94 12·06	3·88	14·00	·903

Single wire without glass.

B – 3½ sep. A – 19½ do.	13·10 17·89	4·79	15·50	1·000
A – 19 sep. less B – 4 do	17·66 13·34	4·32	15·50	

Two wires at 18, repeated.

C – 4 sep. A – 14½ do.	11·07 15·42	4·35	13·24	·854

Th. 52°. S. H. 18. C. H. 4½.

By theory [Art. 152], the proportions should be between those of

	1	·9323	·9053	·8827
and	1	·8926	·8597	·8353

478] Comparison of different substances tried in the usual manner*.

The large tin circle suspended by silk.

				[Article].
B – 4½ sep. B – 3 sep. about $\frac{1}{10}$ A – 19½ do. B – 5 do.	13·82 17·89	4·07	15·85	[1]

The globe suspended on silk.

B – 5 do. A – 19½ do.	13·82 17·89	4·07	15·85	[2]

Coated double glass plate A†.

A – 20 do. B – 5½ do.	18·11 14·06	4·05	16·08	[3]

Double plate B.

B – 5 do. A – 20 do.	13·82 18·11	4·29	15·96	[4]

The large circle supported on waxed glass.

A – 21 B – 6½	18·56 14·51	4·05	16·53	[5]

A tin plate 15·5 square, on do.

B – 5 A – 20	13·82 18·11	4·29	15·96	[6]

* [See Arts. 653, 654, 682.]

† Double plate ground glass A, thickness ·3, diam. coating 1·82, comp. power 11·04.
——————— B, ——— ·31 ——— 1·855, ——— 11·1.

A tin plate 17·9 by 13·4, on do.

				[Art.]
A – 20 B – 5½	18·11 14·06	4·05	16·08	[7]

A tin cylinder 35·9 inches long and 2·53 in diameter, on do.

B – 7 A – 21	14·73 18·56	3·83	16·64	[8]

A tin cylinder 54·2 long and ·73 in diameter, on do.

A – 19½ B – 5	17·89 13·82	4·07	15·85	[9]

Brass wire 72 inches long and ·185 in diameter, on do.

B – 4½ A – 19	13·58 17·66	4·08	15·62	[10]

Th. 50°. S. H. 16½ C. H. – 8*.

479] According to the 5th and 6th article of last page, the quantity of electricity in the square is to that in a circle of the same area as 1·08 to 1, and that in square to that in oblong of the same area as ·991 to 1.

By comparing the 2nd article with the 3 last, the quantity of electricity in {thick cylinder / thin cylinder / wire} may be to that in a globe whose diameter equals the length of $\frac{\text{cylinder}}{\text{wire}}$ as {·939 / ·962 / ·988} to N. log $\frac{\text{length}}{\text{thickness}}$, or {1·184 / 1·116 / 1·103} to N. L. $\frac{2^{ce}\text{ length}}{\text{thickness}}$, or as {1·429 / 1·271 / 1·218} to N. L. $\frac{4\text{ times length}}{\text{thickness}}$.

Therefore, if we suppose that the real quantity of electricity in any cylinder is to that in the globe whose diameter equals the length of the cylinder as $1\frac{1}{7}$ to N. L. $\frac{2^{ce}\text{ length}}{\text{thickness}}$, or as ·4964 to tab. log $\frac{2^{ce}\text{ length}}{\text{thickness}}$, it will agree very well both with theory and experiment.

Or by comparing this with the first article, the quantity of electricity in any cylinder is to that in a circle whose diameter is equal to the length of the cylinder as ·759 to tab. log $\frac{2^{ce}\text{ length}}{\text{thickness}}$.

Comparative charges of bodies tried in the former experiment.

By means of this experiment and that of 1771. [Arts. 455, 456.] If the charge of the globe is called 1, that of the circle will be ·992, therefore, by comparing 6th and 7th articles with 5th, the charges of the square and oblong will be ·957 and ·965.

* [Exp. VII. Art. 281.]

By comparing arts. 1 and 5, the charge of circle on waxed glass is greater than on silk strings in ratio 1·042 to 1, and therefore if charge of cylinders and wire on waxed glass are supposed greater than on strings in the ratio ·1·021 to 1,* the charges of thick cylinder, thin cylinder and wire will be

1·028 ·980 and ·966.

480] Sat. Jan. . Th. 53°. S. H. 23. C. H. $11\frac{1}{2}$.

Comparison of different substances tried in the usual way† except that in the first experiment the touching wire *r*R and the wire RS were of brass ·185 thick.

Tin plate with thick touching wire.

C – 1	9·56	3·78	11·45
B – 4	13·34		

The same plate with the common touching wire.

B – 4	13·34	3·51	11·58
C – $1\frac{1}{2}$	9·83		

Hollow tin plate, 1·01 thick.

C – $2\frac{1}{2}$	10·35	3·71	12·20
B – $5\frac{1}{2}$	14·06		

Glass covered with thick coat of tinfoil.

B – $4\frac{1}{2}$	13·58	3·49	11·83
C – 2	10·09		

——Thin coat of do.

C – 1	9·56	3·78	11·45
B – 4	13·34		

—— gold leaf

B – $4\frac{1}{2}$	13·58	3·62	11·77
C – $1\frac{3}{4}$	9·96		

—— gum.

C – 2	10·09	3·97	12·07
B – $5\frac{1}{2}$	14·06		

——Water with a little gum.

B – $4\frac{1}{2}$	13·58	3·75	11·70
C – $1\frac{1}{2}$	9·83		

The same tin plate as before.

C – 2	10·09	3·97	12·07
B – $5\frac{1}{2}$	14·06		

Th. 55°·5. S. H. 22·5. C. H. 10.

* [The cylinders and wire were supported on waxed glass at one end only.]

† [Exp. IV. Art. 269.]

481] Result of this and Art. 471. [Same as Table, Arts. 269, 270.]

TRIALS OF LEYDEN VIALS.

482] The plates from Nairne made out of the same piece of glass were coated with circles of tinfoil as below*.

Plates.	Thickness.	Diameter of coating.	Computed power.
D	·2057	2·16	22·74
E	·2065	2·16	22·60
F	·2115	2·19	22·68
G	·2022	2·14	22·65
H	·07556	6·79	610·2
I	·07797	2·299	67·78
K	·07712	2·286	67·78
L	·08205	2·358	67·75
M	·07187	2·207	67·78
A	·2112	6·55	203·1
B	·2132	6·586	203·4
C	·2065	6·482	203·4

The old ground glass plates $\frac{A}{B}$ were coated with tinfoil $\frac{2{\cdot}123}{2{\cdot}27}$ square, computed power $\frac{27{\cdot}8}{33{\cdot}66}$, to be used as trial plates.

483] Friday, Jan. . Th. 52°. S. H. 15. C. H. –9.

The plates D, E, F, G of Nairne were compared with the double plates A and B by means of the trial plates A and B and an additional wire sliding on the electrifying wire Mm.

The $\frac{\text{length}}{\text{thickness}}$ of the wire Mm is $\frac{30}{{\cdot}15}$ inches, the additional wire is of the same thickness. The wire Bb is $9\frac{1}{2}$ inches long.

Plates tried.	Additional wire.	Trial plate.	
2 double plates	15	A	sep. near 1 diam. closed soon. Called 1st way.
	18	A	rather more, closed soon, 2nd way.
	3	B	sep. 1 diam. closed much slower, 3rd way.
D	6	B	3rd way.
	15	A	2nd way.
E	15	A	2nd way.
	6	B	3rd
F	3	B	3rd
	18	A	2nd
G	18	A	2nd
	6	B	3rd

* [See Art. 315. The computed power as given in this part of the Journal is the square of the diameter divided by the thickness, which is eight times the computed power as defined in Art. 311, and calculated in Art. 315.]

Plates tried.	Additional wire.	Trial plate.	
Double plates	3	B	sep. full 1 diam. did not close soon.
D	6		do.
E	6		do.
F	6		do.
G	6		do.
G	0		seemed more, but not quite certain.
G	12		seemed pretty certainly less.
G	18	A	sep. about 1 diam. closed faster.
F	18		do.
E	18		do.
D	18		do.
D	12		sep. but certainly less.
Double plates	15		sep. about 1 diameter*.

484] Two of the old ground glass plates H, I, K, L of Nuremberg glass†, were coated with oblong squares to serve for trial plates to the plates I, K, L, M of Nairne, but the observations were found to be so irregular that nothing could be made of them, owing, as was supposed, to the spreading of the electricity on the surface of the glass.

To prevent this, all the four plates H, I, K and L were coated with oblong squares, and cased in cement composed of 2 parts rosin, 1 of bees' wax, and 3 of brick dust†.

In making it the bees' wax was first melted and imperfectly dephlegmated, the rosin was then added and melted with as little heat as possible, and then the brick dust, previously heated so as to be very dry, was added. By this means the cement is more safe and sticky than if more heat is used in making it. In some of the mixtures also a small part of the rosin, never exceeding $\frac{1}{8}$th of the whole, was exchanged for as much pitch, which was added after the rest was melted and mixed.

The plates E, F, G, and I, K, L of Nairne were cased in the same cement, about ·165 thick.

A plate of the same cement was also cast by pouring it out on a tin plate. This was coated with circles about 2·2 in diameter.

485] The spreading of the electricity on the surface of the trial plates seemed not to be prevented by casing them in cement, for putting the plate L of Nairne on the positive side, and the trial plate H on the negative, then if the apparatus was let down and drawn up again immediately‡, the pith balls separated about half an inch negatively, but if the apparatus was suffered to rest at the bottom about half a minute, and then drawn up immediately, they separated considerably more than 1 inch, and if it was suffered to rest at the bottom but a very short time,

* [See Art. 655.] † [See Art. 303.] ‡ [See Art. 302.]

and then kept mid way for $\frac{1}{2}$ minute, and then drawn up, the balls at first separated positively but closed very soon, and after a long time separated negatively.

If a sliding plate containing about $\frac{1}{8}$ part of the electricity in the plate L was put on the positive side as an additional plate, and the apparatus was let down and drawn up immediately, the balls separated about 1 diameter negatively, but if it rested at the bottom $\frac{1}{2}$ a minute and was then drawn up immediately they separated about 1 inch negatively.

486] In order to see how fast the electricity spread on the surface of the glass, the heavy paper cylinders* were placed in the usual place, and the light ones on the wire *Bb*, the wires *Gg* and *Ff* were detached from *Cc* and rested at bottom, and a coated plate on the positive side. The wire *Cc* was suffered to rest on *Aa* and *Bb* while the jars were charging, and the wire *V* drawn up so as not to rest on the coated plate†.

When the heavy cylinders, and *a fortiori* the light ones separated, the wire *V* was let down on the plate and the wire *Cc* immediately drawn up, and the time elapsed till the closing of the light cylinders counted, which was as follows:

G	of Nairne all	20″	D	of Nairne not	20″	L reversed	50″
E	inclosed in	25	M	inclosed in cement.	35	Trial plate H	7
F	cement.	20	M	reversed	36	D° reversed	5
F	reversed	23	L	of N. in cement	50		

487] Another way was taken to try the same thing, namely, the wire *Ff* was taken off and *Gg* placed so as to lye below the wire *Dd* and to be drawn up against it by a string. The coated plate to be tried was placed on the negative side, the wire δ touching its bottom coating. The jars were then charged, the wire *Cc* resting all the while on *Aa* and *Bb*, and *Gg* drawn against *Dd*, and β drawn up so as not to rest on the plate. When the jars were sufficiently charged β was let down on the plate, and the wire *Gg* dropt immediately after, so as to take away the communication between *Dd* and the ground, so that the pith balls which were hung to *D* shewed whether much electricity passed round to under side of plate or spread on the surface.

When the pith balls communicated with the ground they separated about $\frac{3}{10}$ of an inch negatively by the repulsion of the wires on them.

Coated plates.	Balls closed in seconds.	Separated again in	
D of Nairne not in cement	3″ or 4″	30″	in 3′ separated about $\frac{7}{10}$
E of Nairne in cement	25″	2′. 30″	in 4′ separated about $\frac{4}{10}$
F do.	25	2′	
G do.	20	1′. 40″	in 4′ separated about $\frac{3}{10}$
I do.	35		

* [Art. 248.]

† [See Fig. 20, Art. 295.]

Coated plates.	Balls closed in seconds.	Separated again in
K of Nairne in cement	35	
L	10	did not separate again in several minutes
Double plates A	15	1'. 15"
Double plates B	10	1'. 30"
Plate of cement	closed almost instantly, separated again in about 10".	

488] Three sliding coated plates were made, each covered all over with tinfoil on one side, with a slip of tinfoil 1·8 by ·9 in. on the other. Two flat pieces of brass were also prepared, one 1·8 by ·9 and the other 1·8 square.

The tinfoil was divided in breadth into 6 equal parts, and the breadth of the coated surface is expressed in those divisions or in 6th parts of the breadth*.

The first plate was one of exper. rosin ·345 thick.

The 2nd, 2 plates of glass with rosin between.

The 3rd, a bit of the large piece of whitish plate glass.

The rosin sliding plate when the breadth of coating $\left.\begin{matrix} = 24 \\ = 12 \end{matrix}\right\}$ contains
2nd

as much electricity as the double plate A.

The 3rd sliding plate when its breadth $= 10\frac{2}{3}$ contains as much electricity as plate F of Nairne.

Therefore 1 division on $\left\{\begin{matrix} 1^{st} \\ 2^{nd} \\ 3^{rd} \end{matrix}\right.$ sliding plate contains $\left.\begin{matrix} \frac{1}{48} \\ \frac{1}{24} \\ \frac{1}{10\frac{2}{3}} \end{matrix}\right\}$ of the electricity in plate D, E, F, or G of Nairne.

7 inches of additional wire answers to 1 inch of computed power in plates of Nairne.

2 trial plates were made for the plates I, K, L and M of Nairne out of 2 of the ground plates first got from Nairne. The dimensions of the coating of the small one was about 3·3 by 3·1, and that of the large one 3·7 by 3·4, the thickness of glass unknown.

Two trial plates were also made for the plates A, B and C of Nairne out of the old ground plates E and F. F, the smallest, was 5·7 square, and E was 6·3 by 6 nearly.

Two trial plates were made of crown glass for the plate H of Nairne, the small one 5·7 by 5·1, the other 6 by 5·9.

489] Tuesday, Feb. 4 [1772]. Th. 47°. S. H. 17. C. H. −5.

Trial of plates D, E, F and G of Nairne, and of the 2 double plates, the plates E, F and G being cased with cement: tried by means of additional wire†.

* [See Art. 297.]

† [Art. 318.]

Plates tried.	Length of additional wire.	Trial plate.	
D	9	B	If let down and up immediately, sep. neg. about $\frac{1}{10}$ inch. If it rested at bottom 2″ or 3″, rather less.
G	0	B	do.
F	0	B	do.
G	2	B	do.
2 double plates	9	B	do., but closed sooner.
D	9	B	do.
D	16	A	sep. about $\frac{1}{10}$ pos.; much the same if it rested at bottom 2″ or 3″.
2 double plates	18	A	do. if let down and up immediately, rather more if resting at bottom 2 or 3″.
G	9	A	do.
F	7	A	do.
G	7	A	do.
2 double plates	19	A	do.

490] Feb. 4 continued. Comparison of the three plates E, F and G together with the plates I, K, L and M, tried by the two above-mentioned trial plates and the 1st and 2nd sliding plate*.

Plates tried.	Sliding plate and breadth of coating thereon.		Trial plate.	
E, F, G	1st	9	S	sep. pos. about $\frac{1}{10}$. D° if resting 2 or 3″
I	—	24		sep. rather less than $\frac{1}{10}$, rather more if resting 2 or 3″
E, F, G	—	9		same as before
M	—	16		sep. about $\frac{1}{10}$. D° if resting 2 or 3″
K	—	24		D°
L	—	24		D°
L	2nd	17	Large.	Sep. neg. about $\frac{1}{10}$: rather less if resting 2 or 3″, the separation more after a little time than at first, and closed very slow.
K		17		D°
I		15		D°
E, F, G		7		D°
M		9		D°
E, F, G		7		D°

* [Art. 318.]

491] Trials of the same kind as those in [Art. 487].

Large trial plate for the experiments of this page—balls closed in 13″

Small do.	10″ or 15″
Trial plate A	15
B	15
E of Nairne	20
F	20
G	20
D	5
Double plate B	5
A	7
M of Nairne	20
I —	20
K —	35
L —	25

492] Wed. Feb. 5 [1772]. Th. 49°. S. H. 17½. C. H. 4½.

Comparison of I, K, and L together, with A, B and C by means of the trial plates E and F, and of A, B and C together, with H by means of the two crown glass trial plates*.

Plates to be tried.	Sliding plate and breadth of coating.		Trial plates.	
I, K, L	2nd	8	F	sep. about $\frac{1}{10}$ pos. Much the same if resting at bottom 2 or 3″.
B	3rd	8	D°	
A	3rd	8	D°	
C	D°		D°	
I, K, L	2nd	10	D°	
C	3rd	8	D°	
C	3rd	10	E	sep. 1 diam. neg. Much the same if kept at bottom 2 or 3″. Kept increasing for a short time.
A	3rd	10		do.
B	—	11		do.
I, K, L	—	6		the same, only rather less if resting at bottom 2 or 3″.
C	—	11		the same as before.
I, K, L	—	6		the same as before.

* [Art. 318.]

493] Comparison of A, B, C together, with H *.

Plates to be tried.	Sliding plate and breadth of coating.		Trial plates.
A, B, C	3rd	18	Large. Sep. neg. about 1 diam., the same if it rested at bottom 2 or 3″.
H	D°		D°
A, B, C	D		D°
H	D°		D°
H	3rd	6	visibly rather more
—	—	24	very visibly less than last, and seemingly less than former
H	3rd	24	Small—sep. near $\frac{1}{10}$ inch pos. Same if it rested at bottom 2 or 3″
A, B, C	D°		D°
H	D°		D°
—	3	18	sensibly less.

Trials of same kind as those of [Art. 487].

I	closed in 65″	H	in 55″	Trial plate	E in 35″
L	20	B	20		F 7
K	20	A	20	Large trial plate for H	5
		C	35	Small —	5

EXPERIMENTS, 1773 †.

494] *Spreading of electricity on surface of glass plates.*

4 plates of English glass were cut out of same piece and coated with bits of tinfoil of the same size. One of these plates was covered at different times with thick solution of lac, which ran into heaps in drying, another with transparent varnish which also ran into heaps, another with solution of lac and vermillion which lay smooth, and the other left as it was.

They were all done in the end of the summer and suffered to dry in the open air. The spreading of the electricity on their surface was tried in the manner described 1772, p. 22 [Art. 486].

Tu. Oct. 13 [1772]. Th. 63. N. $20\frac{3}{4}$. C. − 7.

D of Nairne, corks closed immediately.
G do. in cem. in 3″ or 4″.
Plate with lac closed immediately, sep. again in 3″ or 4″.
Lac and verm. D°, but much more.
Transparent varnish same as lacquer.
Plate not varnished closed immediately, sep. again in 5″ or 6″.

* [See Art. 658.]

† [This is the heading of this bundle of the Journal, though many of the dates belong to 1772.]

495] *To see whether the machine used for trying Leyden vials* conducted fast.*

The heavy and light paper cylinders being both hung to conducting wire, and globe† turned till heavy ones separated, the light ones closed in 1′. 1″ after the others when the conducting wire communicated with machine, and about 1′. 20″ when it did not.

496] The plate covered with solut. lac was undone, and that and the plate not varnished were lacquered in Nairne's manner, one with vermillion, the other without. Neither of them were dried after the operation.

The old Lac and vermillion and the transparent varnish were dried before fire, heat uncertain.

Frid. Oct. 16 [1772]. Th. 63. N. 20. C. – 10.

The two plates lacquered in Nairne's manner discharged the electricity of the jars presently.

Lac and verm. in 1st manner closed in about 10″.
Transparent varnish uncertain, from 5″ to 20″.

Sat. Oct. 17 [1772]. The last varnished plates being dried before fire. Th. 65. N. 21. C. – 7.

1st lac and vermil. closed in 2″ or 3″, did not sep. in 1′.
2 D° closed rather sooner.
Transparent closed and sep. again in about 4″.

Tu. Oct. 19 [1773‡]. The varnished glasses were baked over stove, the heat being kept for 2 hours at about 170.

Wed. Oct. 20 [1773‡]. Th. 64. N. 22. C. – 5.

Lac & verm. closed in about 15″.
Old lac & verm. D°.
Transparent in 1 or 2″.
D° rubbed with cloth, closed and sep. again in 1 or 2″.

On Wednesday, the varnished plates were baked for above 2 hours, the heat the greatest part of the time about 210, but part of the time the ☿ rose a little way into the ball. I suppose must be at least 235 or 240.

Fr. . Th. 60. N. 22. C. – 5.

Transparent closed and sep. again immed.
Last lac & verm. closed & sep. again almost immed.
1st D° closed and sep. again in about 2″.

497] Glasses for exper. on spreading of elect.

9 plates of English glass & 8 of Nuremberg coated with plates of same size.

Sun. Oct. 24 [1773‡]. Th. 63. Comm. – 6. N. $21\frac{1}{2}$.

Closing of corks.

* [Art. 295.]

† [Of electrical machine. See Arts. 248, 563, 568, 569.]

‡ [Probably 1772. See Note to Art. 502.]

E. 9 closed in about 2″.
8 rather sooner.
7 D°.
6 D°.
5 D° 4, 3, 2, 1 D° 1 did not sep. in 1′.
N 8 closed and sep. again in less than 2″ but did not sep. much.
7 closed in 1 or 2″ but did not sep. again soon.
6 closed presently and sep. again more than 8.
5, 4, 3, 2, 1 D° but seemed to sep. at first, to sep. more before it began to close, it was pos. after closing.

Sun. & Mon. N 1, 2, 3, 4 & 8 were baked with heat from 130 to 200. The tinfoil of the lowest plate was blistered.

Sun. Th. 62. Com. – 15. N. 18.

E 2, 3, 4, 5 closed in about 2″, 2 and 3 sep. again in 30 or 4[0]. E 1 closed not quite so fast.

E 9 not baked closed in about 7″, 7 & 6 in about 4″, & 8 in about 3″.

The 5 baked Nuremberg immediately separated wide, some without closing first, others with.

N 6 closed and sep. almost immediately, 7 closed and sep. almost immed. but did not sep. wide. 5 closed and sep. again in 5″ or 7″. N 8 washed with sp[ts] of wine did not close so soon as before.

N 8 & E 4 were washed with sp wine & a little ros. varnish & then varnished with rosin.

Sun. eve. Th. 60. Com. – 20. N. 16.

E. 4 did not close in 1′.
N. 8 closed in 2 or 3″.
N. 3 closed and sep. again immed.

N. 8 & E. 4 were then cased in soft cement. The plates N. 1 & E. 3 were varnished with lac varnish, & the plates N. 2 & E. 2 with a mixture of 6 parts of varnish & 1 of vermilion, & afterwards baked for about 5 hours with a heat part of the time up to 200, and most of the time above 150, & N. 3 & E. 5 were varnished in the same manner, and then cased in a cement composed of 14 of rosin to 12 of brick dust. N.B. E. 5 was heated in drying the varnish to a great degree, so as to make it smoke violently.

Wed. Nov. 4 [1772]. Th. 59. N. 16½. C. – 18.

E. 4 closed in about 4″.
E. 5 in 3″ or 4″.
N. 8 closed and sep. again in 3 or 4″.
N. 3 seemed to do the same rather sooner.

Fr. Nov. 6 [1772], the plates E. 1 and N. 4 were varnished with rosin and the plates E. 2 & N. 2 varnished with a mixture of 4 parts of rosin varnish to 1 of vermilion & afterwards baked. They were a good deal heated both in varnishing and baking, so as to be somewhat blistered.

The plates E. 3 & N. 1 were also varnished before then and dried before fire.

Sat. Nov. 7 [1772]. N. 5 and E. 8 were varnished with rosin, and N. 6 and E. 6 varnished with 8 parts of solut. rosin & 3 of vermilion, and then baked for about 2 hours with heat which part of the time rose to 146, but commonly did not exceed 130.

Sun. Nov. 7 [1773*]. Th. 63. Com. – 2. N. $22\frac{1}{2}$.

N. 3 closed and sep. wide immed.
N. 8 closed & sep. again in 2″ or 3″.
N. 6 rather slower.
N. 5 closed and sep. wide immed.
N. 4 more so.
N. 2 D°.
N. 1 D°.
E. 6 closed in 3 or 4″.
E. 8 closed and sep. again in 2 or 3″ about 1 inch.
E. 3 closed and sep. wide immed.
E. 2 D°.
E. 1 closed and sep. again in about 1′.
E. 5 closed in about 2″.
E. 4 did not close in $\frac{1}{4}$ min.

498] Order in which the elect. spread.

N	E
6 rosin & verm. last done	4 soft cement
8 in soft cem.	6 rosin and verm. last done
5 & 3 { varnished with ros. last done and hard cem.	5 hard cem.
	8 rosin last done
4, 2, 1 { rosin alone, 2 1st; rosin and verm. 1st	1 rosin alone
	2 & 3 ros. and verm. 1st ros. alone 1st

4·21 of the lac varnish contains 12 gra. of lac.

Th. Nov. Th. 58. Com. – 13. N. 19.

N. 8 in soft cem. closed almost immed. sep. again in 3 or 4″.
E. 4 in D° did not quite close in $\frac{1}{2}$ min.
E. 8 in ros. closed and sep. again almost immed.
E. 6 ros. and verm. closed and sep. wide immed.
N. 5 ros. more so.
N. 6 ros. & verm. same as E. 6.

Th. Nov. Th. 58. Com. – 7. N. 21.

E. 4 at first approached a little nearer, afterwards did not.
N. 8 closed and sep. again in about 3″.
E. 2 Lac and verm. closed and sep. wide immed.
E. 3 Lac. not so soon.
N. 2 Lac and verm. rather quicker.
N. 1 Lac. D°.

* [Probably Nov. 8, 1772. See Note to Art. 502.]

E. 1 not varn. closed in about 8″.
N. 4 closed and sep. again in about 2″.

499] Sat. Nov. E. 8 & N. 5 were varnished with ros. & E. 6 & N. 6 were varnished with 8 parts of solut. rosin & 3 of vermillion. They were afterwards baked over boiling water for about 2 hours, the heat between 115 & 120.

Sun. Nov. E. 1 & N. 4 were varnished with 6·0 of thick solut. lac, 3·3 of verm. & 19 of sp[ts], and E. 2 & N. 1 were varnished with 6·0 of solut. lac, 4·16 of verm. & 9 of sp[ts]. The quantity of this last mixture spread on the glasses was 8·0.

Mon. Nov. 16 [1772]. Th. 52. Com. −11. N. 19½.

N. 1 closed in 1 or 2″, sep. again in about 4.
E. 2 did not quite close in ½ min.
E. 1 nearly the same.
N. 4 nearly the same as N. 1.
E. 6 closed and sep. again immed.
N. 6 do.
N. 5 not quite so soon.
E. 8 closed and sep. again in about 3″.
N. 8 in soft cem. same as N. 1.
N. 2 not covered, closed rather sooner and sep. rather more than N. 1.
E. 3 closed in about 5″, did not sep. in ½ min.

500] Trials of quant. el. in Leyden vials, &c.

The following plates were coated as follows*.

	Mean thick.	Diam. coat.	Comp. power.	Log. D°.	
Thick white	·2115	2·252	23·98	1·3799	
Thin D°	·104	2·234	48	1·6812	
N	·106	2·136	43·04	1·6339	
O	·106	2·522	59·99	1·7781	
P	·127	2·87	64·87	1·8120	
Q	·076	2·082	57·04	1·7562	
G	·1848	3·596	69·97	1·8449	
White plate	·172	3·444	68·98	1·8387	
Crown C	·0659	3·45	180·6	2·2567	
D° A	·0682	3·51	180·6	2·2568	
Thick rosin	·1845	3·760	29·18	1·4651	}
2nd do.	·195	3·374	58·37	1·7662	} [See Art. 514]
3rd do.	·103	4·247	175·1	2·2434	}

Two trial plates were made with two plates of glass with rosin between, for comparing thick rosin with the double plates A and B.

Two trial plates were also made on a piece of the white plate glass for comparing N and thin white with D + E.

* [See Arts. 370, 371.]

501] Sat. Oct. 17 [1772]. Th. 65. N. 21. C. 7.

Plates tried.	Addit. wire.	Trial plate.		Inc. el. answering to addit. wire.
D of Nairne varnished with lac	7	A	sep. a little pos.	1·58
G of N in cem[ent]	7	—	rather more	
——	0		much the same as D or rather less	0
E D°	—	—	D°	
E	10	B	did not sep.	
G	—	—	D°	
D	—	—	scarce separated	
2nd ros. plate	18	A	sep. a little	4·07
D	7	A	D°	1·58
Thick ros.	$3\frac{1}{2}$	small	D°	·79
Doub. plate B	—	—	D°	
Thick ros.	$1\frac{3}{4}$	large	very little	·40
Doub. pl. B	$1\frac{3}{4}$	D°	less	
D	$3\frac{1}{2}$	A	sep.	·79
Thick white	7	A	D°	1·58

Hence it should seem that the plate D contained about 1·6 inc. el. less than the plates E or G, which is nearly conformable to p. 26, 1772 [Art 489].

The thick rosin plate seems to contain just the same as doub. pl. B, and the 2nd rosin plate to contain 2·49 inc. el. less than D. The thick white seemed to contain ·79 inc. el. less than D.

502] Q *and* P *compared with* M *and* K *of Nairne, also green cylinder* L *and white cylinder compared with plates of Nairne by means of sliding trial plates.*

Mon. Oct. 18*. Th. $64\frac{1}{2}$. N. $17\frac{1}{2}$. C. – 15.

[13 observations, Art. 660.]

[Result.] Therefore K seems to contain 5 inc. el. less than M, conformable to 1772, p. 26 [Art. 489].

Q contains 16 inc. less, and P 16 less.

The comp. power of white cyl. = 537·5, and [it] appears to contain 756 inc. el. Therefore inc. el. by comp. power = 1·41.

* [As the records of the actual observations in the following articles are of precisely the same nature as those already given, they will be omitted, and as the author has summed up the results for each day, these statements only will be given, except in cases of more than ordinary importance. According to the day of the week and month the dates for these experiments should belong to 1773, but as the experiments seem continuous with those of dates before and after which are certainly in 1772, I think Cavendish made a mistake in the day of the month which he did not find out till 4th November.]

The comp. power of green cyl. is 318·2, and [it] appears to contain $540 \times \frac{21}{22}$ inc. el., therefore inc. el. by comp. power = 1.62*.

503] 1st *and* 2nd *green and white cylinder and white jar compared with* H *of Nairne in usual manner.*

[12 observations.]

[Result.] Hence it should seem that the white cyl. contained as much el. as H; the 2nd green contained 45 inc. el. more than H; the 1st green uncertain, and the white jar seemed to contain 74 inc. el. more.

504] *Trials of the same cylinders and jar in same manner except that in trying the white jar and* 1st *green cylinder the plate* M *of Nairne was placed on the neg. side as an additional trial plate.*

Sat. Dec. 5 [1772]. Th. 56. N. 20.

[13 observations. Art. 660.]

[Result.] Hence	1st green cylinder should cont.	135 inc. el. more than H
	2nd	56
	white cyl.	7
	white jar	88

By means of this and preceding page, the quant. el., comp. power and quant. el. by comp. power are as follows†:

	Quant. el.	Comp. power.	Quant. el. by comp. power *.
1st green	1170	600·7	1·84
2nd do.	1023	600	1·70
white cyl.	976	684·1	1·43
white jar	1060	680·7	1·56

505] The quant. el. in the 2 coated globes was tried by putting the white cylinder and the 6th sliding plate on neg. side.

[6 observations.]

[Result.] Therefore $\begin{cases}\text{globe 2}\\ \text{globe 3}\end{cases}$ seems to contain $\begin{cases}1782\\ 1555\end{cases}$ [circ.] inc. el.

Trials of jars used in the 1st *sort of experiments:* [*Art.* 240] *tried by putting a sliding plate with or without the white cylinder on neg. side.*

Th. Dec. 3 [1772]. Th. 55. N. 22.

[6 observations.]

[Result.] There seems some mistake in the 3rd exper., therefore if we make use only of those exper. in which they sep. pos. the jar for $\begin{cases}\text{neg. side}\\ \text{pos. side}\end{cases}$ contains $\begin{matrix}162\\ 0\end{matrix}$ more than H;

* [These measures of specific inductive capacity must be multiplied by 5·1. See note to Art. 462.]

† [See Art. 383.]

id est $\frac{1134}{972}$ inc. of el., but if we made use only of the other exper. it would be $\begin{cases}\text{neg. side } 1233\\ \text{pos. side } 1043\end{cases}$.

506] *Trials of the 4 large jars, the jars being placed on the neg. side.*

Fr. Dec. 4 [1772]. Th. 59. N. 19½.

[8 observations.]

[Results.] By mean

jar 1 equals w. cyl. + g. c. 2 + B + $\frac{\text{g. c. } 1 + \text{C}}{2}$ + 5, 6 = 3184

jar 4 w. c. + g. c. 2 + B + $\frac{\text{C}}{2}$ + 5 − 10 = 2675

jar 3 w. c. + g. c. 2 + g. c. 1 + B + 5 − 7½ = 3635

jar 2 w. c. + g. c. 2 + $\frac{\text{g. c. } 1 + \text{B}}{2}$ + 5 − 16 = 3050

507] *Trials of the* 5th *and* 6th *trial plates, the trial plate being placed on neg. side.*

Result. Therefore trial plate 6 − 56 = H.

[Trial plate] 6 − 18 rather more than C′.

By mean 6 − 18 = C, and trial plate 5 − 17 = C.

The trial plate 4 is on same plate as 5, and the area of one division on it is to that of 5 on trial plate 5 as $1{\cdot}8^2$ to 9.

508] *Thick white,* 2nd *rosin,* D *and* F *of Nairne, and the two double plates together, compared together, also thin white with* D *and* E *and* D *and* F.

Sat. Dec. 5 [1772] in evening. Th. 56. N. 28½.

[24 observations.]

N.B. Before these exper. were tried, the plates E and F were freed from cem[ent] and coated afresh with plates of same size. The plate D was also freed from the varnish and coated afresh, and the trial plate B was freed from cement and coated with rather larger plates.

Hence it should seem that thick ros. contained 11 inc. el. less than the doub. plates A or B, *id est* 18·2 inc. el., that the thick white contained same as D, *id est* 36 inc., and that 2nd ros. contained 2·03 inc. less, *id est* 34 inc., which differs very little from p. 12 [Art. 501].

509] *Whitish plate, P, Q, O, old G and thin rosin compared with M.*

Sun. Dec. 6 [1772]. Th. 54. N. 20.

[19 observations. Art. 655.]

[Result.] By these exper. the two double plates should contain about 1·13 inc. el. more than D, that thick white contained same as D, that the 2nd rosin contained 2·26 less, that the thin white contained ·45 less than D + E, and that N contained 1·81 less.

Sunday evening. Th. $57\frac{1}{2}$. N. $17\frac{1}{2}$.

[18 observations.]

510] *Crown* A *and* C *compared with* A, B *and* C *of Nairne.*

Mon. Dec. 7 [1772]. Th. 55. N. $18\frac{1}{2}$.

[10 observations.

21 observations of spreading of electricity on the different plates.]

By the above exper. Crown A and C should contain 15 inc. el. less than A.

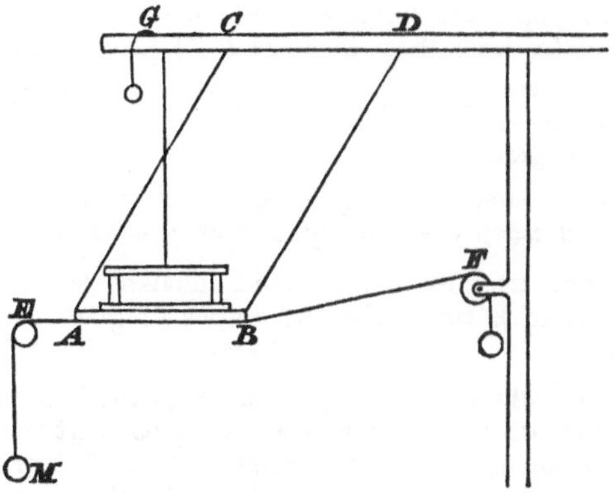

511] *Whether the shock from the plate air was diminished by changing the air between them by moving them horizontally**.

Sat. Dec. 18 [1773 †]. Th. 60°. N. $23\frac{1}{2}$.

It was tried whether shock in charging plate air was sensibly diminished by moving the 2 plates horizontally, and thereby changing the air between them in the manner represented in figure, where *AB* represents the two 8-inch brass plates with sealing wax between them suspended by the silk strings *AC* and *BD*.

AE and *BF* are silk strings fastened to the frame on which lower plate rests, and passing over wire hooks *E* and *F*, and stretched by

* [See Arts. 345, 516.] † [Probably Dec. 19, 1772. See Art. 502.]

weights so that the plates would move from *E* towards *F* and rest in any position.

The electrifying wire was suspended by the string *G* with a counterpoise.

The plates were electrified, holding my finger to bottom plate, they were then moved 24 inches by lifting up the weight *M*, and then discharged by holding my little finger to lower plate and touching upper plate with brass knob held in the other hand. I could feel a small pulse in little finger, having tried this I electrified and discharged the plates in same manner only without moving them first and endeavouring to preserve the same distance of time between the electrification and discharge. I was not able to perceive any difference in the feel. I endeavoured to ascertain the time by the vibrations of a pendulum, but without much success. It seemed needless, however, as I could perceive scarce any difference in the sensation whether I discharged it immediately or waited as long as when the plates were moved. The usual time between the electrification and discharge was about $2\frac{1}{2}''$.

The experiment was also made by Richard, who did not perceive any difference.

The heavy paper electrometer was used. The bits of sealing wax between the plates were those made in 1771 [Art. 457].

512] *Whether globe included within hollow globe is overcharged by electrifying outer globe**.

It was tried whether the globe enclosed within hollow paper globe was overcharged when the outer globe was electrified.

This was first tried by making the 2 hemispheres slide on 2 sticks of glass by means of 2 tin hooks and a stick of glass fixed to the back of the hemisphere.

The wire by which it was elect. was suspended about 4 inches above the hemispheres while the vials were charging. It was then let down, and it was so contrived that the same motion of the hand which lifted up again the elect. wire, lifted up the wire which connected the inner globe with the outer, drew back the hemispheres, and drew up the pith balls fastened to a stick of glass till they touched the inner globe.

It was found that if the elect. of the hemispheres was discharged before they were separated, but after the communication between them and the inner globe was taken away, that the pith balls did not sep., but if they were separated before their elect. was discharged, then the pith balls would at first sep. about an inch or so, but quickly closed, whereas if the inner globe was electrified after the hemispheres were separated, it was found to be a great while before the pith balls closed. It was found that this was owing to the sticks of glass on which the hemispheres slid being electrified thereby, as the same phenomena were produced by electrifying those sticks when the hemispheres were taken off.

* [See Exp. I. Art. 218.]

N.B. These sticks were not covered with sealing wax, and as appeared by this exper. suffered the electric. to run along them pretty readily. The stick of glass run through the globe had all that part without the globe covered with sealing wax.

513] *The same thing tried by a better machine.*

Wed. Dec. 23 [1772]. Th. 52°. N. $18\frac{1}{2}$.

The exper. was tried in a different manner, the hemispheres being fastened by sticks of glass covered with sealing wax within wooden frames turning on hinges.

If the pith balls were made to sep. pos. about 1 inch before the globe was elect. they separated $1\frac{1}{2}$ or two diameters on touching the globe. If they separated only 1 inch before touching, they did not sep. at all on touching the globe. If they separated negatively $1\frac{1}{2}$ or 2 inches before touching, they did not sep. at all after touching.

The event was just the same whether the wires for discharging the elect. of the globes when separated were placed so as to touch the hemispheres as soon as they were sep. an inch from each other, or whether they were placed so as not to touch them till they were separated almost the whole distance.

N.B. Each hemisphere was drawn back about 11 inches from its first situation.

It appears from hence that the inner globe was a small matter overcharged, but not enough so to make the balls sep. unless they were before positively electrified, so that the redundant fluid in it could hardly be $\frac{1}{24}$ of that which it would have received by the same degree of electrification if the outer hemispheres had been taken away, and probably not more than $\frac{1}{2}$ as much.

[*Exper. rosin**.]

514] These rosin plates were made out of a mixture of 4 parts rosin and 1 of bees wax mixed together with a considerable heat, towards the beginning of the year 1771. Towards end of 1772 some round plates were cast out of this by gentle heat, which were pared to a proper size and shape and then pressed out between brass plates heated in wooden box over furnace (the tin lining being not then made), the bits of tinfoil were at first fastened on by just wetting it in a few places with gum water and sticking it on, but as this was found not to do well, the bits of tinfoil were afterwards rubbed with melted wax and fastened on by keeping them some time pressed with slight weights with flannel between them.

515] 1st *and* 2nd *sliding plates compared with double plate* B, *also* P, Q *and* O *and thin rosin. Old* G *and whitish plate compared with* D, E, F, *and* M.

* [See Arts. 337, 373, and 500.]

The 1st and 2nd sliding coated plates were compared with double plate B [6 observations]. The 25 div[isions] on sl[iding] pl[ate] were measured by using brass plate 3 inc. by $1\frac{1}{2}$ and 9 div. of the tinfoil.

2 trial plates were made for plates M &c. out of a white glass hemisphere [8 observations. Art. 656.]

Fr. Dec. 25 [1772]. Th. 49. N. 20.

[19 observations, and 3 on insulation. Art. 656.]

By this exper. with $\left\{\begin{matrix}\text{large}\\ \text{small}\end{matrix}\right.$ trial plate Q contains $\left\{\begin{matrix}0\\ \cdot7\end{matrix}\right.$ inc. el. less than D, E & F, P $\begin{matrix}7\cdot3\\ 11\cdot7\end{matrix}$ less, O $\begin{matrix}4\cdot4\\ 7\cdot3\end{matrix}$ less, old G $\begin{matrix}2\cdot9\\ 7\cdot3\end{matrix}$ less, whitish plate $\begin{matrix}4\cdot4\\ 9\cdot5\end{matrix}$ less, and thin rosin $\begin{matrix}14\cdot7\\ 17\cdot2\end{matrix}$ less.

By mean P contains 9·5 less, O 5·8 less, old G 5·1 less, whitish plate 7 less, thin rosin 16 less, and Q ·3 less.

The wires used in the machine were all cleaned between this experiment and the next.

516] *Whether the charge of plate air is diminished by changing the air between them by lifting up the upper plate* *.

In order to try whether in electrifying plate air the electricity was lodged in the air or in the plates, the two brass 8-inch plates were placed on each other with supports of sealing wax to keep them at about ·4 inc. distance from each other and placed on the machine† with the end *M* of the wire *Mm* resting on it; the uppermost plate being fastened by a stick of waxed glass and 3 pieces of silk to the end of a lever so that it could be lifted up and down. It was also contrived so that in lifting up the plate the wire *Mm* was first lifted up from it about $\frac{1}{2}$ inch, for fear that if *Mm* rested on the plate when lifted from the under one, some electricity might escape from the ends of the wire *Bb*, &c. The third sliding trial plate was put on the negative side.

If the wire *Cc* was let down and up immediately without lifting up the upper plate, the pith balls separated negatively very little with 18 divisions of sliding plate.

If the wire *Cc* was let down and immediately drawn up half way, but not drawn quite up till the upper plate had been drawn up and let down again, the balls separated very little more.

The event was the same also if the trial plate was drawn out so that the balls should separate a little positively.

The upper plate was lifted up 2 or 3 inches. The sliding plate was let out to 12 divisions that the balls should separate positively.

* [See Arts. 344, 511.] † [Art. 295.]

517] *Trials of plate air* 1, 2, 3 *and* 4. [See Arts. 341 and 668.]

Two plates of glass $11\frac{1}{2}$ inches square were coated with tinfoil about 11·4 inches diam. a slip of tinfoil extending from the coating to the other side. These plates were placed upon each other with coated sides to[wards] each other and kept asunder by 3 supports of sealing wax, the supports being placed a little on outside of coated part and tried in the usual manner.

Sun. Dec. 27 [1772]. Th. 50. N. 18.

[4 observations.]

Mon. Dec. 28 [1772]. Th. 53. N. $17\frac{1}{2}$.

The exper. tried in same manner except that only 1 corner of the under plate rested on machine, the rest being supported by 2 wooden pillars, the places where it was supported being nearly under wax supports.

[15 observations. Art. 668.]

By this exper., plate air 1 contains 1 inc. el. more than D, plate air 2, ·1 inc. el. less than D + E, and plate air 3, $10\frac{1}{2}$ inc. less than D + E + F.

Wed. Dec. 30 [1772]. Th. 55. N. 15.

The supports of plate air 3 altered and called plate air 4.

[12 observations.]

One of the pith balls was destroyed by accident, and another put in its room.

The plate air 1 was made to rest intirely on machine.

[3 observations.]

By this exper., plate air 4 contains 1 inc. el. less than D + E + F, plate air 2, 1 inch less than E + F, and plate air 1, 1 inch more than E. It should seem also that the wire *Mm* contained 2 inches less el. when the plate rested intirely on the machine than when it rested on it only by one corner.

The thickness of these plates of air was found by laying these plates on bracket fastened to dividing machine* with or without wax supports between them, and finding the division at which the new machine stood right, the knob of the new machine resting on the middle of upper plate, and the under plate being supported under the wax supports. By this means the thickness of these plates of air were as follows :

plate 1 = ·910
2 = ·420
3 = ·288
4 = ·256

* [Arts. 341, 459, 591.]

Some experiments were made by putting bits of tinfoil between the plates whether the glasses were flat, and consequently whether the measures thus found were true. It seemed as if when the plates lay on each other the middle of the coatings could not want more than ·002 or ·004 of touching, but it did not appear that it wanted so much, and it seemed as if the outside did not want anything of touching, so that the above measures seem pretty just.

The diameter of the coating was 11·4.

The above coatings were taken off from these plates of glass, and coatings 6·254 in diameter put in their room, these with the small wax supports placed between them is called plate air 5.

N.B. 8 folds of the tinfoil used for these coatings was found to be $\frac{1}{100}$ inch thinner than the same number of folds of that used for former coatings, so that this plate air is about ·003 thicker than plate air 4.

The coatings were also taken from thin rosin and coatings 4·525 put in their room.

A plate of pure lac was also pressed out ·125 thick, and the coatings used before for thin rosin put on, which were found at a medium 4·23 in diameter.

Two plates of dephlegmated* bees wax pressed out the year before were also coated.

N.B. The bees wax was heated very hot in dephlegmating, and melted with gentle heat when cast into plates.

The thickness of the $\frac{\text{thinnest}}{\text{thickest}}$ of these plates was $\frac{\cdot 064}{\cdot 303}$, and the diameter of their coatings $\frac{2\cdot 74}{3\cdot 78}$.

518] *Lac plate and 4th rosin compared with* D + E + F; *also thin wax with* E + F; *also thick wax and plate air 5 with* D.

Mon. Jan. 4 [1773]. Th. 51. N. 17.

[23 observations.]

By these exper. Lac plate contains $1\frac{1}{2}$ inc. el. more than D + E + F; 4th rosin from 3 inc. more to 2 inc. less, by mean much the same as D + E + F; thin wax 4 inc. less than E + F; thick wax $2\frac{1}{2}$ less than D; plate air 5 $\frac{1}{4}$ more, and 1st made rosin $\frac{1}{4}$ more.

In the preceding experiments the plates of rosin &c. were exposed to the heat of the fire during trials, which seemed to cause an irregularity. To avoid that, in the following days' experiments the plates were laid on table at same distance from fire as the machine for some time before they were tried, and a screen was placed between all of them (except plate air) and fire while trying.

519] *Lac and 4th rosin with* D + E + F; *also thin wax with* D + E; *also thick wax, 2nd rosin and 1st made rosin and plate air 5 with* F.

* [Art. 375.]

Tu. Jan. 5 [1773]. Th. 50. N. 17.

[22 observations.]

By these exper. 4th rosin contains from 2 inc. more to $2\frac{1}{2}$ less than D + E + F; by mean, the same as D + E + F.

Lac from 4 more to $\frac{1}{2}$ less, by mean $2\frac{1}{4}$ more than D + E + F.

Thin wax from 4 less to 1 less, by mean $2\frac{1}{2}$ less than D + E.

Thick wax 3 less than F.

2nd rosin, $1\frac{3}{4}$ less, plate air 5 $\frac{1}{4}$ more, and 1st made rosin same as F.

N.B. The 1st made rosin was made of the same proportion of rosin and bees wax as the others, but not of the same parcel: it is uncertain how much it was heated in making the mixture.

Result of the exper. on plate air.

	Diam.	Thick.	Comp. pow.	Inc. el.	Inc. el. by comp. pow.	Do.* $\times\frac{8\times 12\cdot 1}{18\cdot 8}$	Inc. el. / diam.	Last col. into excess preceding above unity.
1st plate	11·4	·910	143	37	·259	1·34	3·25	1·11
2nd		·42	310	71	·229	1·18	6·23	1·12
3rd		·288	451	$97\frac{1}{2}$	·216	1·11	8·55	·95
4		·256	508	107	·211	1·09	9·39	·84
5	6 254	·259	151	$36\frac{1}{4}$	·240	1·24	5·80	1·39

520] *Breaking of electricity through thin plates of lac, exper. rosin and dephleg. bees wax.*

Thin plates were pressed out of lac, experimental rosin and dephlegmated bees wax, very thin at onesend and thicker at the other. The tinfoil was stript from one side of the e plates but the other left on, and was fastened to a piece of glass with gum water, and a piece of tinfoil fastened to the under side of glass communicating with the other.

These plates were placed on [the] negative side of the machine with wire δ bearing against bottom and a flat piece of brass at top on which wire β was suffered to rest. The machine was electrified in usual degree, and the bit of brass shifted from thicker to thinner part, till the electricity broke through the plate and discharged the jars.

A piece of the plate with the tinfoil under it was then cut out of the size of the brass plate, as near as possible to the place where the electricity broke through, and the thickness of the plate found by weighing it and also the tinfoil after the plate was separated from it.

The thickness of the plates thus found was as follows†, the specific

* [See Art. 343. The inches of electricity are circular inches, and to reduce them to globular inches must be multiplied by 12·1, the diameter of globe, and divided by 18·8, the diameter of a circle which has the same charge. The computed power here is the square of the diameter divided by the thickness, and this must be multiplied by 8 to get the computed power as defined in Art. 311.]

† [With the "usual degree of electrification" Lane's electrometer discharged at ·04 inch. See Art. 329. The electric strength of wax, rosin, and lac is therefore about three times that of air.]

gravity of {wax, rosin, lac} being supposed {·955, 1·06, 1·14},

wax at 1st place		·0130
	2nd	·0123
rosin		·0131
lac		·0143

521] *The quantity of electricity in a Florence flask tried with and without a magazine.*

The quant. el. in a Florence flask was tried by putting it on negative side, and some of the jars &c. on the other, the battery of 6 Florence flasks being used instead of the jars.

With the 1st, 2nd, 3rd jar with sliding plate 6 – 40 sep. neg. rather more than 1 diam.

4 jars + white cyl.	sep. a little pos.
1, 2, & 3 jars + 6 – 48	D° neg.

Sat. Jan. Th. 56. N. 19.

The same thing tried again in same manner

with the	4 jars and white cyl.	sep. about 1 diam.	pos.
with	1, 2, & 3 jar	D°	neg.
	4 jars + white cyl.	D°	pos.

Tried without mag.

With 4 jars and white cyl. sep. at 1st about 1 diam. but soon closed,
with 1, 2, & 3 jar sep. a good deal neg.
with 1, 2, & 3 jar + wh. cyl. + gr. cyl. 2 after a time sep. near 1 diam.
With 4 jars and the 2 cyl., sep at 1st a good deal, after a time sep. about 1 diam.

Sun. Jan. Th. 56. N. 21.

A coating of tinfoil to a part of the Florence flask out of water.

With 4 jars + wh. cyl. + gr. cyl. 2 sep. rather more than 1 diam.

The case was much the same whether wire was suffered to rest at bottom 2″ or 3″, or less than 1″.

With 4 jars + wh. cyl.	+ 6 – 16	sep. less than 1 diam.	
With 1, 2, & 3 jars	+ 6 – 16	D°	neg.

Without mag.

With 4 jars sep. a little neg., increased after a time to full 1 diam.

By the 1st night's experiments the flask contains	12126 inc. el.
by the 2nd	11694
and by the 3rd	11495
Without magazine by 2nd night it contained	13205 inc. el.
The true quantity is supposed	11700

522] *Computed power of above flask.*

The diameter of the flask at the surface of the water in tin pan on Saturday was 1·7; the height of that part above the bottom 5·1; the height of top of tinfoil coating above bottom 6·55; and the diameter of that place, ·68; and the circumference at the widest part 13.

The weight of that part under water was 1 .. 2 .. 7*, and that of the part between that and the top of coating was 2 .. 4.

If the spheroid *agdm* does not differ much from a sphere, and *ab* does not differ much from *ad*, the surface *afea* is nearly equal to the circumference

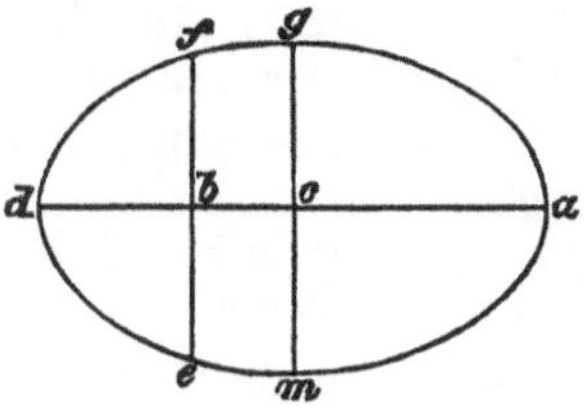

of $gm \times ab \times \frac{2ab+gm}{3ab}$, the $\left\{\begin{array}{l}\text{surface}\\ \text{thickness}\\ \text{comp. pow.}\end{array}\right.$ of the part under water was $\left\{\begin{array}{l}62\\ \cdot 0127,\\ 6200\end{array}\right.$ and that of the part above $\left\{\begin{array}{l}5\cdot 3\\ \cdot 0179,\\ 375\end{array}\right.$ and the comp. power of the whole part below top of coating 6575, the specific gravity of the glass being supposed 2·68.

Therefore inc. el. by comp. pow. = 1·78.

523] As it appears from the above experiments that the Florence flask contains more electricity when it continues charged for a good while than when charged and discharged immediately, it was tried whether the white globes would do the same.

This was done by putting the globe 3 on positive side and the white cylinder and trial plate 6 on negative side, and first charging and discharging them in the common manner, and then discharging the magazine and charging it again, while the end *c* of the wire *Cc* rested on *Bb*, while the end *C* was prevented from resting on *Aa* by a silk string. When the magazine was charged, and had continued so for a little time, the end *C* was let down on *Aa* and the wire *Cc* immediately drawn up again so as to discharge the globe &c. The event was as follows,

in common way,	wh. cyl.	+ 6 − 20	sep. near 1 diam. pos.
globe elect. first,		+ 6 − 24	D°.
in common way,		+ 6 − 47	D° neg.
globe elect. first,		+ 6 − 48	D°.

By these experiments the globe contains 45 inc. el. or about $\frac{1}{34}$ less when electrified in the common way than when charged before the rest, which is as much as is contained in 1 inch in length of the uncoated part of the neck (the whole neck being 1½ inches), so that supposing the experiment exact it seems as if the globe contained rather more elec-

* [Troy weight.]

tricity when it continued charged a considerable time than when charged and discharged immediately*.

524] *Diminution of shock by passing through different liquors*†. Tried in November [1772].

The electricity was made to pass through 42 inches of a saturated solution of sea salt in a thermometer tube of a wide bore, and the two jars charged in such manner as that a slight shock should be felt in [the] elbows: it was then made to pass through rain water in a tube of a rather greater capacity, and the electricity made rather stronger. The wires were obliged to be placed within ·18 of each other in order to feel the shock in the same degree. Therefore the electricity meets with more than 230 times the resistance in passing through rain water than salt.

The above jars were electrified till light paper cylinders began to separate, and the shock made to pass through a tube filled with rain water. The wires were obliged to be brought within ·48 inches of each other in order that the shock should be just felt in the elbows.

When the same tube was filled with saturated solution of sea salt diluted with 29 its bulk of rain water, a much greater shock was felt when the wires were at $16\frac{1}{2}$ inches from each other.

Therefore electricity meets with much more than 34 times the resistance from rain water than from a saturated solution of sea salt with 29 of rain water.

When the same tube was filled with kitchen salt in 1000 of rain water, the wires must be brought within 4·4 inches; with pump water within 2 inches, and with spirit of wine almost close; therefore the resistance of

$$\left.\begin{array}{c}\text{Pump water}\\ \text{S. salt in 1000 of rain water}\end{array}\right\}\text{ is }\left\{\begin{array}{l}4\frac{1}{6}\\ 9\end{array}\right.\text{ less than that of rain water.}$$

Mon. Nov. 16 [1772] with straw electrometer. With sea water a shock was felt when the wires were $19\frac{1}{2}$ inches distant; with rain water when they were at about ·19 inches distant. Therefore resistance of sea water is about 100 times less than that of rain water.

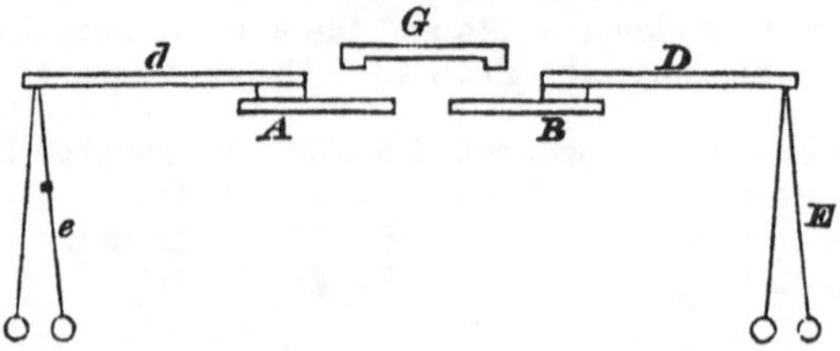

525] Exper. *Whether force with which two bodies repel is as square of redundant fluid in them*‡.

* [These phenomena are connected with the 'residual charge.' A careful investigation of them has been made by Dr Hopkinson, *Phil. Trans.*, Vol. 167 (1877), p. 599.]

† [This is the first experiment on electric resistance.]

‡ [Arts. 386, 563, 567.]

Tried by pith balls hung by threads.

A and *B* are the coated plates A and B, the bottoms of which communicate with the ground, *D* and *d* are two bits of wood resting on them, supporting the pith balls *E* and *e*. *G* is a bit of wood for making a communication between them. The wire for electrifying the plates rests on *B*, and is so contrived that when that is lifted up the wood *G* is let fall on the plates.

The pith balls *E* had bits of wire made to run into them in order to increase their weight.

A paper with divisions was placed 6 inches behind the pith balls and a guide for the eye 30 inches before them.

Tu. Oct. 26 [1773]. Th. 60. Com. $-6\frac{1}{2}$. N. $21\frac{1}{2}$.

One of the balls *E* with its string weighed ·5 gr. and its wire 1·4, the other ·6 gr. and its wire 1·7.

The two balls and strings together—the weight of one of the strings weighed 1·05 gr., the weight of the string about ·05, the weight of the two wires together was 3·2.

When the wires were taken out of the balls *E*, and a communication made between the two plates, while the electrifying wire rested on *B*, then when balls *e* sep. $\begin{cases}1{\cdot}2\\1{\cdot}08\end{cases}$, balls *E* sep. $\begin{cases}1{\cdot}25\\1{\cdot}14\end{cases}$.

The wires were put into balls *E* and the jars electrified while the electrifying wire rested on *B*. When the balls *E* sep. 1·3 inc. the electrifying wire was lifted up and the electricity of the plates taken away, immediately after which the electrifying wire was let down and immediately drawn up again when the balls *e* separated to 1·44. The electrifying wire being then let fall on *B* and suffered to remain, the balls *E* separated to 1·14.

The jars were charged, and the electricity diminished by alternately drawing up and down the electrifying wire and discharging the electricity of the plates till the balls *e* separated to 1·2; then letting the electrifying wire rest on *B*, the balls *E* separated to 1·08.

Wed. morn. The new heavy electrometer made with large wood ball and pith ball separates when the balls *E* separate to 1·52, and new light electrometer separates when the balls *e* separate to 1·44.

When balls *e* separate to 1·44, balls *E* separate to ·96.

The new heavy electrometer above mentioned separates about $\frac{1}{4}$ or $\frac{1}{2}$ inch when old light cylinder electrometer just separates.

Result of these experiments.

Balls *E* without weight separate $\frac{1}{24}$ farther than balls *e* with the same degree of electrification.

If balls separate $\frac{1\cdot22}{1\cdot08}$ with 1 part of redundant fluid, balls of $\frac{1}{4}$ weight separate $\frac{1\cdot5}{1\cdot25}$ with $\frac{1}{2}$ part of redundant of fluid.

If balls of given weight separate 1·5 with given degree of electrification, balls of 4 times weight separate ·96, therefore if balls of given size are electrified in given degree, the distance to which they separate is inversely as $\frac{1}{3\cdot1}$ power of their weight.

Therefore, in last paragraph, if balls of given weight separate $\frac{1\cdot5}{1\cdot25}$, balls of $\frac{1\cdot9}{1\cdot57}$ their weight will separate to $\frac{1\cdot22}{1\cdot08}$, therefore if balls of given weight with given quantity of redundant fluid separate to given distance balls of $\frac{\cdot475}{\cdot392}$ that weight separate to same distance with half that quantity of redundant fluid.

526] *Whether the charge of plate* E *bears the same proportion to that of another body whether the electrification is strong or weak: tried by machine for Leyden vials.*

Wed. Oct. 27 [1773]. Th. 61. Com. 8½. N. 20¼.

Plate E of Nairne on neg. side against sliding tin plates placed at end of long wire [20 observations].

Result. Therefore with light electrom. the plate E is balanced

$$\text{by a square of } \frac{14\cdot03+8\cdot17}{2}+x=11\cdot1+x,$$

$$\text{with heavy el. by } \frac{10\cdot09+12\cdot33}{2}+x=11\cdot21+x,$$

$$\frac{10\cdot59+11\cdot23}{2}+x=10\cdot91+x.$$

The plate E is balanced by 37 inc. el.

527] *Plain wax and* 3rd *dephlegmated wax with* E + F *and* 5th *rosin with double plate* A *and* B. *Also small ground crown with* D + E + F, *and large do. with* C.

The coatings were taken off from 4th rosin, and coatings 1·79 inc. diam. put in their room. This is called 5th rosin.

A plate of dephlegmated bees wax was also made [·120]* inc. thick and coatings put on 3·525 inc. in diam. This is called 3 dephlegmated bees wax.

A plate of plain bees wax was also pressed out [·119]* inc. thick and coatings put on 3·475 inc. diam.

* [These measures are left blank in the Journal. I have supplied them from Art. 371.]

A piece of thick crown glass was procured from Nairne about ·26 thick and ground down equally on both sides to about ·07 inc. thick. Two circular coatings were put on, one 3·54 inc. in diam. the other 2·035.

Wed. Jan. Th. 56. N. 27½ [16 observations].

528] K, L *and* M *compared with* D + E + F *at distance and close together; also large ground crown with* C *and small one with* D + E + F; *also* 3rd *dephlegmated wax and plain wax with* E + F; *also* 5th *rosin with double* B.

Friday, Jan. 29 [1773]. Th. 3½. N. 16½.

Tried with middle sized cork balls and a new white large trial plate [18 observations. Art. 656].

529] K + L + M *compared with* A, B, *and* C; *also* A + B + C *with* H.

Sat. Jan. 30 [1773]. Th. 50½. N. 16½. [20 observations. Art. 657.]

530] K + L + M *compared with* B *with electrification of different strengths.*

Sun. Jan. 31 [1773]. Trial plate F enlarged.

[14 observations with light and heavy electrometer alternately. Art. 656, 658.]

531] K + L + M *with* A, B, *and* C; *also* D + E + F *with* K, L, *and* M; *also small ground crown with* K, L, *and* M; *and* D + E + F *and large ground crown with* A, B *and* C *and* K + L + M.

Mon. Feb. 1 [1773]. Th. 48. N. 16.

[20 observations. Art. 657.]

532] *On light visible round edges of coated plates on charging them*.*

Mon. Feb. 1 [1773]. Th. 48. N. 16.

Some coated glass plates were placed on pos. side and electrified in usual manner in dark room in order to see whether any light was visible round their edges. With the plates M and L of Nairne and with the small ground crown glass a light was visible round the edges when the light electrometer was used, and nearly equally so with the large ground crown glass. The light seemed of the 2 rather stronger with the plate F and A of Nairne; no light was visible when light electrometer was used, but it was with the heavy electrometer.

533] *Crown* A *and* C *and large ground crown with* C; *also* 3rd *dephlegmated wax, plain wax and sliding plate* 3 *with* E + F; *also* 2 *double plates with* E, F, *and* D.

* [Art. 307.]

Mon. evening. Th. 53. N. 15. [25 observations. Art. 655.]

534] *Charge of the triple plate—the three plates* A, B *and* C *placed over each other, with bits of lead between coatings**.

The three plates A, B and C were placed over one another with the coatings nearly perpendicularly over each other, with bits of lead between them, so as to keep them at the distance of † inches from each other. This compound plate was tried in the usual manner.

Tu. Feb. 2 [1773]. Th. 50. N. 17½. [9 observations. Art. 677.]

535] *Whether the charge of plate* D *bears the same proportion to that of another body whether the charge is strong or weak: tried with machine for Leyden vials*‡. [Art. 664.]

Th. Feb. 4 [1773]. Th. 48½. N. 13½.

Tried with smallest cork balls and the light straw balls as electrometers.

The plate D placed on the neg. side and the sliding tin plates at 23½ inc. dist. from wire.

Div. on el[ectrometer]§.	Div. on sliding plate.		Side square equiv. to trial plate.	Diff.	Sum.
1 + 3	2 – 2	sep. neg.	10·09	6·61	26·79
	4 17	D° pos.	16·70		
Tin plates at 17½ inches dist.					
1 + 3	4 18½	D° pos.	17·42	5·88	28·96
	2 5	D° neg.	11·54		
3 + 1	3 1½	D° neg.	12·06	3·62	27·74
	4 15	D° pos.	15·68		
1 + 3	4 18½	D° pos.	17·42	5·88	28·96
	2 5	D° neg.	11·54		
3 + 1	3 1½	D°	12·06	3·88	28·00
	4 15½	D° pos.	15·94		
The same repeated with neg. elect.					
3 + 1	4 – 15	D°	15·68	3·62	27·74
	3 1½	D° neg.	12·06		
1 + 3	2 4½	D°	11·31	5·63	28·25
	4 17½	D° pos.	16·94		

* [Art. 380.]
† [So in MS.]
‡ [Art. 356, 664.]
§ [Divisions and quarter divisions.]

536] H *with slits and a crown glass with oblong coating compared with white cylinder; also* A *and* C *with slits compared with* B.

The coatings were taken off from the plates A, C and H, and oblong coatings with slits put in their room; an oblong coating without slits was also put to a piece of crown glass, vide Measures [Art. 593].

Tu. Feb. 9 [1773]. Th. 50. N. $12\frac{1}{2}$. 50 observations. Art. 660.

These plates were tried with middle cork balls.

Spreading of el. on surf.

A & C. Balls at first sep. wider. Closed in about 10″.

B. D°, but rather sooner. As it was supposed that this proceeded from the wires not conducting ready enough, the machine was moved slower, there was then but little of this and B was a great while before it closed, C about 5″, H a great while.

It was suspected that this increase of separation of the balls before they closed was owing to the wire designed to carry off el. to earth* not conducting fast enough. To try this, the next evening a long wire was insulated, and the cork balls hung to it. It was electrified sufficiently to make them sep. about an inch. They closed instantly on touching the wire with a bit of iron either communicating with wire for carrying off el. to ground, or whether it was only held in the hand. The air was as dry as the night before.

537] *Crown with slits and* H *with* D° *compared with white cylinder; and* A *and* C *with oblongs compared with* B†.

The coatings were taken off from the plates A and C and oblong coatings without slits put in their room. The coatings were also taken from the crown glass, and oblong coatings with slits like those put to C put in their room.

Fr. Feb. 12 [1773]. Th. 49. N.

[34 observations, Art. 660.]

538] *Experiment of p.* 61 [Art. 535] *tried with small ball blown to the end of a thermometer tube.*

A ball rather less than $\frac{1}{2}$ inch diam. was blown at end of glass tube and was coated on outside with tinfoil, the inside being filled with ☿. This was used instead of plate D in exper. to see whether charge of Leyden vial bore the same proportion to that of another body whatever force it was electrified with. It was found that 12 inches of this tube when coated contained as much el. as $K + \frac{8}{30}$ D, and therefore the spreading of the el. $\frac{2}{10}$ inch on surface of this tube increases its charge by $\frac{1}{18}$ D, whereas the spreading of el. $\frac{1}{10}$ on surface of D increases its charge by $\frac{1}{5\frac{1}{2}}$ D.

* [Art. 258.]

† [Art. 321.]

Th. 49. N. $13\frac{1}{2}$. Tin plates at 17 inches from wire.

With electrometer at 1 + 3.

Div. on sliding plate.			Equiv. to trial plate.	Diff.	Sum.
4	22	sep. about 1 diam. pos.	19	5·9	32·1
3	$3\frac{1}{2}$	D° neg.	13·1		

With electrometer at 3 + 1.

3	6	D°	14·3	3·6	32·2
4	$19\frac{1}{2}$	D°	17·9		

*Fringed rings on plate of crown glass &c.**

Sat. Feb. 13 [1773].

It was found on looking at the plate of crown glass that there were narrow fringed rings of dirt all round the edges of the coatings, the space between these rings and the coating being clean. This was supposed to be done by the explosions.

The distance of these rings from the edge of the coating seemed nearly the same both within the slits and without, but of the 2 seemed less within the slits. The mean distance seemed about ·105 inc. which seems to shew that the electricity spreads pretty nearly the same both within the slits and without.

Something of this kind has been frequently observed in the sliding trial plate 1 and sometimes I believe in some of the coated glass plates.

Sun. Feb. 14 [1773]. Th. 49. N. 17. Last exper. repeated.

[At 1 + 3, Sum = 29·6, at 3 + 1, Sum = 28·3. Plate D gave 26 and 27·5 respectively. See Art. 664.]

539] *Experiment to determine whether the charge of a Leyden vial bears the same proportion to that of another body when elect. is very weak as when it is strong*†.

AB is a tin cylinder 14 feet 8·7 inches long and 17·1 inches in circumference. *DC* is a brass wire 37·1 inches long and ·15 in diameter; both supported by non conductors; with the middle sized cork balls hung at *D*.

FE communicates with the prime conductor and is charged till light paper electrometer separates. A brass wire is suspended by silk, so as to be made alternately to touch *E* and *DC*.

Mon. Feb. 15 [1773]. Th. 55. N. 22.

The cylinder *AB* and wire *DC* were electrified negatively till the balls separated about 1 diameter. On touching *DC* twice with the wire, the corks separated about as much positively.

The wire was 27·6 inches long and ·15 in diameter.

* [Art. 308.] † [See Arts. 358, 666, and Note 25.]

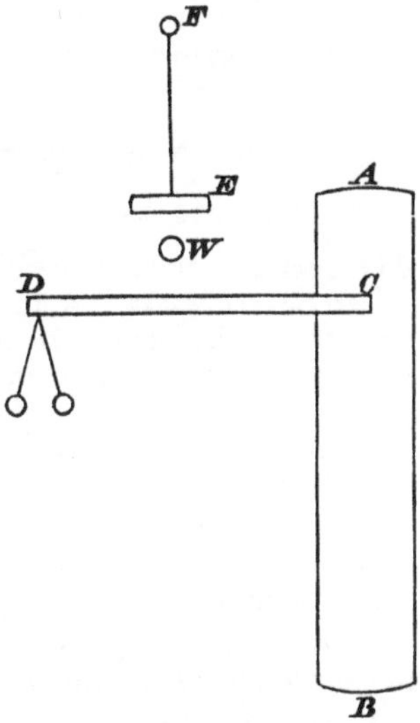

The cylinder AB was then taken away and the plates D and E placed under the wire DC. The wire was obliged to be changed for one 20·8 inches long to exhibit the same phenomenon. [See Art. 666.]

Tu. Feb. 16 [1773]. Th. 57. N. 20.

Same exper. repeated.

Cylinder touched twice with wire 31 inches long; changed from about 1 diam. neg. to D° pos.

D and E with wire 24 inches D°.
cyl. with wire 31 D°.
cyl. with 27½ did not. [See Art. 666.]

540] *Lane's electrometer compared with straw and paper electrometers.*

In the afternoon. Th. 56½. N. 19. I tried the distance to which the spark would fly by Lane's electrometer.

Divisions on electrometer*.	Distance. [Lane.]		
0 + 5	Knobs touched		
0 + 48	·027	Straw elect. sep.	1 + 3
1 + 5	·038		2 + 1½
1 + 15	·044		2 + 2½
1 + 20	·047		2 + 3
1 + 25	·051		3 + 0½
1 + 27½	·053		3 + 1
1 + 5	·038	light paper elect. just sep.	
25 + 28	knobs at ·965 inc. dist.		

* [Revolutions and 60th parts of a revolution. One revolution = ·038 inch.]

541] *Crown and H with slits compared with white cylinder; also on the excitation of electricity by separating a brass plate from a glass one.*

Wed. Feb. 17 [1773]. Th. 55. N. 21. [6 observations, Art. 660.]

Fr. Feb. 19 [1773]. Th. $53\frac{1}{2}$. N. $18\frac{1}{2}$.

A plate of glass $11\frac{1}{2}$ inches square, coated with tinfoil 8 inc. in diameter, was supported on waxed glass. A brass plate 8 inc. in diameter was supported over it by silk strings in such manner as to lye on the plate perpendicularly over the tinfoil, and to be drawn up till it touched a piece of wire supported on waxed glass with the middle sized cork balls suspended from it. This was done in order to see how much of the charge of the plate was contained in the coating.

It was found that though the plate was not electrified, yet on lifting up the brass plate the balls separated some inches if the tinfoil communicated with the ground, but if it did not communicate, the balls, as well as I remember, separated considerably less. Some bits of thin silk thread were placed between the glass and brass plate.

In the afternoon. Th. 54. N. $17\frac{1}{2}$.

The experiment repeated with bits of card between the glass plate and brass.

When tinfoil $\begin{Bmatrix}\text{commun.}\\ \text{did not commun.}\end{Bmatrix}$ with ground, balls sep. about $\begin{Bmatrix}1\\ \frac{7}{8}\end{Bmatrix}$ inch.

When there was nothing between the glass and brass plate, they sep. 1·4 inc. whether the tinfoil communicated with the ground or not.

In all these cases the brass plate was negative.

The glass plate was found to be pos. if the tinfoil did not communicate with the ground, but I could not perceive it to be at all electrified if it did communicate.

The next morning the experiment was repeated, but the balls separated much less than before. The temper. of the air was much the same.

542] It was tried whether when three tin plates 1 foot square were placed near to and parallel to each other, the line joining their centers being perpendicular to their planes, the middle plate would receive much electricity on electrifying the plates*.

The experiment was tried with the same apparatus and nearly in the same manner as the experiment with the globe†, except that the two outer plates were suspended by two sticks of waxed glass turning on hinges. The wire too by which the plates were electrified was made so as to touch all three plates at the same time. Four bits of sealing

* [Exp. VIII., Art. 288 and Note 23.] † [Art. 218.]

wax were stuck to the middle plate, two on each side, to prevent the outer plates coming too near.

Sun. Feb. 21. Th. supposed about 55. N. $20\frac{1}{2}$.

If the bits of sealing wax were of such size that the distances of the outer plates were about $\begin{cases} 1\cdot15 \\ 1\cdot65 \end{cases}$, the middle-sized cork balls separated about $\begin{cases} \frac{5}{8} \\ \frac{3}{4} \end{cases}$.

The light paper electrometer was used in this experiment. If the globe 2 was electrified in the same degree, and its electricity communicated to $\begin{cases} \text{the 4 jars} \\ \text{1, 2 \& 4 jars} \end{cases}$ and the middle tin plate electrified by one of these jars (the two outer being drawn aside) and the cork balls then drawn up against the plate, they separated about $\begin{cases} \frac{5}{8} \\ \frac{7}{8} \end{cases}$.

In the $\begin{matrix} 1^{st} \\ 2^{nd} \end{matrix}$ case the electricity of the globe was diminished $\begin{cases} 8 \\ 6 \end{cases}$ [times], and therefore when the outside plates were at $\begin{cases} 1\cdot15 \\ 1\cdot65 \end{cases}$, the quantity of electricity in the middle plate was about $\begin{cases} \frac{1}{8} \\ \frac{1}{7} \end{cases}$ of what it would have received by the same degree of electrification if placed by itself.

543] *Charge of* A, B, *and* C *laid on each other without any coatings between; also charge of* 1st *thermometer tube.*

The coatings were taken from the 3 plates A, B and C of Nairne, and the plates cleaned and placed one on the other without anything between them, and stuck together by dropping some melted wax on the edges. The outside surfaces were then coated with circles 6·6 inc. diam. This is called Triple Plate*.

A thermometer tube was coated with coatings 11 inch long, the inside being filled with ☿, with wire let into one end, and the ends stopped with cement. The tube was 12·7 inc. long; weighed 1 .. 3 .. 0, and the bore held 22 gra. of water, the specific gravity of a piece of the same tube weighed twice over was $\frac{11\cdot7}{3\cdot15\frac{1}{2}} = 3\cdot1$.

N.B. The comp. pow. of this tube is about $90\frac{1}{2}$.

This is called Tube 1†.

Mon. Feb. 22 [1773]. Th. $53\frac{1}{2}$. N. $20\frac{1}{2}$. [8 observations. Art. 675.]

544] *Lane's electrometer compared with straw and paper electrometer; also charge of plate rosin with brass coating made to prevent spreading of electricity.*

* [Art. 380.] † [Art. 382.]

Lane's elect.	Rev[olutions.]	Div[isions.]
Light paper just sep. when [Lane's] el. at	1	13
Straw at 1 + 3	0	54
3 + 1	1	36
Knobs touched at	0	$7\frac{1}{2}$

A plate of rosin and bees wax of the same proportions as for exper. rosin was cast of the shape of figure, *ABDC* and *abcd* being brass plates

2·45 in. diam. their distance before the rosin was poured in being about ·12 inc.

Tu. Feb. 23 [1773] in afternoon, the rosin plate being cast that morning, the hygrom. as well as I remember being about 22.

[4 observations, comparison with E.]

Wed. Feb. 24 [1773]. Th. 54. N. 20. [4 observations.]

Spreading of electricity on surface.

Rosin closed in about 7″, sep. again in 35.

E was irregular.

545] *Second thermometer tube; also comparison of charge of cylinder used in* [Art. 539] *with* D + E.

A thermometer tube whose length was 22·1 inc., weight = 2, 17, 21 and weight water which filled bore 14 gra. was coated with tinfoil 15·5 long, conseq. comp. power = * the spec. gra. of a part of the same tube being 3·243.

Fr. Feb. 26 [1773]. Th. 52. N. $20\frac{1}{2}$.

The cyl. used in [Art. 539] compared with the plates D and E, the wire *Mm* of machine being drawn out to $39\frac{1}{2}$ inches, and resting on the cylinder as in that experiment. A sliding trial plate on neg. side.

[6 observations. See Art. 666.]

546] *Charge of second thermometer tube; also that of rosin plate with brass coating; also that of* A, B, *and* C *laid on each other without coatings between.* [10 observations. Art. 675.]

The same things were tried the day before, Th. 55, N. $17\frac{1}{2}$, but the wire for making communication between machine and ground was forgot to be fixed. [14 observations. See Art. 666.]

547]† *The quantity of electricity in Plate* D *compared with that in a tin circle of* 36 *and another of* 30 *inches diameter by means of*

* [So in MS.]

† [Arts. 356, 664.]

the machine used for comparing simple plates, the trial plate being a tin cylinder inches long, and † in circumference, fastened to the end of the usual sliding trial plates, with another cylinder of the same size sliding within it.*

Tried with elect. of usual strength and with the middle sized corks.

Also the double plate A compared with the circle of 18½ inc.

Wed. March 3 [1773]. Th. 56. N. 21. [16 observations.]

548] *Charge of plate of experimental rosin designed for compound plate of glass and rosin; tried both when warm and when cold*‡.

A plate of experimental rosin near 8 inches square was pressed out between two glass plates with tinfoil coatings fastened on by oil, the heat being such that it required very little weight to press out the rosin.

The thickness of the plate was much less toward one end than the other, varying in different parts of the coated plate from ·137 to ·108, but the mean thickness was ·122.

It was coated with circles of tinfoil 6·61 in. diameter.

Its charge was compared with that of the plates K, D and E of Nairne by means of a sliding trial plate made of the plate † of Nairne.

Sat. March 6 [1773]. Th. 55½. N. 17.

	Tr. sl. pl.	
K + D + E	24	sep. neg.
	19	D° pos.
ros. plate	19⅓	D°.
	24⅓	D° neg.

In the afternoon. Th. 57½. N. 16.

ros. plate	19½	sep. pos.
	25½	D° neg.
K + D + E	25	D°.
	19	D°.

The rosin plate was then warmed before the fire between two glass plates with flannel between them and the rosin till it would not support its own weight without bending. As soon as it was strong enough to bear its own weight it was compared as before.

rosin plate	20½	sep. pos.
	26	did not sep.

In about 2 or 3 hours after, when it was quite cold it was tried again:

rosin	25	sep. neg.
	19	D° pos.
	18¾	D°.
	24½	D° neg.

* [Art. 240.] † [So in MS.] ‡ [Arts. 381, 678.]

549] *Whether charge of glass plate is the same when warm as when cold.*

The same afternoon the charge of a glass plate when hot and cold was compared together in the same manner.

The glass was $11\frac{1}{2}$ inches square, used for Æpinus experiment*, coated on one side with a circle of 8 inc. diameter, and a brass plate of same diameter used for the other coating.

The glass and brass plate were both heated before the fire till almost as hot as I could bear my hand on, and then tried by the help of the 6th sliding plate, when the breadth of the sliding plate was required to be 37 in order that it should sep. pos.

After the plate was cold it was tried again, the breadth of the sliding plate was obliged to be 36.

Hence it would seem as if the charge both of glass and of rosin plate was the same when hot as when cold, the small difference between them being most likely owing to the electricity spreading more on the surface of the warm plate than of the cold one.

550] *Crown with slit coatings and* H *with oblong compared with white cylinder; also second thermometer tube with* D + E + F.

The slit coatings were taken from H and plain coatings, 6·03 square, put in their room.

Sun. March 7 [1773]. Th. 56. N. 15.

Straw elect. at 2 + 3, which is equivalent to light paper electrometer [4 obs.]. Elect. at 3 + 1 [6 obs.]. Elect. at 1 + 3 [7 obs. Art. 660].

Mon. Mar 8 [1773]. Th. 54. N. $14\frac{1}{2}$. [4 obs.].

551] *Quantity of electricity in plate* D *and rosin with brass coatings compared with that of tin circle of* 36″ *and one of* 30″ *by machine for trying simple plates*†; *with different degrees of electrification*‡.

Tu. Mar. 9 [1773]. Th. 51. N. 15.

The exper. of p. 78 [Art. 547] repeated, only using square tin plates of different sizes made to fasten on to sliding cylinder instead of the sliding trial plates.

The tin circles and the square plates both supported on silk.

Straw el. at inner marks N° 2.

* [Arts. 134, 341, 517.] † [Art. 240.] ‡ [Art. 664.]

	Square plate.	Inc. cyl. drawn out.	
circ. 36″	5	24	sep. a little neg.
—	3	11	D° pos.
E	1	26	D°.
	5	9	D° neg.
rosin	3	17	D°.
	1	4	D° pos.
circ. 30″	1	20	D° pos.
	3	28	D° neg.
El[ectrometer] at outer marks.			
circ. 30	3	20	D°.
	1	29	D° pos.
circ. 36	4	34	D° neg.
	2	28	D° pos.
E	4	21	D° neg.
	2	24	D°.

The electricity was found to break through the rosin plate when electrified with this strength, but there was no hole made in the plate, as it was found not to break through after that with the weak degree of electrification. It seemed not to pass over the surface, as no light was perceived.

552] *Charge of compound plate of glass and rosin.*

The rosin plate of p. 79 [Art. 548] without its coatings was included between the plates B and H of Nairne and the outside surfaces coated with circles 6·6 inc. diam. and is called Compound Plate.

Th. Mar. 11 [1773]. Th. 53. N. 17. [4 comparisons with K.]

Fr. Mar. 12 [1773]. Th. 53½. N. 13½. [20 observations, Art. 664.]

553] *Circle of* 18½″ *compared with double plates, also plate D, plate air and the two double plates compared with circles of* 36″ *and* 30″.

A sliding trial plate was made of deal, with an additional piece to fit on, the breadth was 31 inches, the length when not drawn out, and without the additional piece, was 15, and the additional piece increased the length 10½ inches.

The number in the 2nd column shows the number of inches by which the sliding piece is drawn out.

March 3.					Inc. el.	Diff.	Mean.
Circle 30 on silk	12·5	1½ 30	27·3 34·3	− ·8	26·5 33·5	7·0	30·0
D° on glass	14·7	4 36	27·9 35·6	+ 1·0	28·9 36·6	7·7	32·7
D° on silk	11·3	5½ 36	24·6 32·9	+ 1·7	26·3 34·6	8·3	30·4
E	15·7	1½ 38	27·3 36·1	+ 1·9	29·2 38	8·8	33·6
Circle 30″ on silk	18·7	8	33·9	0	33·9		

Tu. Mar. 9. [Electrometer] At inner marks. At outer marks.

	Diff.	Mean.	Diff.	Mean.
Circle 36″	11·6	35·6	5·8	36·3
E	12·2	32·5	6·9	33·2
Circle 30″	11·4	30·4	7	30·8

March 12. [14 observations.]

Sat. Mar. 13. Th. 55. N. 12. [12 observations. See Art. 649.]

Mon. Mar. 15. Th. 54. N. 14. [15 observations. See Arts. 649, 655.]

554] *The same with addit. four small rosin plates.*

Four plates of rosin and bees wax were cast 4 inches square and about ·22 thick and coated with circles 1·8 in. diam. A tin trial plate was also made 6 inches long and 5 broad. It is called N. The plates of rosin were connected by bits of brass wire like that used for connecting the two double plates.

Fr. Mar. 19. [20 obs. Arts. 649, 651.]

Tu. Mar. 23. [21 obs. Art. 649.]

Wed. Mar. 24. [22 obs. Art. 649.]

555] Sun. Mar. 21st [1773]. Th. about 55. N. about 15.

It was tried whether the 4 rosin plates contained the same quantity of electricity whether they were placed close together or at a distance, and what is to be allowed for the connecting wires, &c.

This was tried with the usual machine*, the rosin plates being placed on the positive side and sliding plate 3 on the negative side, the sliding plate remaining always at the same division, the small variations of the charge being found by the additional wire.

They were tried in 5 different ways.

1st way. The plates placed close together near the end *m*, the usual wires *V* resting on the plates, with the connecting wires put on the plates.

2nd way. D° without the connecting wires.

3rd way. The connecting wires suffered to remain, and also one of the wires *V*, but the 3 others removed towards end *M*, placed at 4 inches

* [Art. 295. See Art. 337.]

distance from each other, and supported in their usual situation by silk strings.

4th way. The same, except that the 3 wires *V* were taken quite away.

5th way. The rosin plates placed at as great a distance from each other as possible *id est* * inches with the usual wires *V*, but without the connecting wires.

	Inc. el. on addit. wire.	Sliding plate.		Inc. el. on addit. wire.	Sliding plate.	
3rd way, with connect. wires, usual ones removed to end	0	3...14	sep. pos.	1	3...12¾	D° neg.
4th way, without usual wires	2½		D°	2		D°
1st way, with connecting wires and usual also	1		D°	2		D°
2nd way, without connecting wire	2		D°	2½		D°
5th way, removed to distance	1½		D°	2½		D°

556] *Whether charge of white glass thermometer tube is the same when hot as when cold*†.

Sun. Mar. 21 [1773] afternoon. Th. about 55. N. about 15.

A ball about 1 inch in diameter was blown at the end of a thermometer tube with a bulb 4·3 inches above. This was filled with ☿ sufficient to rise into the bulb. The tube was coated 3·4 inches from the ball with gummed paper dipped in salt water and bound on with iron wire. This ball was placed in a glass of ☿ surrounded with iron filings and placed on machine near *M*, and heated by a spirit lamp, the ☿ in which the ball was immersed being made to communicate with the ground, and a bit of iron wire bound round the wire *Mm* being dipped into the ☿ in bulb.

The crown glass plate * and the plate A of Nairne, which was coated as a sliding plate, being put on negative side.

The 1st column being the number of square inches which it was necessary to give to the coating of the sliding plate in order that the balls might sep. pos. and the 2nd column that they might sep. neg. The charge of the crown glass plate being equal to that of the sliding plate when its coated surface is 33 square inches.

* [So in MS.] † [See Art. 366 and Note 26.]

Sep. pos.	Sep. neg.	Heat of ☿ in which ball was immersed.
12·9	29·4	cold
15	33	170
21		210
26·6		270 [fast.
29·4	———	300 elect. passed through glass pretty
31·8		305 passed through much faster.
29·4		290
19·6		235
17·2		160
14·8	33	145
12·9	30·6	cold

Tu. Mar. 23 and Wed. Mar. 24. [43 obs. Art. 649.]

557] *Allowance for connecting wires in p.* 86. [Art. 554.]

The allowance to be made for the charge of the connecting wires was endeavoured to be found by suspending the two circles of 9·3 inc. horizontally by silk lines at 11 inches distance from each other and finding their charge by means of the forked electrifying wire as in 1772 p. 7 [Art. 472], both when the plates were connected by a wire similar to that used for connecting the rosin plates, and without any connexion.

The event was as follows.

Fr. Mar. 26th [1773. 8 obs. See Art. 647.]

Therefore the plates contain about $\frac{2\frac{1}{2}}{2}$ square inc., or 1·41 inc. el. more with the connecting wire than without *.

Sat. Mar. 27 [1773].

It was tried by usual machine whether the 4 rosin plates contained more el. when at a distance than near. The trial plate B *id est* the largest trial plate used for D &a being placed on neg. side.

With a quantity of additional wire = to $9\frac{1}{2}$ inc. el. the balls sep. pos. when the plates were at as great a distance as possible. When they were placed close together they seemed to require rather more additional wire, and as well as I could judge, a quantity = about $\frac{1}{2}$ inc. el.

558] *Excitation of electricity by separating brass plate from glass one.*

Sat. afternoon. Th. 60. N. 9.

The experiment of p. 71 [Art. 541] was repeated. It was found that the brass plate was electrified on lifting up as before, though the plate was not electrified before. But if the plate was first charged and discharged again before the plate was lifted up, it was found to be stronger electrified.

I then took a piece of tinfoil of the same size as the brass plate, with a silk string fastened to it near the edge, and laid it on the glass and

* [See Art. 647.]

lifted it up gently by the silk string. The tinfoil was found to be electrified thereby.

559] *Comparison of Henly's, Lane's, and straw electrometer.*

Sun. Mar. 28 [1773]. Th. about 58. N. about 8.

The two conductors of Nairne were placed end to end, and Henly's electrometer placed on that furthest from globe* parallel to conductor and the cork pointing from globe. The four jars were also joined to the usual wire with the straw electrometer hung to it, the wire and jars being placed at such a distance from the conductors that the electricity was found not to flow sensibly from them to the jars.

The globe 3† was then applied to that conductor nearest the globe and electrified till Henly's electrometer stood at 90°. The globe 3 was then removed from the conductors and its electricity communicated to the jars‡.

The straw electrometer separated to $2 + \frac{1}{2}$.

The experiment was repeated several times and was found to agree together pretty well.

The jars were then electrified, they and the straw electrometer standing in the same place, and it was found that Lane's electrometer fastened to one of them discharged at $0{\cdot}53\frac{1}{2}$ with that degree of electrification, the same jar being applied to the conductor and electrified till Henly's electrometer stood at 90°, Lane's discharged at 12·15.

The conductors being then taken away and the jars and straw electrometer placed in usual position, Lane's discharged at 1·17 when straw stood at 2 + 3, and at 1 + 2 when light paper electrometer just separated. The knobs touched at 0·4.

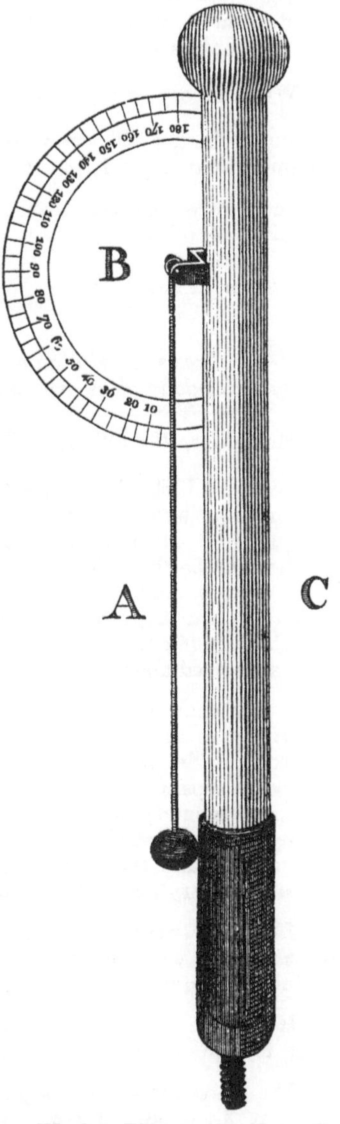

[Henly's Electrometer, from the original figure, *Phil. Trans.* 1772, p. 359.]

* [Of Nairne's electrical machine.]

† [Globes 2 and 3 are glass globes coated as Leyden jars. See Art. 505 for their charges.]

‡ [For the charges of these jars see Art. 506.]

Sun. eve. Th. 58. N. 8.

The globe 3 electrified till Henly stood at 90°, and its electricity communicated to 1, 2, and 3 jars, straw electrometer separated to $2 + 1\frac{3}{4}$. Lane's with that degree of electrification discharged at 1·7.

When Henly's stood at 90°, Lane's discharged at 12·20.

Jar 2 charged till straw electrometer separated to 4, and electricity communicated to jar 1, straw separated to $2 + \frac{1}{4}$.

When straw electrometer separated to	4	Lane's discharged at	2·0
	$2 + \frac{1}{4}$		·52
	$2 + 3$		1·19
light paper just separated			1·1

560] *Excess of redundant fluid on positive side above deficient fluid on negative side in glass plate and plate air &c.**

Mon. Mar. 29th [1773]. Th. 58. N. 7.

The $11\frac{1}{2}$ inch plate coated with circles of 8 inches diameter was supported on waxed glass. I charged this by touching the top with a vial charged till the straw electrometer separated to $2 + 3$ while I touched the bottom with a wire. At the same time an assistant stood ready with a bent wire in his hand ready to discharge it as soon as I took the jar away, the wire was fastened to a stick of waxed glass and had the pair of cork balls commonly made use of hanging to it, the cork balls separated about 1 inch.

I then charged the jar 4 to the same degree and communicated its electricity to the jars 1 & 2 and touched the upper side of the plate with one of the jars, but without touching the bottom with the wire. The corks separated very nearly the same as before, but of the 2 rather more. I then charged the jar till the straw electrometer separated to $2 + 2$ and diminished its electricity as before, the corks now separated rather less than the first time. The experiment was repeated several times with very nearly the same event.

I could perceive no difference in the separation of the cork balls whether the wire of the jar with which I touched the plate was 17 inches long or only $2\frac{3}{4}$.

If the four jars were charged to $2 + 3$ and its electricity communicated to globe 3, it was diminished to $2 + 2$.

The plate air 4 was charged by jar charged till straw electrometer stood at $2 + 0$, and if jar 4 was charged to the same degree and its electricity communicated to jar 2, the corks separated the same if bottom was not touched.

With plate air 1 the charge was obliged to be reduced by communi-

* [See Note 30.]

cating jar 2 to jar 4 to make the same separation when bottom was not touched as when it was.

Tu. Mar. 30 [1773]. Th. 56. N. 8.

The same experiment was repeated, only putting a piece of sealing wax with marks on it supported by glass about 2 inches below the corks to serve by way of comparison.

Compound plate of [Art. 552]	2 + 1	Jar 1 commun. to 2 + 4	2 + $1\frac{1}{2}$
Plate air 1	2 + $\frac{1}{2}$	jar 1 to 2	2 + 1
Plate air 4	2 + 1	1 to 2	2

The second column is the distance to which the straw electrometer separated in charging jar with [which] the plate was electrified when the bottom was touched in order that the cork balls should separate equal to marks on wax. The third column is the ratio in which the electricity of the jar was diminished when the bottom was not touched, and the fourth column shews the degree in which the jar was electrified (as expressed by distance to which the electrometer separated) in order that the balls should separate to the required distance.

N.B. The paper of divisions used for the electrometer was different from that used before, but the divisions nearly of same strength. The marks on sealing wax used for compound plate were nearer than those for plate air.

The jars 1, 3 & 4 being charged till straws separated to 3 + 0 and the electricity communicated to jar 2, they separated to 2 + 1, and the electricity of jar 2 being destroyed and the electricity of the others again communicated to it, they separated to 1 + 3*.

Therefore diminishing the electricity in ratio of 95 to 126 diminishes distance to which the balls separate in ratio of 126 to 165, or diminishing the electricity in ratio 1·33 to 1 diminishes distance in ratio 1·31 to 1.

Result.

On Monday the excess of redundant fluid on the positive side above deficient fluid on negative side in

$11\frac{1}{2}$ inch plate with 8 inch coating		$\frac{1}{3·56}$
Plate air 1	is	$\frac{1}{1·88}$
Plate air 4		$\frac{1}{2·14}$

* The smaller divisions are equal to $\frac{1}{4}$ of large ones.

of the quantity of electricity which is given to it with the same degree of electrification if the bottom plate is not touched.

	Compound plate	$\frac{1}{2{\cdot}66}$
On Tuesday D^o excess in	Plate air 1	$\frac{1}{1{\cdot}86}$
	Plate air 4	$\frac{1}{2{\cdot}16}$

561] Fr. Apr. 2 [1773]. Th. about 55. N. about 10.

It was tried whether a parallelepiped box included within another box of the same shape and communicating with it would receive any electricity on electrifying the outer box*.

The experiment was tried just in the same manner as that with the globe in p. 26 [Art. 513]. The inner box was 12 inches square and 2 thick. The outer box was 14 inches square and 4 thick on the outside, and 13 square and 3·4 thick within.

The boxes were made of wainscot and well salted. I could not perceive that the inner box was at all over or undercharged, for if I previously electrified the cork balls positively sufficiently to make them separate in touching the inner box, they would separate as much if I previously electrified them negatively in the same degree.

Globe within hollow globe tried again†.

562] Sun. Apr. 4 [1773]. Th. 58. N. 11.

The globe included between the 2 hemispheres was tried again in the same manner, except that the hemispheres were coated with tinfoil and were made to shut closer.

I could not perceive the inner globe to be at all electrified either way.

In order to see how small a degree of electricity I could perceive this way, I separated the two hemispheres as far as in the experiment, and electrified the 2nd thermometer tube with the same strength of electricity as was used in the experiment, and communicated its electricity to the jars 1 and 2, then touched the inner globe with one of those jars and drew up the cork balls, previously positively electrified, against the globe. I found them to separate very visibly.

I then repeated the experiment in the same manner except that the balls were negatively electrified in the same degree.

The elect. of the thermometer tube was diminished by communicating to the 2 jars in the ratio of 105 to 6339‡, or of 1 to 60, so that if

* [Exp. II. Art. 235.] † [Exp. I. Art. 218.]

‡ [Charge of 2nd thermometer tube = 80·7 glob. inc. = 124·3 circ. inc., by Art. 675, jar 1 + jar 2 = 6234, by Art. 506.]

the redundant fluid in the globe had been so much as $\frac{1}{60}$ of that in the hemispheres, I must have perceived it.

As it might be suspected that in the principal experiment the neighbourhood of the hemispheres communicating with [the] ground would enable the globe to hold more than it would otherwise do, and that therefore the cork balls would not separate so much as they would do if the hemispheres were taken away and the quantity of redundant fluid in the globe was the same, and consequently that the above computation of the quantity I could perceive is not just, I took away the hemispheres, made the corks touch the globe, and electrified it till they separated, then holding the hemispheres in my hands as near the globe as in the experiment, I did not perceive any alteration in the separation of the corks.

The outside diameter of the hemispheres was 13·3 inches.

563] *Experiment to see whether the force with which two bodies repel is as the square of the redundant fluid in them* : tried with straw electrometer and glass globes.*

The two electrometers were hung at opposite ends of a horizontal stick of wood 43 inches long, supported on sticks of waxed glass and communicating near the middle with one of the globes†. The same string also which lifted up the electrifying wire let down a piece of wood for making a communication between the two globes. The board with divisions was placed 6 inches behind the electrometers, and the guide for the eye 30 inches before it.

The electricity of the globes wasted very slowly, so that it could not be sensibly diminished in the time between reading off divisions to heavy electrometer and those to light one.

The electrifying wire rested on horizontal wood while globe† was turned, two jars being used as a magazine to prevent the globe Leyden vial from charging too fast. The globe‡ was turned till the heavy electrometer separated to rather more than the intended division, after which I waited till it came right, when by the string I lifted up the electrifying wire and made the communication between the two globes and looked at the division of light electrometer. The electricity of the magazine was discharged as soon as the electrifying wire was lifted up.

564] One of the straws used for the heavy electrometer was black in some places, and is called "blighted," the other is called "fair."

* [See Art. 386.]

† [The coated globes 2 and 3. Their charges are given in Art. 505 as 1782 and 1555 circ. inc., or 1159 and 1009 glob. inc., the sum of which, 2168, agrees with Art. 391.]

‡ [Of Nairne's electrical machine.]

Sun. Jan. 24 [1773].

	Weight.	Length.	Cent. gr. from needle.	Wire length.	Added weight.	Weight with addit. tried immed. after.
Blighted straw	7	11·1	5·1	2·2	10·7	17·8
Fair straw	6	11·1	4·99	1·8	8·8	14·8

The additional wire was run into straw very easily, and was fastened by putting a little wax on the end, which by heat was pressed quite smooth against the end of the straw.

Mon. morn. Jan. 25. Th. 55. N. 20½.

The globe 3 was made to communicate with horizontal wood, then if heavy electrometer separated to {8, 9, 10} divisions, the light electrometer separated, on communicating electricity to globe 2, to the same number of divisions.

565] *Trials of time in which the electricity of jar* 1 *was diminished by these straws from degree in which the heavy electrometer used in former experiments of this kind to that in which the pith balls began to close.*

			Weight.
Light electrometer	1	30″	7·8
	2	35	6·8
	fair	27	14·8
Heavy electrometer	blighted	15	17·8

In the afternoon the blighted straw was by mistake for the other covered for an hour with paper soaked in salt water. After standing about an hour to dry, it was found that when heavy electrometer was made to separate 10 divisions, light separated 11½.

The blighted/fair straw discharged the electricity as before in 5″/22, weight of blighted straw 17·8.

The fair straw was then kept in salted paper in the same manner for about 3 hours.

Tu. morning. Th. 57. N. 23.

If heavy electrometer separates to 15, 10, 9, 8, light electrometer separates about ¼ less.

Fair straw discharged the electricity almost immediately, blighted in about 5″.

When fair straw rested on cork ball it was about 30″, the blighted was much longer.

Weight of fair straw 14·95.

Ball of fair straw moistened with salt water.

Heavy electrometer at $\frac{15}{10}$, light at $\frac{14}{10}$.

The blighted straw was kept in salted paper for $1\frac{1}{2}$ hours, and then set to dry till the afternoon. In the afternoon its ball was moistened with salt water.

When heavy el. at $\begin{cases}15\\10\end{cases}$, light at $\begin{cases}14+\\9+\end{cases}$.

$\left.\begin{array}{l}\text{Fair}\\\text{Blighted}\end{array}\right\}$ straw closed in about $\begin{array}{c}2''\\2''\end{array}$ when resting on straw, and about 5″ or 7″ when resting on cork ball.

In about $1\frac{1}{2}$ hours after they were tried again without any alteration having been made.

When heavy at $\begin{cases}15\\10\end{cases}$, light at $\begin{cases}14\\9\frac{1}{4}\end{cases}$.

$\left.\begin{array}{l}\text{fair}\\\text{blighted}\end{array}\right\}$ closed $\begin{cases}2\frac{1}{2}\\2\frac{1}{2}\end{cases}$ on straw $\begin{cases}8\\10\end{cases}$ on ball.

The light straw N° 1 was soaked in salt paper at night for $3\frac{1}{2}$ hours.

Wed. Jan. 27. Th. 57. N. 23.

When heavy sep. to 15, light at 14, but increased after a time to near 15. As it was suspected that this increase might be owing to the air being electrified, I tried and found the air to be much electrified in all parts of the room.

N° 1 resting on $\begin{cases}\text{straw}\\\text{ball}\end{cases}$ closed in $\begin{array}{c}2''\text{ or }3''\\90\end{array}$.

N° 2 on $\begin{array}{c}\text{straw}\\\text{ball}\end{array}$ closed in $\begin{array}{c}20''\\\text{very slow.}\end{array}$

The ball of N° 1 was then moistened with salt water.

Heavy sep. to 15, light to 13, but increased to near 14.

In order to avoid in some measure the inconvenience from electrifying the air, Richard turned the globe, by which means the electricity was not made so strong.

heavy to $\frac{15}{10}$, light to $\frac{14}{10}$.

N° 1 closed in about 4″ whether resting on ball or straw.

N° 2 was soaked in salt water for $2\frac{1}{2}$ hours till 3 in afternoon, about 5 or 6 it was tried.

heavy to $\frac{15}{10}$, light to $\frac{15}{9\frac{1}{2}}$ 1st time, for several times after to 14.

	On straw.	On ball.
N° 2	4″	4
1	3 or 4	3 or 4
fair	2 or 3	8
blighted	2 or 3	9

Wed. Feb. 3. Th. 46½. N. 12.

As it was found the preceding day that the straws conducted ill, they were kept about 3 or 4 hours in the morning in salted paper, at about 3 they were taken out of the paper and hung up to dry. In the afternoon they were tried, a screen being placed to keep them from the fire.

Globe 3 elect.		Globe 2 elect.	
Heavy.	Light.	Heavy.	Light.
15	16¼	15	17½
15	16	15	17¼
12	12	15	17½
12	12¼	12	14¼
10	10¼	12	13½
10	10½	12	13
8	8¼	10	11½
8	8	10	11½
8	8½	8	9½
8	8¼	8	9½
10	9¼	10	11½
10	10	10	11½
10	10	12	14
12	13	12	13½
12	12	15	17½
12	12	15	17½
15	16		
15	15½		
15	15½		

	On straw.	On ball.
The blighted heavy straw closed in	25″	1′.30″
fair	10	not near closed in 2′

	Weight with addition.	Without.	Distance of pin from cent. grav.
Fair	14·85	6·05	5·07
Blighted	17·85	7·05	5·155

566] After the additional wire had been taken from the heavy electrometer, the two electrometers were electrified and compared to gether without the process of communicating the electricity from one globe to the other, when they stood as follows.

Heavy.	Light.			Heavy straw electrometer without additions.
8	9	Heavy paper	$\frac{2}{10}$	17
10	$10\frac{3}{4}$	electrometer	very little	15
12	$12\frac{1}{2}$	Light do.	$\frac{2}{10}$	13
15	$15\frac{1}{2}$		very little	12
12	$12\frac{1}{3}$			
10	$10\frac{1}{2}$			
8	$8\frac{1}{2}$			

If globe 2 was electrified till D° electrometer separated to 17, on communicating electricity to globe 3 it separated to $9\frac{1}{4}$.

The light straw electrometer was then placed instead of the paper electrometer, and a paper with divisions placed behind it. It was found that when heavy straw electrometer separated to $\frac{17}{9\frac{1}{4}}$ divisions, the light straw electrometer separated to

Large divisions.	Small divisions.
3	1
1	3

567] As the straws seemed not to conduct well enough, they were gilt. The gilding was not perfect in several places, but it was sufficient to conduct the shock of a jar very weakly electrified.

		Weight.	Cent. grav. from pin.	Wire length.	Added weight.
Heavy electrometer	N. 1	7·55	5·25	2·45	12·
	N. 2	6·55	5·17	2·01	10·1

Tu. Apr. 13.

The globe 2 electrified and communicated to globe 3.

Heavy el.	Light.
13	$15\frac{1}{2}$
12	14
10	12
8	$10\frac{1}{2}$

Wed. Apr. 14. Th. 51. N. 13.

Globe 2 communicated to 3.		Globe 3 communicated to 2.	
8	$9\frac{1}{2}$	8	$8\frac{1}{2}$
10	$11\frac{3}{4}$	10	$10\frac{1}{2}$
12	14	12	13
13	$15\frac{1}{2}$	13	14

It was found that some electricity ran from the electrifying wire to the knob of the globe to which electricity was to be communicated, on which the knob was removed to such a distance that no sensible electricity ran from one to the other.

Globe 3 communicated to 2.		Globe 2 communicated to 3.	
13	$13\frac{1}{4}$	13	$15\frac{1}{4}$
12	$12\frac{1}{4}$	12	14
10	$10\frac{1}{2}$		
8	$8\frac{1}{2}$		

N.B. The holes where the wires were put in were gilt over.

N. 2 } of heavy electrometer were found to weigh 16·65
N. 1 } 19·65.

The wires were then taken out, the holes stopped up with wax and gilt over. It was then found on electrifying the globe without communicating its electricity to the other, that when the heavy electrometer stood at

8		9
10	the light stood at	$11\frac{1}{4}$
12		$13\frac{1}{4}$
13		$14\frac{1}{4}$

	Weight.	Cent. grav. from pin.
N. 1	7·6	5·36
N. 2	6·65	5·285

	N. 1	N. 2
Force requisite to separate straws without wires	40·8	35·1
with wires	159·9	136·1

Therefore force required to separate heavy electrometer falls short of four times force required to separate light electrometer in the ratio of 296 to 303·6, or of 1 to 1·027.

568] *Separation of Henly's electrometer by different strengths of electrification.*

Nairne's jar being tried against the two trial plates for plate H, the pith balls separated a little after a short time the same way as the two trial plates. Therefore Nairne's jar is supposed to contain about $\frac{1}{9}^{6}$ of plate H, or 16 times as much as plate M.

The two conductors of Nairne were set end to end with [Henly's] electrometer on furthest, and the jar applied to the same, the furthest conductor being without any point, and the plate M was placed near it, set on a conductor communicating with the ground. When the electrometer was raised a little above 90°, the nearest conductor was removed and the electricity of globe taken away. Then as soon as the electrometer was sunk to 90° a communication was made between conductor and plate M and immediately taken away again, and the figure to which the electrometer sunk wrote down and the electricity of plate M discharged, after which a communication was again made between the conductor and plate M.

The results of the experiments are contained in the following Table, where the first column is the number of times that a communication has been made between the conductor and M.

The second column shows the quantity of electricity in the jar, which must diminish each time in the ratio of 15 to 16, and the other column is the number which the electrometer stood at in the different experiments.

Number of times.	Elect. in jar.	Numbers on electrometer. 1st	2	3	4	5	6
1	·938	70	73	73	79	77	80
2	·879	32	36	44	63	58	63
3	·824	20	21	23	31	35	30
4	·773	16	16	17	19	18	18½
5	·725	14	14	15	16	14½	15½
6	·679	12	12	13	14	13	12½
7	·637	10	11	12	12	11½	11½
8	·597	9	10	10½	11	10½	10½
9	·560	8	9	10	10	10	9½
10	·525	7½	8				8½
11	·492	7	7½				8
12	·461	6	6½				7
13	·433	5½	6				6½
14	·406	5	5½				5½
15	·380		5				
16	·357		4½				
17	·334		4				
18	·313		4				
19	·294		3½				
20	·276		3				

Number of times.	Diff.	Supposed true. Elect. in jar.	Number on electr.	Diff.	2nd diff. by first.
1	063	1·000	90	10	160
2	59	·938	80	17	290
3	55	·879	63	31	564
4	51	·824	32	13·5	262
5	48	·773	18·5	3	62
6	46	·725	15·5	2·5	55
7	42	·679	13	1·5	35
8	40	·637	11·5	1	
9	37	·597	10·5	1	26
10	35	·560	9·5	1	
11	33	·525	8·5	·5	
12	31	·492	8	1	
13	28	·461	7	·5	25
14	27	·433	6·5	1	
15		·406	5·5		

The above experiment is supposed to have been made in the autumn of 1772.

569] *Separation of Henly's electrometer when fixed in the usual way and on an upright rod.*

Aug. 13, 1773. Th. about 78.

Henly's electrometer was stuck on a thin wooden rod 25 inches long, the end of which was fixed into the hole made in the conductor for receiving the electrometer, being parallel to the conductor as usual. The conductor to which this was fixed was connected to the other conductor which received the electricity from the machine by a brass wire about 10 inches long, and a jar with Lane's electrometer fastened to it was made to communicate with this last conductor, so that the rod to which the electrometer was fastened was about *inches from the globe and *inches from the jar.

Henly's electrometer was then compared with Lane's while in this situation, and when this was done the wooden rod was taken away and

* [So in MS.]

Henly's placed on the conductor in the usual manner, everything else being the same as before, and compared with Lane's as before.

N.B. In both trials the cork ball of Henly's was turned from the globe*.

The result was as follows :—

Lane.	Henly.	
Rev. div.	On rod.	Usual way.
4·30	21	5
6·30	37	10
8·30	38	18
10·30	40	32

Hence it appears that when Henly's [electrometer] is fixed on the rod it is more sensible towards the beginning of its motion than afterwards, whereas when put in the usual way it is the contrary.

570] Result of P. 70, 75, & 95 [Arts. 540, 544, 559], being a comparison of the different electrometers.

Straw electrometer at	P. 70. Th. 56½. N. 19.	P. 75. Th. 53½. N. 20½.	P. 94. Th. 58. N. 8.	P. 95. Th. 58. N. 8.
1 + 3	43	46½		
2 + ¼				48
2 + ½			50	
2 + 1½	60			
2 + 1¾				63
2 + 2½	70			
2 + 3	75		73	75
3 + ½	80			
3 + 1	82½	88½		
4				116
Light paper elect. just sep.	60	64½	58	
Henly's at 90°			731	736

The three last columns are the distances at which Lane's electrometer discharged, expressed in divisions, or 60th parts of a revolution of the screw.

By $\frac{\text{P. } 94}{\text{P. } 95}$ [Art. 551] the distance at which Lane's discharges is as the $\begin{cases} 1{\cdot}228 \\ 1{\cdot}226 \end{cases}$ power of the quantity of electricity in the jar, and the quantity of electricity when the straw electrometer is at 2 + 3, *id est* the usual charge is to that when Henly's is at 90° as 1 to $\begin{cases} 6{\cdot}53 \\ 6{\cdot}38 \end{cases}$.

* [Of the electrical machine.]

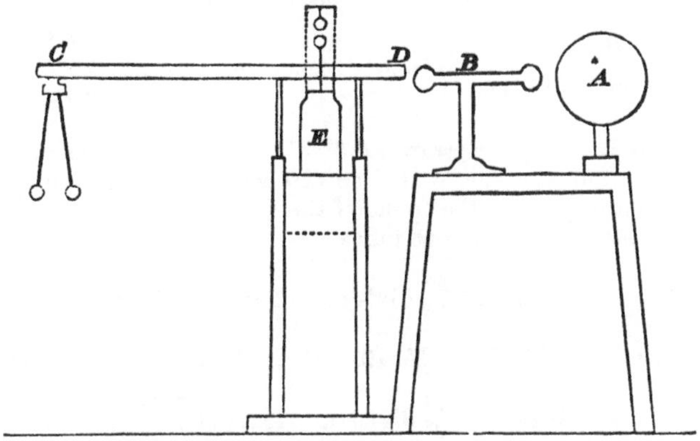

571] *Comparison of Lane's electrometer with light straw electrometer in different weather.*

Lane's electrometer was compared with the light straw electrometer by the apparatus represented above. *A* being the globe, *B* a conductor, *CD* a wooden rod supported on two waxed glass pillars, having a pin at *D* almost in contact with the conductor, the straw electrometer being hung to *C*. *E* is a jar with Lane's electrometer fastened to it, supported on a bracket fixed to glass pillars, the wire of which touches *CD*.

The distance of *C* from the globe is $54\frac{1}{2}$ inches and from the nearest glass pillar 32 inches. The height of the pith balls above the floor is $36\frac{3}{4}$ inches.

A small board with divisions on it, not represented in the figure, supported on an upright wooden rod, is placed behind the straw electrometer 25 inches from it, and a bit of tin with a narrow notch in it for an eye sight is placed at the same distance before the electrometer.

The outward divisions on the board, or those called the 4th, are at 5 inches asunder, the 3rd at 4 inches, the 2nd at 3 inches, and the 1st at 2.

As I found it impracticable looking attentively at both balls of the electrometer, I looked only at one, which, as my eye was guided by a narrow slit, was sufficient, and when I had made the experiment looking at one ball I repeated it looking at the other, so that the mean would be right though the slit was not right placed.

A wire was continued from the coating of the jar to the earth.

Wed. Aug. 18, 1773.

Th. 63°. N. 19. Bar. 29·64.

With two more jars communicating with *E* by wire.

Knobs of Lane's electrometer touched at 0·29.

$$\text{Straw elect. at} \begin{cases} 2^{nd} \\ 3^{rd} \\ 4^{th} \end{cases} \text{division; Lane discharged at} \begin{cases} 1{\cdot}43 \\ 2{\cdot}27 \\ 3{\cdot}1\tfrac{1}{2} \end{cases}.$$

With only one jar; straw at 3rd division, Lane discharged at 2·27. A slip of tinfoil was then pasted on *CD* the whole length so as to touch the wire of the jar and the frame of the straw electrometer. The result with only one jar was then as follows.

Straw at $\genfrac{}{}{0pt}{}{3^{rd}}{4^{th}}$ division. Lane at $\genfrac{}{}{0pt}{}{2{\cdot}26}{3{\cdot}1}$.

Th. Sept. 2. Th. 65°. N. 19. Bar. 29·865.

Straw at $\genfrac{}{}{0pt}{}{3^{rd}}{4^{th}}$ division. Lane at $\genfrac{}{}{0pt}{}{1{\cdot}58\frac{1}{2}}{2{\cdot}41}$.

Wed. Sep. 8th. Th. 62°½. N. 19½. C. 18. Bar. 29·235.

Straw at $\genfrac{}{}{0pt}{}{3^{rd}}{4^{th}}$ division. Lane at $\genfrac{}{}{0pt}{}{2{\cdot}22}{3{\cdot}1}$.

In the afternoon. Th. 62°. N. 19. C. 17. Bar. supp. 29·37.

Straw at $\genfrac{}{}{0pt}{}{3^{rd}}{4^{th}}$ division. Lane at $\genfrac{}{}{0pt}{}{2{\cdot}33}{3{\cdot}0}$.

Fr. Sept. 17. Th. 58°½. N. 28½. C. 29. Bar. 29·61.

Straw at $\genfrac{}{}{0pt}{}{3}{4}$ Lane at $\genfrac{}{}{0pt}{}{2{\cdot}24}{2{\cdot}59}$.

572] *Comparison of strength of shocks by points and blunt bodies.*

The wooden rod used in P. 118 [Art. 571] was supported on waxed glass with the straw electrometer at the end, and some tinfoil wound round part of the rod. The white glass cylinder was put in contact with it, electrified in such a degree that I felt a slight shock in discharging it with a piece of brass wire with a round knob at the end. If it was then electrified in [the] same degree, and discharged [with] a like brass wire with a needle fastened to the end, I could perceive no shock, and but a very slight sensation, even though the point was approached pretty quick. The distance to which the straw electrometer separated was about 1·8 inches.

The white cylinder was then changed for one of the large jars, the shock was not very different whether it was discharged by the knob or point unless the point was approached very slow. The distance to which the electrometer separated was about ·9 inch.

The wooden rod was taken away, and the white glass cylinder made to rest on the conductor with Henly's electrometer on it, and electrified till it stood at 90°, and to prevent the shock being too strong it had its choice whether it would pass through my body or some salt water, the wires in the salt water being brought within such a distance that the shock was weak when taken by the blunt body. I then found that if I took it with the point I could scarce perceive any spark.

The experiment was tried in the same manner with a large jar. The shock was very sensibly less though the point was approached almost as fast as I could.

573] *Whether shock of one jar is greater or less than that of twice that quantity of fluid spread on four jars**.

It was found that if the jars 3 and 4 were electrified in a given degree, and their electricity communicated to the jars 1 and 2, the shock produced by discharging them was nearly the same, or of the two rather more, than that produced by discharging the jar 1 or 2 by itself. The shock of the jar 3 was found to be very sensibly greater than that of jar 4.

It was tried with the wooden rod, the jars to be electrified being placed in contact with the tinfoil thereon, and when they were sufficiently electrified those to which the electricity was to be communicated being approached till they touched the rod, all four standing on the same tin plate. The jars were electrified till the straws separated ·9 inch.

N.B. The jars 1 and 2 contain pretty nearly the same quantity of electricity and their sum is nearly equal to the sum of jars 3 and 4. The quantity of electricity in jar 3 exceeds that in jar 4 in the ratio of 37 to 27, or of 4 to 3 nearly †.

574] *Comparison of the diminution which the shock receives by passing through water in tubes of different bores, and whether it is as much diminished in passing through* 9 *small tubes as through the same length of one large tube the area of whose bore is equal to that of the* 9 *small ones* ‡.

Nov. 1773. It was tried whether a shock was as much diminished by passing through a glass tube filled with water, 37 inches of which held 250 grains of water, as in passing through 9 tubes, 37 inches of all which together held 258 grains of water, the length of water which it passed through being the same in both cases, namely about 40 inches.

Two jars were used, and charged till the straw electrometer separated to 3 + 0. The water in the tubes was mixed with a very little salt, and the shock just enough to be perceived.

I could not be certain that there was any difference, but if any, that with the single tube seemed greatest. The shock was then made to pass through 7 of the small tubes, 37 inches of which hold about 200 grains of water. The shock was then sensibly less than with the large tube.

* [Art. 406 and Note 31.] † [Art. 685.] ‡ [See Art. 506.]

It was afterwards tried through what length of a tube, 37 inches of which held 44 grains, the shock must pass, so as to be as much diminished as in passing through 44¼ of the large one.

It was found that when it passed through 5·2 inches the shock was sensibly greater, and when it passed through 8·4 sensibly less than with the large one, so that it is supposed it would be equal if it passed through 6·8.

$$\frac{6{\cdot}8}{44\frac{1}{4}} = \overline{\frac{44}{250}}\Big|^{1{\cdot}08}$$

so that the resistance should seem as the 1·08 power of the velocity.

N.B. The quantity of water which the tubes held was not measured very exactly.

575] The tubes used in p. 123 [Art. 574] were measured by ☿ and are as follows:

N°	Length col. ☿.	Weight. [Troy] [oz. pwt. gr.]	Length of same column when near. Straight end.	Bent end.	Weight of 37 inches in grains.
1	37·1	16 . 12	20·9	20·2	395
2	37·3	14 . 2	23·65	22·9	335
3	38·4	1 . 0 . 8	24·5	21·8	470
4	38	11 . 6	24·5	24·7	263
5	37·7	17 . 17	21·7	23·3	417
6	36·8	14 . 10	20·4	20·3	348
7	38·8	15 . 22	26·6	22·3	364
8	38·6	17 . 17	24·8	21·6	407
9	39·8	16 . 18	26·3	22·2	374
10	37·8	1 . 10 . 0	16·9	17·4	705
11	37·3	1 . 3 . 20	22·8	22·2	567
large	44·7	8 . 15 . 8	...	...	3480

37 inches of the 9 first tubes, which are what was used in p. 123 [Art. 574], held together 3373 grains, therefore the shock was very nearly the same, but if anything rather greater when it passed through one tube, 37 inches of which held 3480 grains of ☿, than when it passed through 9 tubes, 37 inches of all which together held 3373 grains.

By p. 124 the shock is as much diminished in passing through 6·8 inches of a tube, 37 inches of which hold 567 grains, as through 44¼ of one, 37 inches of which hold 3480, so that resistance should seem as the 1·03 power of the velocity.

576] *Comparison of diminution of shock by passing through iron wire or through salt water**.

In order to compare the conducting power of iron wire and salt water, the shock of two jars had its choice whether it would pass through

* [Art. 398 and Note 32.]

2540 inches of nealed iron wire, 12 feet of which weighed 14·2 grains, or through my body, each end of the iron wire being fastened to a pretty thick piece of brass wire which I grasped tight, one in one hand and the other in the other, and with them discharged the jars.

It was found that when the straw electrometer separated to 1 + 0, I just felt a shock in my wrists, and when it separated to 2 + 0, I felt a pretty brisk one in them but not higher up.

I then gave the shock its choice whether it would pass through my body, or 5·1 inches of a column of a saturated solution of sea salt contained in a glass tube, 1 inch of which holds 9·12 grains of fresh water, the wires running into the salt water being fastened to brass wires as before.

I found the shock to be just the same as before, and found too that increasing the length of the column of salt water not more than $\frac{1}{4}$ of an inch made a sensible difference in the strength of the shock.

Therefore the electricity meets with the same resistance in passing through 2540 inches of wire whose base is $\frac{142}{78 \times 144} = \frac{1}{79}$ as through 5·1 inches of salt water whose base is 9·12.

Therefore, if the resistance is as the 1·08 power of the velocity, the resistance of iron wire is 607000 times less than that of a column of salt water of the same diameter *.

577] *Comparison of conducting powers of saturated solution of sea salt and distilled water.*

The shock of 1 jar charged till the straw electrometer separated to $1 + 0\frac{1}{2}$, discharged through a column of $\begin{cases} \cdot 8 \\ 1\cdot 0 \end{cases}$ inches of a mixture of saturated solution of sea salt with 99 of distilled water in tube 6, was $\begin{cases} \text{greater} \\ \text{less} \end{cases}$ than when it was discharged through $35\frac{1}{2}$ inches of saturated solution of sea salt in tube 2.

By a former experiment, the shock passed through $\left.\begin{matrix} \cdot 87 \\ 1\cdot 35 \end{matrix}\right\}$ of the mixed water was $\begin{cases} \text{greater} \\ \text{less} \end{cases}$ than through $40\frac{1}{2}$ of saturated solution.

By a mean, the resistance of one inch of the mixed water is equal to that of 38 of the saturated solution, therefore allowing for the different bases of the tubes, the resistance of the mixed water is 39 times greater than that of the saturated solution.

The shock of two jars, charged to 4 + 0, and discharged through $\begin{matrix} \cdot 55 \\ 1\cdot 8 \end{matrix}$

* [If the resistance is as the velocity, resistance of saturated solution of salt is 855400 times that of iron wire. By Matthiessen and Kohlrausch it should be about 502500. See Note 32.]

of distilled water in tube 5, was $\left\{\begin{matrix}\text{greater}\\ \text{less}\end{matrix}\right.$ than when it was discharged through $23\frac{1}{2}$ of the above-mentioned mixed water in tube 8.

By a former experiment, the shock passed through $\left\{\begin{matrix}\cdot 8\\ 2\cdot 0\end{matrix}\right.$ of distilled water was $\left\{\begin{matrix}\text{greater}\\ \text{less}\end{matrix}\right.$ than through $23\frac{1}{2}$ of the mixed.

By the mean, the resistance of 1·3 of distilled water = that of $23\frac{1}{2}$ of mixed.

10·9 inches of tube 5 in the place where used holds 120 grains of ☿, or 37 inches holds 408 grains, which is the same as tube 8: therefore the resistance of distilled water is 18 times greater than that of mixed, or 702 times greater than that of a saturated solution of sea salt.

578] *Whether the electricity is resisted in passing out of one medium into another in perfect contact with it.*

The 9th tube of P. 126 [Art. 575] was filled with 8* columns of saturated solution of sea salt inclosed between columns of ☿, the end columns being ☿. The tube 7 was filled with one short column of ☿ at the bent end, and a long column of saturated solution of sea salt.

It was found that the shock of one jar, charged till the straw electrometer separated to $1{\cdot}0\frac{1}{2}$, passed through a column of the salt water in tube 7, $\left.\begin{matrix}27\cdot 7\\ 21\cdot 2\end{matrix}\right\}$ inches long, was rather $\left\{\begin{matrix}\text{more}\\ \text{less}\end{matrix}\right.$ diminished than in passing through the mixed column in tube 9, the wires used in tube 9 being immersed in the end columns of ☿, and those used in tube 7 being immersed one in the short column of ☿ at the end and the other in the column of salt water.

The length of the mixed column in tube 9 was 43·5 inches, its weight was 10·5, the weight of a column of ☿ of the same length was 18·10, therefore the sum of the lengths of all the columns of salt water was 21·8 inches, and by the experiment the shock was as much diminished by passing through 24·4 inches of salt water in tube 7 as through this. But as the bore of tube 7 in that part which was used was greater than tube 9 in the ratio $\frac{24\cdot 4}{22\cdot 3}\times\frac{36}{37\cdot 4}=1\cdot 06$ to 1 nearly, the shock should be as much diminished in passing through a column 22·94 long in tube 9 as through one of 24·4 in this. Therefore the shock is as much diminished in passing through a mixed column, in which the length of salt water is 21·8 inches, as through a single column of the same size whose length is 22·94 inches.

The difference is much less than what might proceed from the error of the experiment.

579] A slip of tin was made consisting of 40 bits soldered together, all $\frac{1}{10}$ inch broad and all about $\frac{1}{6}$ inch long. They were made to lap

* [8_0 in MS. Perhaps 80.]

about $\frac{1}{20}$ inch over each other in soldering. I could not perceive that the shock of a jar was sensibly less when received through this than through a slip of tin of same length and breadth of one single piece.

If the jar were charged pretty high and a double circuit made for it, namely through this piece of tin and my body, I could not perceive the least sensation.

580] *Made at Nairne's with his large machine.*

A long conductor was applied to the electrical machine and a smaller conductor to its end, a Henly's electrometer was placed on the middle of the long conductor, and a small jar with a Lane's electrometer fastened to it was made to touch the short one. When Henly's stood at

30	Lane's discharged at	17 + 35 = ·668	
55		17 + 50 = ·678	inch.
70		19 + 30 = ·741	

The jar was then changed for one of rather more coated surface and a much smaller knob. When Henly's stood at 30 or 35, Lane's discharged at 17·7 = ·650, so that Lane's discharged at nearly the same distance with the same charge, whichever jar was used.

Henly's electrometer was then placed on an upright rod, touching the long conductor near the furthest end, Lane's electrometer with the first jar being placed as before.

Henly then rose to 55 or 60 before Lane discharged at 17·55 = ·681 inch. Henly being then lifted higher it rose to 65, Lane remaining as before. It was then lifted still higher, when it rose to

65	before Lane's discharged at	17·55 = ·631
50		9·55 = ·377
35 or 40		6·55 = ·263

Lane's being then separated to 27·55 = 1·060, the jar once discharged over surface of glass and once to the electrometer, but there seemed reason to think that Henly's rose no higher than before, namely 65.

My Henly's electrometer usually rose to 90 when Lane's discharged at 12·20 = ·467 inches.

Therefore the distance at which Lane's discharges, answering to different numbers on Henly's, is as follows:

		[Lane]
Henly on highest rod	65	1·060
	65	·681
	50	·377
	35 or 40	·263
Henly on conductor	70	·741
	55	·678
	30	·668
My Henly on conductor	90	·469

The distance at which Lane's discharges with a given jar is nearly proportional to the quantity of electricity in the jar, for if a jar is charged to a degree at which Lane is found to discharge at a given distance, and its electricity is communicated to another jar of the same size, so as to contain only $\frac{1}{2}$ as much electricity as before, Lane will then discharge at nearly $\frac{1}{2}$ the former distance.

M[EASURES]*.

581] M. 1. *Comparative charges of jars and battery*†.

If jar 1 is electrified till straw electrometer separates to $1\frac{1}{2}$, and its electricity is communicated to jars 2 + 4, pith electrometer separates $5\frac{3}{4}$. Therefore charge required to make pith balls separate $5\frac{3}{4}$ is to that required to make straw electrometer separate $1\frac{1}{2}$ as 3184 to 8909, and that to make pith separate $5\frac{1}{4}$ to that to make straw separate $1\frac{1}{2}$ as 2920 to 8909.

Jars 1 and 2 being electrified by wire and jar $\left\{\begin{matrix}5\\6\\7\end{matrix}\right.$ by coating till pith electrometer separated $1\frac{1}{2}$ and a communication being then made between them in the manner used for trying Leyden vials, pith balls separate $\begin{matrix}5\frac{3}{4}\\5\frac{1}{2}\\5\frac{1}{4}\end{matrix}$ negative, therefore charge of jar $\begin{matrix}5\\6\\7\end{matrix}$ should be $\left\{\begin{matrix}1316\ddagger\\1273.\\1231\end{matrix}\right.$

Charge of 1 + 2 + 3 + 4 = 12544.

Jars 1 + 2 + 3 + 4 being compared in the same manner with jar $\left\{\begin{matrix}5\\6\\7\end{matrix}\right.$ the pith balls did not separate at all.

M. 2. If the charge of jars 1 + 2 + 3 + 4 is called 4

jar 1 or 2 is nearly	=	1
5, 6, or 7	=	4
1 row of battery	=	22
whole battery	=	154

Jar 8 being electrified it was found that it must be touched $7\frac{1}{2}$ times by white cyl. to reduce the quantity of electricity to $\frac{1}{2}$. The 4 jars must be touched $8\frac{1}{2}$ times by do. Therefore charge of jar 8 = $3\frac{1}{2}$.

A piece of crown glass 1 foot square of which weighed 10·12 was coated with tinfoil about 10 inches square.

* [These "Measures" are on a set of loose sheets of different sizes marked M. 1 to M. 21, and another set marked M. 1 to M. 12.]

† [Art. 411.]

‡ [These numbers are given as in the MS. They should be each multiplied by 10. See also Art. 585, where the numbers seem to be deduced from some other experiment.]

M. 3. The charge of each row of the battery was found by charging to a given degree by electrometer and touching it repeatedly with jar 4 till the separation of electrometer was reduced to that answering to ½ the charge.

*The 1st, 2nd, 3rd, 4th, 5th, 6th, 7th row required to be touched 18, 19, 17, 18½, 17, 17, 18 times, therefore charge

1st row	= 26	charge of jar 4
2nd	= 27·4	
3rd	= 24·6	
4th	= 26·7	
5th	= 24·6	
6th	= 24·6	
7th	= 26	

and charge of whole battery = 180 times that of jar 4
and real charge = 321000
and if real charge by computed of white glass = 7·5,
computed charge = 42800
which answers to 187 square feet of glass whose thickness = $\frac{1}{10}$.

Therefore charge of jar 4 answers to 1·04 square feet of D° thickness. The coating is about $\frac{7}{12}$ of a square foot, and therefore the mean thickness = ·058.

582] M. 4. Let jar be touched n times† by jar which is to first as x to 1, it will be reduced in ratio of 1 to $(1+x)^n$, therefore if it is reduced to ½ thereby

$$(1+x)^n = 2.$$

Therefore let $\quad$ N. L $2 = a$ and N. L $(1+x) = px$,

$$pxn = a,$$

and
$$\frac{1}{x} = \frac{pn}{a};$$

but
$$\text{N. L } (1+x) = x - \frac{x^2}{2} \text{ nearly}, = x\left(1 - \frac{x}{2}\right),$$

therefore
$$p = 1 - \frac{x}{2} \text{ nearly } = 1 - \frac{a}{2pn} \text{ nearly},$$

therefore
$$\frac{1}{x} = \frac{n}{a}\left(1 - \frac{a}{2n}\right) \text{ nearly},$$

$$= \frac{n}{a} - \tfrac{1}{2} \text{ nearly},$$

whence we have the following

Rule for finding ratio of charge of 2 jars, supposing the charge of first to be reduced to ½ by touching n times by 2nd.

Charge of 1st is to that of 2nd :: $1·444n - \frac{1}{2}$ to 1.

* N.B. The left-hand row is supposed to be called the 1st row. [If Jar 4 = 2675 circ. inc. (See Art. 506) whole battery = 481500 circ. inc. or 321000 glob. inc., counting 1 glob. inc. = 1·5 circ. inc., as Cavendish seems to do here.]

† [Art. 413.]

583] M. 5.

$$\begin{aligned} \text{jar } 1 &= 3184 \text{ [circ. inc.]} \\ 2 &= 3050 \\ 3 &= 3635 \\ 4 &= 2675 \\ 5 &= 11816 \\ 6 &= 12544 \\ 7 &= 11816 \\ 8 &= 10761 \\ 1^{st} \text{ row} &= 64538^* \end{aligned}$$

Quantity of electricity communicated to whole battery† by

$$\begin{aligned} B + 2A &= 3{\cdot}61 + 2A \\ 2B + 2A &= 7{\cdot}07 + 2A \\ 3B + 2A &= 10{\cdot}36 + 2A \\ 3B + C + 2A &= 13{\cdot}16 + 2A \\ R &= 20{\cdot}58 \\ R + B &= 23{\cdot}66 \qquad D = \tfrac{1}{2} \\ R + 2B &= 26{\cdot}74 \\ R + 3B &= 29{\cdot}83 \end{aligned}$$

Quantity of electricity communicated to 1st row by

$$\begin{aligned} A &= {\cdot}95 \\ 2A &= 1{\cdot}8 \\ 3A &= 2{\cdot}6 \\ 4A &= 3{\cdot}3 \end{aligned} \qquad \left.\begin{aligned} B + 2A &= 2{\cdot}56 \\ 2B + \quad &= 4{\cdot}58 \\ 3B + \quad &= 6{\cdot}17 \\ 3B + \quad &= 7{\cdot}54 \end{aligned}\right\} + 2A$$

Charge of 1st battery of Nairne.

584] M. 6. Electricity of 1st row of old battery was reduced to $\frac{1}{2}$ by touching $11\frac{1}{4}$ times by crown glass of 10 inches square. Therefore charge of 1st row to that of crown glass as $15\frac{3}{4}$ to 1. The first row of new battery appeared by that means to contain 10·7, the 2nd row 11, and the 3rd 11·4 times the charge of the same plate.

The mean area of the convex coating of each jar seemed to be $14 \times 12\frac{1}{2} = 175$ inches, to which adding 5, *id est* $\frac{5}{12}$ of area of bottom, whole coating may be estimated at 180 square inches of same thickness as sides.

Elect. $\begin{cases} 1^{st} \\ 2 \\ 3 \end{cases}$ row of new battery was reduced to $\frac{1}{2}$ by touching $\begin{cases} 10 \\ 10\frac{1}{2} \\ 10 \end{cases}$ times by jar 1, therefore charge = jar 1 × $\begin{cases} 13{\cdot}94 \\ 14{\cdot}66, \\ 13{\cdot}94 \end{cases}$ and charge mean row = jar 1 × 14·18 = 45149 inc. el.

* [See Art. 506.]

† [Here A seems to be the charge of one of the first 4 jars taken as unit, B that of one of the others taken as 4, and R that of the row taken as 22, the battery being 154, as in M. 2.]

By top leaf its charge should be $\frac{64538 \times 11{\cdot}33}{15{\cdot}75} = 46390$ inc. el., therefore its computed charge $= \frac{45150}{1{\cdot}5} = 30100$, and thickness of glass should seem $= \frac{180 \times 6 \times \frac{4}{3} \times \frac{21}{20}}{30100} = \frac{1}{20}$ inc.

585] M. 7. *Whether shock of battery is sensibly diminished by imperfect conduction of the salt water in the jars.*

An uncoated glass jar like the coated ones was filled with fresh water and put into a glass jar of fresh water, a brass wire with knob being put into it, and a slip of tinfoil into the outer jar, it was charged till straw electrometer separated to 8 and tried by shock melter* filled [with] sea water, wires about 3 inc. dist.

The water in inner jar was then changed for sat. sol. s. s.† and that in outer for about equal parts of D° and fresh water, and tried in the same manner. The shock seemed rather greater, but was plainly less when electrometer was at 7.

When shock was taken without shock melter* it was as strong with el. at 5 as with D° at 8. Jar 2 being charged to 8 and its electricity communicated to jar, the electrometer separated to 4½.

586] M. 8. Feb. 28, 1775.

Specific gravity bottle filled with salt water from torpedo trough weighed 8 . 4 . 18 by ingraved weights. Th. at 49. Specific gravity = 1·0254.

Being mixed with $\frac{707}{2006} = {\cdot}3525$ its weight of rain water, specific gravity bottle weighed 8 . 4 . 1, Th. at 49½, specific gravity 1·0190.

Excess of specific gravity above unity of stronger is to that of weaker as 1·335 to 1. The quantity of salt in them is as 1·3524 [to 1].

Therefore the excess of specific gravity above 1 differs pretty nearly, but not quite, in as great a ratio as the quantity of salt in them.

M. 9. April 1. D° Specific gravity bottle with water from torpedo trough weighed 8 . 4 . 22 by D° weights.

April 29. Torpedo trough filled with water to within 1 inch of top, and 58 oz. salt added.

Specific gravity bottle filled therewith, Th. at 70°, weighed 8 . 4 . 12. At 54½ same water weighed 8 . 4 . 16½.

One bottle of sea water weighed 8 . 4 . 11, Th. at 67. Another bottle weighed 8 . 4 . 19½, Th. D°.

Specific gravity bottle with rain water weighs 8 . 1 . 22½.

[M. 10 blank].

* [This word occurs also in Arts. 622 and 637. See facsimile at Art. 622.]
† [Saturated solution of sea salt.]

M. 11. *Rule for finding the quantity of salt in water by its specific gravity.*

Let the specific gravity of the solution at $46\frac{1}{2} = \mathrm{S}$, and $\dfrac{\text{quantity of salt}}{\text{solution}} = x$. If S is above

$$\left.\begin{matrix}1{\cdot}0675\\ 1{\cdot}0261\\ 1{\cdot}\end{matrix}\right\} \frac{1}{x} = \left\{\begin{matrix}\dfrac{\cdot 779}{\mathrm{S}-1+\cdot 0033}\\ \dfrac{\cdot 719}{\mathrm{S}-1+\cdot 0022}\\ \dfrac{\cdot 789}{\mathrm{S}-1}\end{matrix}\right.$$

$$\begin{aligned}\cdot 779 &= 9{\cdot}8917\\ \mathrm{L}\ \cdot 719 &= 9{\cdot}8568\\ \cdot 784 &= 9{\cdot}8942\end{aligned}$$

vide Heat P. 98*

587. In 2nd Lane's electrometer or 1st detached do.

40 threads screw = $1\frac{1}{2}$ inches, or 1 division of plate = $\frac{1}{1600}$ inch.

For 3rd Lane D°.

588] M. 12. June 11.

	For salting.†	Not salting.
Mahogany	2 . 19 . 5	3 . 15 . 15
Wainscot	2 . 16 . 10	3 . 8 . 21
Beech	3 . 11 . 1	4 . 5 . 8
Ash	3 . 14 . 10	4 . 7 . 9
Alder	2 . 4 . 10	2 . 13 . 12
Lime	2 . 11 . 8	2 . 18 . 13
Deal	2 . 12 . 14	3 . 4 . 22

Weight of the unsalted ones on June 18, and number of vibrations of a pendulum ‡ inches long, in which the electricity of 1 row of the battery was reduced from 2 to 1 by pith balls by touching with them, the ends being wrapt round with tinfoil fastened on with gum.

	Weight.	Number of vibrations.	Loss of weight.
Mahog.	3 . 14 . 12	34	1 . 3
Wainscot	3 . 7 . 20	19	1 . 1
Beech	3 . 15 . 5	36	10 . 3
Ash	4 . 0 . 16	6	6 . 17
Alder	2 . 11 . 18	200	1 . 18
Lime	2 . 15 . 14	22	2 . 23
Deal	3 . 3 . 12	60	1 . 10

* [Mr Vernon Harcourt, in his Address to the British Association (*B.A. Report*, 1839, p. 48), has given extracts from Cavendish's MS. on Heat, p. 1 to p. 50, but he does not mention any page 98.]

† [Art. 609.]

‡ [So in MS.

M. 13. The salted ones taken out of water two or three hours weighed

Mahogany	3 . 15 . 14
Wainscot	3 . 9 . 0
Beech	4 . 8 . 0
Ash	4 . 14 . 12
Alder	3 . 9 . 22
Lime	3 . 17 . 11
Deal	3 . 1 . 4

June 19th. Bits of tinfoil were fastened round the ends of these pieces of wood with gum.

2B being electrified to $1\frac{1}{2}$ and its electricity communicated to the whole battery gave a slight shock when received through the Lime, Alder, Ash and Beech, but most through the Lime.

1R + 3B through wainscot and 2R + 3B through deal gave much the same shock, and 3R was just sensible through Mahogany.

589] M. 14. Dimensions of coatings made to pieces of glass D, E, F, G; A, B, C, I, K, L, M, H. [See Art. 324.]

M. 15. Bluish-green ground glass from Nairne called R and S.

M. 16. Logarithms for calculations of these plates.

M. 17. Do.

M. 18. Straight piece of elect[rifying] wire, thickness ·15, length 30.

Increase of quantity of electricity in wire of that thickness by increasing its length from 33 to 53 inches = 4·53 inches, therefore increase of quantity of electricity in wire = increase length × ·226.

The two trial plates of white plate glass, which contain together 66 inches of computed power, or 66 × 1·6 inches of electricity, were balanced by twice the sum of the double plates A and B + 48 of additional wire = 73·4 + 10·85 = 84·2 inc. el., therefore 10·7 inc. el. is to be allowed for the usual length of the wire.

[*Rules for making trial plates.*]

M. 19. By P. 21 [Art. 459] it should seem that the difference between two trial plates ought to be to their sum as L + S + 20 inc. of wire to L + S, or as $3{\cdot}6 + \dfrac{20 \times {\cdot}226}{1{\cdot}6}$ to 66, or as 32 : 330.

The plates $\begin{cases} D \\ E \\ F \end{cases}$ of Nairne are to be coated with circles $\begin{cases} 2{\cdot}16 \\ 2{\cdot}16 \\ 2{\cdot}19 \end{cases}$ in diameter.

The mean quantity of electricity in the trial plates should be 47·4 inc. if nothing is allowed for additional wire, therefore if this is increased by $\frac{1}{20}$ to allow for uncertainty, and the plate $\frac{A}{B}$ is used for $\begin{matrix}\text{small}\\\text{large}\end{matrix}$ trial plate, the computed power should be $\frac{27\cdot84}{33\cdot68}$, and [M. 20] the coating should be a square whose side $= \frac{2\cdot123}{2\cdot27}$.

Quantity of electricity in small thin plates to be 110·1, computed power = 67·8, diameter of circles to be

$$\begin{aligned} I &= 2\cdot299\\ K &= 2\cdot286\\ L &= 2\cdot358\\ M &= 2\cdot207 \end{aligned}$$

The trial plates to them to be made of plates I and L $\frac{\text{quant. el.}}{\text{comp. power}}$ of that glass supposed = 1·95.

Computed power of mean between the two plates to be $\frac{120\cdot8}{1\cdot95} \times \frac{11}{10}$ = 68 = a square of 1·933, the thickness of the glass being ·07, therefore $\left.\begin{matrix}\text{Small} = \text{L}\\\text{Large} = \text{I}\end{matrix}\right\} = 1\cdot933$ long and $\left\{\begin{matrix}2\cdot126\\1\cdot\ 74\end{matrix}\right.$ broad.

Oblong coatings 1·82 long and $\left\{\begin{matrix}1\cdot62\\1\cdot74\\2\cdot02\\2\cdot14\end{matrix}\right.$ were made to old ground plates $\left\{\begin{matrix}\text{L}\\\text{I}\\\text{H}\\\text{K}\end{matrix}\right.$

M. 21. $\left.\begin{matrix}\text{Small}\\\text{Large}\end{matrix}\right\}$ trial plate for large thick plates $\left\{\begin{matrix}\text{F}\\\text{E}\end{matrix}\right.$ to be coated with oblong square $\left\{\begin{matrix}5\cdot7\\6\cdot6\end{matrix}\right.$ by 6.

Dimensions of trial plates.

	Length.	Breadth.
A	16·4	11·8
B	13	9·7
C	10·4	7·8
D	8·5	6·4

Sliding Plates.

Value of 1 division in

Number of Plate.	Inc. el.	Parts of D.
1st	·75	·0208
2nd	1·50	·0416
3rd	3·37	·0940
4th	8·57	·238
5th	19·0	·530
6th	18·0	·500
1 inch additional wire	·226	·0063

[From this table it appears that D is supposed to contain 36 "inc. el." Now, by Art. 655, D contains 26·3 "globular inches," which is equal to 41 "circular inches," or 36·7 "square inches".]

[Specimen of Measurements of thickness by dividing engine.]

590] 2nd rosin plate, [Art. 371, 500] mean diameter 5·6.

M. 23. At 2·110 knobs coincided,

2·306
05 } center 2·3057
06

2·306
05
04
04 } one inch [from center] 2·3045
04
04

2·110 coinc.

2·110 coinc.

2·306
04
06
05
00
02 } $1\frac{1}{2}$ inc. from center 2·303
02
00
02
02
02
02
01

2·110 coinc.

Mean thickness = ·195.

591] M. 32. Measures of thickness of crown glass [Arts. 370, 500] measured in the middle of each 16th square part, the numbers being placed in the same situation as the squares.

C. At 1·881 the knobs coincided.

1·944	48	51	48
48	52	52	44
53	53	46	38
53	46	40	34

mean 1·947, thickness ·066.

592] M. 1*. *List of Plate glasses.*

First got.

	[Thickness.]	[Side.]	
A	·205	4·02	Made into trial plates and cased in cem[ent] of a greenish colour inclining to blue with a great want of transparency.
B	·193	3·99	
C	·162	4·01	not used.
D	·16	3·98	coated, comp. pow. = 46.
E	·19	8·06	Same sort of glass as A and B, but from their greater length and breadth I could not so well judge of their colour. Made up into trial plates.
F	·204	8·03	
G	·184	8·04	Same kind as C and D. Remains not approp.
H	·062	4·35	It is marked on side with single scratch of file, I with 2, and so on to L. Made into trial plates and broke except L. More transparent than the thick plates.
I	·066	4·36	
K	·063	4·34	
L	·068	4·37	
M	·055	8·68	not used.

The thickness was measured in the middle of each side, beginning with that side to right of letter, the letter being held towards eye.

		Thickness.	Diam. coating.	Comp. power.
Double plate ground glass	A	·3	1·82	11·04
	B	·31	1·855	11·1

1st got from Nairne.

2 thickish plates of 8 inc. made into trial plates for 2nd sort, called S and L.

Another cut into 4 pieces for trying effect of different varnishes.

A 4th not used.

2 thin ones of bluish glass coated in order to serve for trial plates to largest plate, and called L and S, but not used.

2 thick plates and 1 thin one rough.

2 white glass plates from Nairne.

4 pieces out of same piece with different sorts of surface.

A large piece of whitish plate glass divided into 3 pieces, one used for sliding trial plate.

4 irregular shaped pieces called N, O, P, Q. [Art. 459.]

2 thinnish pieces 5 inches square with very thin plate rosin between.

M. 2. 1 piece of much the same kind fastened to piece of crown glass with cement between, used for sliding plate.

* [Here follows another series of measures on loose leaves of different sizes.]

593] M. 3. *Nairne's plates of same piece**.

	Out of water.	Loss in water.			
4 thin pieces	7. 1.14	2.12.21	7·079	2·644	2·6774
4 thick do.	19. 3.11	7. 3. 4	19·173	7·158	2·6785
large thin piece	7. 0. 2	2.12. 7	7·004	2·615	2·6785
A	19. 6.17	7. 4.10	19·335	7·221	2·6776
B	19.12.11	7. 6.13	19·623	7·327	2·6782
	19. 1.16	7. 2.13	19·083	7·127	2·6775
					Mean 2·6779

1 cub. inc. water = ·5278 oz.
1 oz. of glass = 9·84973 cub. inc.

[The weights in 1st & 2nd column are in ounces, pennyweights and grains (Troy), in the 3rd and 4th in decimals of an ounce, and the 5th is the specific gravity. The number 9·84973 is the logarithm of the number of cubic inches in an ounce of glass.]

M. 4. [Gives 1st the length of each side of each piece of glass, and the distance between the middles of opposite sides to hundredths of an inch.

2nd the thickness at each corner and middle of each side to thousandths of an inch.

3rd specific gravity and mean thickness deduced from it for plates A to M of Nairne.

The results are given in M. 5. The thicknesses are as follows:

	Thickness calculated.	Measured.	Diff.
A	·2112	·2095	·0017
B	·2132	·2109	23
C	·2065	·2057	8
D	·2057	·2047	10
E	·2065	·2055	10
F	·2115	·2101	14
G	·2022	·2103	9
H	·07556	·0735	21
I	·07797	·0759	21
K	·07712	·0755	16
L	·08205	·0804	16
M	·07187	·0707	12

[The thicknesses given in Art. 324 are those calculated from the weight in and out of water and the measurement of the sides. They are greater than the measured thicknesses in every case.]

594] M. 6. *Measures of thickness &a of green glass cylinder.*

Longest cylinder: a mark made with file near middle.

The 1st column is the distance in inches of the point to which cylinder is immersed in water from the scratch.

* [Art. 314.]

The 2nd column is weight required to balance it in that position.

The 3rd is the same thing in the 2nd trial.

The 4th is the difference of these numbers, or bulk of intermediate portion of glass.

The 5th is the same thing in 2nd trial, and

The 6th is the mean between them.

The 7th is the circumference in the middle of that space.

Towards wide end.	1st tri[al].	2nd tri.	Bulk int. space by 1st	Bulk int. space by 2nd	Mean.	Circum.
13	11.12.15	12.22				
			1.13	1.12	1.12.5	3·595
12	11.11. 2	11.10				
			2.22	2.22	2.22	3·435
10	11. 8. 4	8.12				
			2.18	2.17	2.17.5	3·265
8	11. 5.10	5.19				
			2.20	2.21	2.20.5	3·140
6	11. 2.14	2.22				
			2.16	2.18	2.17	3·020
4	10.19.22	17.10				
			2.18	2.18	2.18	2·940
2	10.17. 4	17.10				
			2.23	2.21	2.22	2·905
0	10.14. 5	14.13				

[after this a table for the narrower half of 1st cylinder, and in M. 7 for 2nd, 3rd and 4th cylinders. M. 8 and M. 9 is a table of 11 columns.

1st column, distance from mark.

2nd Mean loss [of weight] for 2 inches.

3rd Supp[osed] mean circumference.

4th Log. loss.

5th Log. supp. circ[umference].

6th Log. thick[ness] × p.

7 Thick. × p [p = ratio of circumference to diameter].

8 True mean circ.*

9 Log. do.

10 Log. comp. power of 1 inch.

11 Comp. power of 1 inch.]

M. 10. Measures of the circumference and substance of glass in jars and cylinder.

Marks with file are made at the extremities of the whole space, and the numbers begin with the space marked with double mark.

* By mean circumference is meant the mean between the inside and outside circumference.

The circumference was measured by a slip of tinfoil put round, and the intersection marked with knife.

The substance of glass was found by hanging it to end of sliding ruler fastened to one end of balance, and weighing it in water; and by sliding the ruler I made more or less of it to be immersed, and knew the difference of the space immersed.

M. 11. *Specific gravity of different pieces of white glass.*

Large jar	3·253	3·253
small D°	3·256	3·257
long cylinder	3·281	3·279
thick flat glass	3·280	3·279
thin do.	3·280	3·284

The small jar being broke, a 2nd was measured.

Thickness measured by calipers in 4 different rows parallel to axis and in 5 different places in each row, beginning at a scratch with a file near bottom.

[Here follow the measures.]

The thickness was then tried in 4 different parts of circumference at 4·4 inc. distance from scratch.

It was then weighed in water in the same manner as the others.

The jar was dried before each trial, and before the 3rd was rubbed with solut. p. ash*, which made the water stick less to side, for which reason it is supposed most exact.

The circumference was measured in two parts of the middle space, and they came out both the same.

595] M. 12. *Measures of coatings to jars and cylind.*†

A coating made to 2nd small jar extending to 4·4 inches from scratch. Comp. power = 680·7.

Coating to white cylinder extends 9·86 inches from double mark. Comp. power = 684·1.

A coating made to 4th green cyl. extending 7 inches from mark. Comp. power = 318·2.

A mark was made on wide part, extending 7·16 inc. from new mark. Comp. power 600·7.

M. 13. A mark made on 2nd green cylinder 11 inches from first towards thick end, and the tube cut off about 1 inch from 1st mark.

A coating made to the thick part extending 8·55 inches from 2nd mark. Comp. power = 600.

* [Pearl ash.]

† [See Art. 383. The computed power here is 8 times the true value, and there is no correction for spreading of electricity.]

EXPERIMENTS WITH THE ARTIFICIAL TORPEDO.

596] Torp. 1 in water touching sides*.

3 rows $1\frac{1}{2}$†	felt plain shock	in hands.
4 —	more brisk	in D°.
7 —	more violent	in D°.
2	plain	in D°.
1	sensible.	

+ 2 + 3‡

4 + 1 + 5 + 6 + 7—but just sensible.

Out of water.

4 + 1	uncertain.
4 + 1 + 2	sensible.
4 + 1 + 2 + 3	D°.
5 + 4 + 1	sensible in elbows.
5 + 6	gentle in elbows.
5 + 6 + 7 + 1 + 2 + 3 + 4	strong in elbows.
1 row	more violent.

Uncoated, out of water.

4	scarce percept.
4 + 1	sensible.
4 + 1 + 2	gentle.

In water.

5 + 6 + 7	perceptible.

Without any torped. jar 4 was perceptible.

I could not perceive any sensible difference in the conducting power of the water I used & of sea water, but the difference caused by mixing $\frac{1}{11}$ part of rain water with the sea water was scarcely perceptible.

§ N.B. resistance of $\left\{\begin{array}{l}\text{distilled water}\\ \text{sat. sol. in } \frac{99}{29}\\ \text{sat. sol.}\end{array}\right\}$ are to each other nearly as $\left\{\begin{array}{l}1\\ 18\\ 100\\ 702\end{array}\right.$

so that there seems no reason to think that the resistance of water about the saltness of sea water varies in a quicker ratio than that of the quantity of salt in it.

* [Art. 415.]

† [3 rows of battery electrified till the electrometer separated to $1\frac{1}{2}$.]

‡ [These numbers are those of the jars of the first row of the battery. See Art. 583.

§ [This should be conducting power, instead of resistance. The numbers then agree with those in Art. 684.]

Without torpedo jar 1 + 2 + 4 was very sensible in elbows, but 1 + 2 was felt only in wrists.

597] Let a given charge be passed by double circuit through your body and another circuit; let the quantity of electricity which passes along the second circuit be to that which passes through your body as x to 1; the rapidity with which the fluid passes through your body is the same whatever is the value of x, and the quantity which passes through your body is* as $1 + x$.

If the resistance which the electricity meets with before it comes to the double circuit is to that which it would meet with in passing through your body alone as a to 1, the force required to drive electricity through the whole circuit in given time is as $a + \frac{1}{1+x}$, and therefore the time in which it is discharged $= \dfrac{1}{a + \frac{1}{1+x}} = \dfrac{1+x}{1+a+ax}$, and the velocity with which it passes through your body is as $\frac{1}{1+a+ax}$, and the strength of shock is as $\frac{1}{(1+x)(1+a+ax)}$.

In trying resistance of liquors by double circuit, if the quantity of electricity which passes through the liquor is to that which passes through your body as x to [1], the quantity of electricity which passes through your body is as $\frac{1}{1+x}$, and the rapidity with which it passes through your body is given.

In trying it with single circuit, if resistance el. in passing through liquor is to that in passing through your body as x to 1, velocity of electricity is as $\frac{1}{1+x}$, and the quantity is given, therefore in both ways of trying it, the greater x is, the more exact will be the method, and both methods will be equally exact if x is given or very great, supposing the strength of the shock to be as the quantity of electricity into its velocity †.

598] Shock produced by charge $\begin{cases}16\\22\\44\end{cases}$ in water bears the proper proportion to that caused by charge $\begin{cases}6\\8\\16\end{cases}$ out of water.

* [Should be *inversely* as $1+x$. The rest is a correct statement of the strength of derived currents according to the law afterwards published by Ohm. See Art. 417.]

† [The "velocity" is what is now called the strength of the current. The strength of the shock is assumed to be proportional to the energy of the discharge. See Arts. 406, 573, 610, and Note 31.]

It is supposed that it required about $2\frac{3}{4}$ the charge to give a proper shock in water as [it does] out, or it is supposed to require 5 times quant. el. It is supposed too that it requires 2[ce] charge of 3 times quant. el. to give same shock with torp. out of water as without torp.

Let quant. el. which passes through $\begin{cases}\text{body}\\ \text{torp.}\\ \text{water}\end{cases}$ be as $\begin{cases}1\\ 2\\ a\end{cases}$, if quant. el. which passes through torp. is increased to $2n$, quant. el. which passes through your body $\begin{cases}\text{in wat.}\\ \text{out wat.}\end{cases}$ will be $\begin{cases}\dfrac{1}{2n+1}\\[2ex] \dfrac{1}{a+2n+1}\end{cases}$, therefore

$$\frac{a+2n+1}{2n+1} \text{ must} = \frac{a+2+1}{3\times 5}, \text{ or}$$

$$15a+30n+15 = 2na+6n+a+3,$$

$$n(2a+6-30) = 14a+12,$$

$$n = \frac{14a+12}{2a-24}, \text{ which if } a = 60, \text{ is}$$

$$= \frac{14a}{2a\times\frac{4}{5}} = \frac{7\times 5}{4} = 9,$$

and therefore it should require about 9 times quant. el., or about $5\frac{1}{4}$ times the charge to give the same shock out of water as at present.

599] Tu. Mar. First leather Torpedo *.

Out of water.

1 row jars el. to $1\frac{1}{2}$ by straw el. and commun. to rest, a shock just sensible in elbows.

1 + 2 + 3 + 4 + 5 + 6 + 7 : just sensible in hands.

D° + 1 row : stronger than N° 1.

In water.

2 rows : plain in hands,
1 row : just sensible,
3 rows : rather stronger than 2 D° out of water.

600] Tu. Apr. 4 [1775] 2nd leather Torpedo.

Out of water,

5 + 6 + 7 : very slight in fingers.
2 rows : only in hands, there seemed to be something wrong.
4 rows : brisk in elbows.
2 rows : briskish in elbows.

* [Art. 416.]

In water.

2 rows just sensible in hands.
3 rows stronger.
4 rows pretty strong D°.

1st leather Torpedo in water.

4 rows nearly same, but I believe not so strong as last.
2 rows very slight.

Out of water.

2 rows slight in elbows.
4 rows strong in elbows.

601] Sat. May 27 [1775] with 2nd leather Torpedo under water, 3 rows charged to $1\frac{1}{2}$ on electrom.*

Shock with one hand to one person seemed stronger, to another weaker than with both.

Communication being made with metals instead of the hands, no shock was felt, but when all the rows were charged to 3, Mr Ronayne felt a small shock.

With wooden Torpedo, 1 row to $1\frac{1}{2}$; shock passed across 27 links of heavy chain with light. It also passed across 4 links of small chain with light, but not across 6.

Without Torpedo, $5 + 1 + 2$ to $1\frac{1}{2}$; shock passed with light through electrom., no candle in room; also with torp. charged as in trial.

On a former night in trying wooden Torp., charged I believe much the same as this time, no light was perceived, though Mr Hunter felt a shock, but very weak.

One candle in room, hid as well as possible behind screen.

With Gymnotus, all rows charged to $3\frac{1}{2}$.
Doubtful.

Dr Priest[ley and Mr Lane touching with 1 hand at same time, Dr Priestly felt shock extend to elbow.

A former night, 3 rows charged to $1\frac{1}{2}$, Mr N. thought the shock extended to elbow; no one else thought so.

Sat. May 27 [1775].

602] Old Torp. out of water 2B + A (8·1)†, tried with metals, weak shock.

New torp. B + A (4·6) as strong as former.

The old Torp. tried with one hand holding metal against bottom side, in other hand holding bright link.

* [Art. 419. The (artificial) Gymnotus is not elsewhere mentioned.]
† [The numbers in brackets are the charge communicated to the battery or the row. See Art. 583.]

3A (3) no shock, with B (3·6) a very slight shock when torp. was just wetted, none else.

With long link, not bright, 3A, (3), sometimes felt it, not always; with 2A never.

With wire of same size bright without link, seemed not to feel it so well.

With small link not bright no shock with B + A (4·6), but there was with B + 2A (5·6); with bright wire without link felt shock with B + A (4·6), but not with B. (3·6).

With dirty link 2B + A (8·1), sometimes a small shock, not always; with 2B + 2A (9·1), certain.

603] Tried with Lane's electrom.; dirt unaltered.

Rows of batt. to which el. is comm.	Jars el.	Equiv.*	
7	R + 2B	26·7	shock.
	R + B	23·7	none.
1	B + 2A	4·56	small shock.
	B + A	3·56	none.
7	R + B	23·7	shock.
	R	20·6	none.
7	R + 2B	26·7	shock.
	R + B	23·7	none.
1	B + 3A	5·6	shock.
	B + 2A	4·6	none.
7	R + 2B	26·7	shock.
	R + B	23·7	none.
1	B + 3A	5·6	shock.
	B + 2A	4·6	none.

Tu. May 30 [1775].

604] It was tried whether distance on Lane's electrom. at which jars discharged was the same at the same separation of straw & pith ball electrom. whether number of jars was great or small†.

This was tried first by laying small knob'd Lane on wire while jars were charging, and afterwards by charging jars, without Lane lying on wire, to a little greater and little less degree by electrom. than what it was before found that they discharged at; then touching them with Lane, I could not perceive that the number of jars made any difference.

It was tried by comparing 1 & 4 jars with straw el. at 2 and by comparing 1 and 7 rows of battery with pith balls at 1.

* [See Art. 583.] † [Art. 402.]

It was also tried whether the number of jars electrified affected the separation of straw el.: by connecting 4 jars to the wire & then withdrawing 2 of them. It was not found to be at all affected.

Tu. May 30 [1775].

605] *Charge required to force el. through 4 links of small chain, and also through 2 loops of machine*, 5 links of chain in each loop.*

Rows of Batt.	Jars el.	Equiv.		
7	R	20·6	passed through 4 links	old torp.
	3B + C + 4A	17·1	did not	
	R	20·6	passed through 2 loops	
	3B + C + 4A	17·1	did not	
	3B + C + 4A	17·1	passed through 4 links	new torp.
	2B + 4A	11·1	did not	
	R	20·6	passed through 2 loops	
	3B + C + 4A	17·1	did once, failed once	
	R	20·6	passed through 4 links	2nd leather torp.
	3B + C + 4A	17·1	did not	
	R + 3B + C	33	passed through 2 loops	
	R	20·6	did once, failed once	
	3B + C + 4A	17·1	did not	

leather torp.	R	= 20·6	gave same shock.
old	2B + A	= 8·1	
new	B + 3A	= 6·6	
without torp.	A + D	= 1·5	

Tried with new Torpedo.

Rows Batt.	Jars el.	Equiv.	
7	B + A	4·6	gave same shock.
1	3A + D	3	
7	2B + 2A	9·1	gave same shock.
	B + 3A	5·4	

Trial of charge required to pass through 4 links of chain.

1	3A	2·6	sometimes passed, sometimes not.
7	3B + C + 4A	17·2	D°.
1	4A	3·3	passed.
7	R	20·6	passed.
7	3B + C + 4A	17·2	did not.
1	3A	2·6	

* [Art. 433.]

Tried with 2 loops of machine*.

[Rows.	Jars el.	Equiv.]	
7	3B + C + 4A	17·2	did not.
1	4A	3·3	
1	B + A	3·6	did.
7	R	20·6	
1	B + 2A	4·6	did not.
1	B + 3A	5·6	did.
7	R	20·6	did.
1	B + A	3·5	did not.
1	B + 2A	4·5	did.
1	B + A	3·5	did not.
1	B + 2A	4·5	did.

3B + C + 2A = (9·2) commun. to 1 row, 2^co passed through 2 loops, once missed, once did not pass through 3, never through 5.

R communicated to 7 rows = 20·6 was tried 3 times without ever passing through 3 rows.

Wed. May 31 [1775].

606] 1 jar was elect. and commun. to 1 row of battery, and shock taken without torpedo. There seemed a little difference in the strength of the shock according to which row it was communicated to, but hardly more than was observed at different times from the same row.

Result of exp. May 30.

607] By mean, quant. el. req. to give same shock with 7 rows is to that with 1 :: 18·3 : 11·5 :: 1·6 : 1†.

Charge req. to force through $\begin{cases}\text{4 links}\\\text{2 loops}\end{cases}$ with 7 rows is to that with 1 in rat. between $\begin{cases}6{\cdot}6 \text{ to } 1\\5{\cdot}7 \text{ to } 1\end{cases}$ & $\begin{cases}6.2 \text{ to } 1\\3{\cdot}7 \text{ to } 1\end{cases}$, by mean as $\begin{cases}6{\cdot}4 \text{ to } 1\\4{\cdot}8 \text{ to } 1\end{cases}$.

Tu. June 6 [1775].

608] The 2nd leather Torp. was tried in sand‡ wetted with salt water. The Torp. lay flat on sand and was covered by it all but pos. elect. parts & middle of back. With 3 rows charged to $1\frac{1}{2}$, felt a shock whether I laid bare hands on torp. & on sand 16 inc. dist. from nearest part of D°. or whether I touched torp. with metals. In latter case shock seemed much the same as shock 10 inch plate crown glass § received through Lane's el. at $\frac{9\frac{1}{2}}{1600}$ inc.

If I laid pieces of sole leather|| which had been soaked in salt water for a week and then pressed between paper with $\frac{1}{2}$ hundred weight for

* [Arts. 433, 605.] † [Arts. 406, 573, 610, and Note 31.]
‡ [Art. 422.] § [Arts. 411, 430.] || [Art. 423.]

½ day to drain out moisture on torp. and on sand, and received shock with metal that way, shock was about equal to that of 10 inc. plate with Lane at $\frac{6\frac{1}{2}}{1600}$.

The torp. taken out of sand and tried with metals in usual way gave shock about equal to D° plate, Lane at 18½.

Being tried in same manner with 1 row, shock was weaker than in sand through leathers, & with 2 rows stronger than without leathers.

The spe. gra. bottle with water which came from sand weighed 8 . 4 . 11. Th. at 69, so that the water with which it was moistened appears to be of right strength.

609] Bits of beech, wainscot & deal* about ¾ inch square were soaked in salt water for 3 or 4 days, then taken out and wiped and exposed to the air in dry room for about 6 hours.

The shock of the Torp. was received touching pos. el. part with metal and neg. with one of these bits, the end which touched the torp. and that part which I held in hand being bound round with tin foil.

With 6 rows elect. to 1½, I felt slight shock through wainscot: dist. tinfoils 2 inc.

With D° charge through deal, tinfoils at 1 inc., none.

With 3 rows to 1½; received shock through beech, tinfoils at 4½ inc. dist., about as strong as with 1½ rows when touched with metals on both sides.

With D° charge through 4½ inches of dry deal dipt in salt water and tried the instant it was taken out, none.

Taking hold of tail in one hand & touching pos. side with metal, brisk shock. When touching neg. side with metal much slighter, the exper. tried with each pos. and each neg. part.

Mon. June 12 [1775].

610] Jar 1 elect. to 2½ by pith el. seemed to give shock of same strength as B + 2A comm. to whole battery; it was weaker than 2B and stronger than B commun. to D°, but as there is a good deal of difference between the sensations of the 2, it is not easy comparing them.

According to this exp. the numb. jars which el. should be divided amongst in order to produce given shock should be as the $2\frac{7}{8}$ power of quant. el., and therefore el. 2 jars should be comm. to $5\frac{1}{3}$ more in order to produce same shock as 1 jar †.

* [Art. 588.] † [Arts. 406, 573, and Note 31.]

Mon. June 18 [1776 *].

1R + 3B + C + 2A comm. to 7 rows = ($34\frac{1}{2}$), & el. to a given mark on pith el. gives shock equal or rather greater than 1 row el. to same degree and not commun. to rest.

1R + 3B + C + 2A elect. to $1\frac{1}{4}$ on straw el. and comm. to rest always passed through 1 loop of machine. The same elect. to 1 sometimes passed, sometimes failed.

1 row charged to 1 and not commun. to any more passed 3 times through 5 loops without once failing.

1R + 3B + C + 2A el. to 1 and comm. to rest would never pass through 2 loops.

611] 2nd Leather torpedo tried under water with metals with glass tubes on them, all rows charged to 4 gave briskish shock, which was much greater than shock out of water with 1 row to $1\frac{1}{2}$, but rather less than with 2 rows to D°.

The shock received in same manner with 1 row not communicated to rest was less when el. to $1\frac{1}{2}$, and about equal when el. to $2\frac{1}{2}$.

With 7 rows el. to $1\frac{1}{2}$ shock of D° Torp. when received through the salted lime tree wood gave slight shock about equal to 3A passed through same wood without torpedo.

Charge of 7 rows el. to 4 is to that of $1\frac{2}{3}$ row el. to $1\frac{1}{2}$ as $\frac{7 \times 2{\cdot}8}{1\frac{2}{3}} : 1 :: 12$ to 1.

612] Tu. July 4 [1775]. 2nd leather torp., the wire belonging to convex side fastened to outside of battery and inside of battery touched by wire of flat side.

3 rows of battery charged to $1\frac{1}{2}$ and comm. to remainder. Under water no sensible diff. whether I touched convex or flat side with one hand.

Out of water, touching tail with one hand and one side of one elect. organ with metal, a much greater shock if I touched convex side than flat side. The event was the same if it was elect. by neg. elect.

Touching convex side of both organs with one hand only, standing on electrical stool, a shock in that hand, but I think scarcely so strong as under water.

Touching flat side in same way, much the same.

Laying 1 finger on convex surface of one organ & another finger of same hand on the middle of the convex surface, a very slight shock.

* [Probably June 19, 1775.]

Laying one finger on convex surface of one organ & the other on the nearest edge of the torpedo, a considerably greater shock, but not strong.

Laying one finger on convex and another on flat side of same organ, a considerably greater shock, but do not know how to compare it in point of strength with that taken the usual way.

Tried without any torpedo *.

613] 3B being comm. to 7 rows and passed through 1 loop of 26 links of small chain. If the chain was not stretched by any additional weight, the shock did not pass. If the middle link was stretched by a weight of 7 pwt. it passed, & the light was visible in a few links. If it was stretched by a weight of $13\frac{1}{2}$ pwt. no light was seen. There was no remarkable difference in the strength of the shock, whether it was received through chain tended by $13\frac{1}{2}$ pw. or without chain.

The chain was fastened to the same machine that was used in a former experiment, it was 7·9 inc. long and the distance of the supports 5·1.

The room was quite dark, it being tried at night without any candle in the room.

3 rows of battery were elect. till pith el. sep. to 1, its el. was then comm. to the rest of the battery, & I received the shock of 1 row, the elect. having its choice whether it would pass through my body or through some salt water. I then elect. 1 row of battery till pith el. sep. to same degree, and commun. its elect. to rest of battery and received the shock of 5 rows of it in same manner. The shock seemed to be nearly of same strength, perhaps rather less.

Therefore shock of 5 rows elect. to a given degree seems about equal or perhaps rather less than that of 1 row el. to 3 times that degree.

614] The mean thickness of the section of the elect. organ in the section given in Mr H. † paper, in which the breadth is 10·3 inches, that is, the same as my torpedo's, is 1·3 inc.; the area of one organ is $2{\cdot}5 \times 5\frac{1}{4} \times \frac{12\frac{1}{4}}{15\frac{1}{2}} = 9\frac{1}{2}$ sq. inc., as found by cutting out a piece of paper of that size and weighing it.

And according to Mr H. there are about 150 partitions in 1 inch, therefore comp. charge both organs reckoned in old way is

$$19 \times 1{\cdot}3 \times 150 \times \tfrac{4}{3} \times 150 = 748000,$$

and the real charge is 1122000 inches of el. supposing the par-

* [Art. 437.]

† [Anatomical observations on the torpedo. By John Hunter, F.R.S., [*Phil. Trans.* 1773]. Art. 436.]

titions to consist of plates of white glass $\frac{1}{150}$ inc. thick, which is about $2\frac{1}{2}$ times as much as my battery, that being = 451000 inc. el.

615] Tried with the 2nd leather torpedo, new covered, in large trough full of water, the torpedo laid flat as in figure, the electrical organ being (as supposed) 3 inches under water.

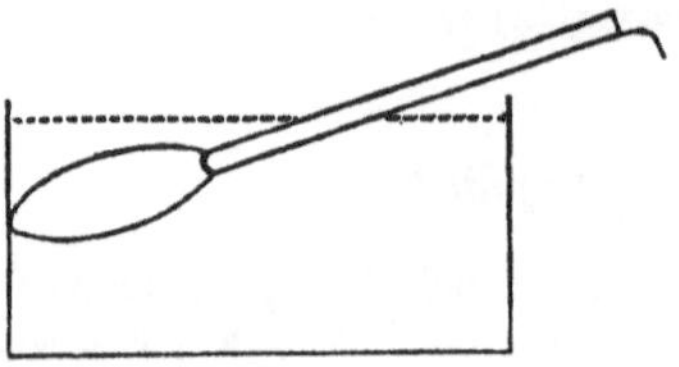

If torpedo was tried out of water with 1 row to $1\frac{1}{2}$ comm. to 7 and touched with hands in usual manner, the shock was just felt in hands, and if touched with metals, was just sensible in elbow.

Tried under water in above-mentioned manner with 7 rows el. to 4, the upper surface being touched with the pestle of a mortar held in one hand, the other hand dipt into water as far as wrist, a shock in the wrist of the hand in the water I believe full as strong as the former.

The place where the hand was dipt into water was about 11 inc. from the front of the fish, and conseq. about 14 from elect. organ.

Tried in the same manner as before, except that the fish lay in an open wicker basket*, just big enough to receive it, and which had been soaked for some days in salt water. The shock seemed much the same.

Holding hand in water in same manner as before without touching torpedo—no sensation.

With three rows to $1\frac{1}{2}$ out of water, the shock was stronger if I touched convex side with one hand laid flat on elect. organs than if I touched flat side in same manner, but the difference was not great.

Charge of 7 rows el. to 4 is to that of 1 row el. to $1\frac{1}{2}$ as 19·6 to 1.

The water appeared by its spe. gra. to contain $\frac{1}{31}$ of salt.

* [Art. 421.]

RESISTANCE TO ELECTRICITY.

616] *Comparison of conducting power of salt and fresh water in the latter end of March and beginning of April,* 1776.

Tried with Nairne's last battery, 6 jars being chose, each of which held very nearly the same quantity of electricity; the wires run into the bent ends of the tubes being made to communicate with the outside of the battery, and the wires run into the straight ends being fastened to separate pieces of tinfoil.

The six jars were all charged by the same conductor: the communication with that and each other was then taken away, and the jars discharged through the tubes, one after the other, by touching the abovementioned bit of tinfoil by metal held in one hand, and the wire of the jar by metal held in the other hand, the shock being received alternately through each tube.

617] Exp. 1.

Distance of wires In tube 15	14	Sat. sol. S.S.* in tube 14, salt in 69 of water in tube 15.	
6·5 inc.	40·7	very sensibly less	in short tube than in long one.
5·8		sensibly less	
3·5		sensibly greater	
4·2		scarce sensibly	
5·3		just sensibly less	

Straw electrom. = 4. Th. = 57. [Resistance = 390000 Ohms.†]

Resistance of 4·7 inches in tube 15 supposed equal to 40·7 in 14. Therefore sat. sol. conducts 8·6 times better than salt in 69 of water.

Exp. 2. *The same solution tried in tubes* 22 *and* 23.

Tube 23 22 electrom. at $1\frac{1}{4}$. Th. = 58. [R. = 118000].†

Tube 23	22	
3·3	41	sensibly greater.
5·5		less in same proportion.

4·4 inches in tube 23 = 41 in tube 22.

Therefore sat. sol. conducts 8·94 times better than salt in 69 of water.

* [Saturated solution of sea salt.]

† [The resistance of the saturated solution in Ohms. calculated from the measurements in Art. 635 by Kohlrausch's data, is given for each tube within brackets to indicate the absolute value of the resistances compared.]

Exp. 3. *A new saturated solution and solution in* 69 *of water made and tried in tubes* 15 *and* 14. [R. = 390000.]

3·5	40·7	just sensibly greater.	Electrom. = 3.
3·1		very plain.	Th. = 57.
5·5		sensibly less.	
5·0		just sensibly less.	

Therefore new saturated solution conducts 9·61 times better than new solution in 69.

Exp. 4. *Salt in* 69 *of water compared with salt in* 999 *of water in tubes* 22 *and* 23.

[R. = 1230000.]
Electrometer = $1\frac{3}{4}$.
Th. = 57.

23	22	
3·1	41·1	sensibly greater.
3·5		scarce sens.
4·3		scarce sensibly less.
4·9		just sensib.
5·3		very sensib.

Resist. 4·1 in tube 23 supposed equal to 41·1 in 22.

Therefore salt in 69 conducts 9·57 times better than salt in 999.

Exp. 5. *Salt in* 999 *compared with distilled water in tubes* 12 *and* 20.

[R. = 462000.]
Electrom. = 3.
Th. = 58.

20	12	
·78	43·5	sensib. greater.
1·2		scarce sensib. less.
1·4		diff. more sensib. than in 1st trial.
1·05		supposed right.

Therefore salt in 999 conducts 36·3 times better than distilled water.

The distilled water changed for rain water.

1·9		sensib. greater. Electrom. = 3. Th. = 58°.
3·3		less, rather more sensib. than former.
2·55		supposed right.

Therefore rain water conducts 2·4 times better than distilled water, or 15·2 times worse than salt in 999.

The rain water changed for distilled water with $\frac{1}{20000}$ of salt in it.

5·3		sensib. less.
2·7		about as much greater.

Therefore salt in 20000 of distilled water conducts [3·67] times better than distilled water, or 9·92 worse than salt in 999.

[6] *Saturated solution and salt in* 69 *of water* (*the new solutions*) *compared in the same manner, only using the jars* 1 *and* 2 *instead of the battery; with the tubes* 5 *and* 17.

17	5	[R. = 25100.]
2·6	41·1	sensib. greater.
5·5		sensib. less.

Therefore saturated solution conducts 10·05 times better than salt in 69 of water.

618] The electricity of the 6 jars was found to be as much diminished by being communicated to 3 rows of the battery as that of 1 row is by being communicated to 4 rows, therefore quantity of electricity in the 6 jars is to that in one row as 3 to 4.

Exp. 7. *Saturated solution and salt in* 69 (*the new solutions*) *tried in the same manner with battery;* 1 *row being electrified to* 2, *and its electricity communicated to remaining rows, and one row used at a time.*

Tried in tube 5 and 17. [R. = 25300.]

17	5	
5·4	41·4	plainly less.
2·6		about as much greater.
3·0		scarce sensibly greater.
5·0		just sensib. less.

3·95 supposed right. Therefore sat sol. conducts 10·31 times better than salt in 69.

Exp. 8. *Saturated solution compared with salt in* 29 *in tubes* 22 *&* 23 *with Nairne's jars.*

23	22	Electrometer 1. Th. 63. [R. = 65000.]
8·7	24·9	just sensibly less.
4·8		about as much greater.

The bore of that part of tube 23 which was used is supposed $\frac{3}{80}$ greater than that of whole tube together. Therefore sat. sol. conducts 3·51 times better than salt in 29 of water.

Exp. 9. *The solution in* 29 *diluted with* $1\frac{1}{3}$ *of water*, id est, *solution of salt in* 69, *compared with sat. solution in same tubes.*

23	22	Electrom. = 1. Th. = 63. [R. = 65000.]
2·0	24·9	greater.
3·8		about as much less.

Therefore sat. sol. conducts 7·79 times better than the diluted solution, and the diluted solution conducts 2·2 times worse than solution in 29.

Exp. 10. *Saturated solution compared with salt in* 69 *in same tubes.*

23	22	[R. = 65000.]
3·1	24·9	sensib. less. Electrom. = 1. Th. = 63.
1·9		as much greater.

Therefore sat. solut. conducts 9·02 better than the solution in 69.

619] *Examination whether salt in* 69 *conducts better when warm than when cold.*

Salt in 69 in tube 17 placed in water; solution in 29 in tube 23 out of water, the distance of wires in tube 17 being not measured, but remaining always the same.

Electrometer = $\frac{3}{4}$.

23		
8·1	sensib. less	heat of water = $58\frac{1}{2}$.
5	about as much greater	
4	plainly less	heat of water = 105.
2·6	as much greater	

Therefore salt in 69 conducts 1·97 times better in heat 105 than in that of $58\frac{1}{2}$*.

620] *Examination whether the proportion which conducting power of sat. sol. and salt in* 999 *bear to each other is altered by heat.*

Sat. sol. in tube 15, salt in 999 in tube 19, both in water; distance of wires in tube 15 not altered.

19	Electrom. = $1\frac{1}{4}$.	
3·25	sensib. less	heat of water 50.
2·15	sensib. greater	
2·25	just sensib. greater	electrom. = 1.
3·5	rather more sensib. less	heat of water 95.

Therefore the proportion seems very little altered by heat†.

621] Jan. 1, 1777. *Salt in* 2999 *of water compared with water distilled in preceding summer in tubes* 12 *and* 20.

20	12	Electrom. = $4\frac{1}{2}$.
1·5	43·5	rather greater. Column of 1·6 in tube 20.
1·95		plainly less. Supposed equal to 43·5 in tube 12.
1·4		greater.

Therefore allowing for different bores of tubes, salt in 2999 conducts 24 times better than distilled water.

* [By the experiments of Kohlrausch, this ratio would be 1·59. See Art. 691 and Note 33.]

† [This agrees with the results of Kohlrausch.]

Jan. 2 M. Same experiment repeated with the same water, which had been left in the tubes all night.

20	12	
1·4		plainly greater.
1·9		seemingly less.
2·1		plainly less.

Therefore salt in 2999 conducts 22 times better than distilled water.

The same experiment repeated, only the water in the tubes was changed for fresh by pouring out the old and putting in fresh by small funnel, without taking out the wires.

1·1		plainly less.
·35		plainly greater.
·7		plainly less.

Therefore salt in 2999 conducts 72 times better than distilled water.

The same experiment repeated, only the distilled water changed for that used in the preceding year.

·4		considerably greater.
1·1		plainly less. ·8 supposed equal.

Therefore salt in 2999 conducts 47 times better than distilled water.

Jan. 3. Experiment repeated with the same water left in.

·8		plainly greater.
1·2		plainly less.

Therefore salt in 2999 conducts 38 times better than the distilled water.

The distilled water changed for the new distilled water.

·28		plainly greater.
·6		plainly less.

Therefore salt in 2999 conducts 86 times better than distilled water*.

Salt in 2999 *compared with salt in* 150,000 *in same tubes.*

1·2	43·5	sensib. greater. Electrom. = 4½.
1·7		sensib. less.

Therefore salt in 2999 conducts 26 times better than salt in 150,000.

The experiment repeated with the same waters, only the wires in tube 12 brought nearer.

·3	12·5	sensib. greater. Electrom. = 1½.
·45	—	sensib. less.

* [See Art. 690.]

622] *Examination whether comparative resistance of salt in* 2999 *and salt in* 150,000 *was the same when tried in the above-mentioned manner, or when passed through* 2 *wires in glass of water, as in fig.**

Jan. 6. The tubes 12 and 20 filled with salt in about 105 of water: salt in 150,000 of water in glass. 2 jars electrified to $1\frac{3}{4}$ and communicated to the rest.

If the distance of wires in tube 12 was $\begin{matrix}33\cdot5\\18\cdot5\end{matrix}$ the shock was sensibly $\left\{\begin{matrix}\text{less}\\\text{greater}\end{matrix}\right.$ than that through the wires in glass.

The same tried as before, only with the jars electrified to 2 and the shock received with shock melter*.

If the distance of wires in tube was $\left\{\begin{matrix}31\cdot5\\20\cdot3\end{matrix}\right.$ shock was plainly $\left\{\begin{matrix}\text{less}\\\text{greater}\end{matrix}\right.$ than through wires in glass.

The glass filled with salt in 2999 and the shock compared with that through tube 20 with same solution of salt in 105.

The jar electrified to 2 and received with shock melter†.

If dist. wires in tube 20 was $\left\{\begin{matrix}\cdot5\\1\cdot\end{matrix}\right.$ shock was $\left\{\begin{matrix}\text{greater}\\\text{less}\end{matrix}\right.$ than through wires in glass.

N.B. Great irregularity was found in trying this last experiment, the cause of which I am unacquainted with.

Therefore salt in 2999 conducts 31·5 times better than salt in 150,000.

The same salt in 150,000 which was used in this experiment was saved and compared with salt in 2999 in the usual manner with tubes 12 and 20, electrometer at $4\frac{1}{2}$.

If distance of wires in tube 20 was $\left\{\begin{matrix}1\cdot2\\1\cdot85\end{matrix}\right.$ shock was plainly $\left\{\begin{matrix}\text{greater}\\\text{less}\end{matrix}\right.$ than through tube 12 with wires at 42·4 inches distant.

Therefore salt in 2999 conducts 24·6 times better than salt in 150,000.

The thermometer in all the foregoing experiments of this year supposed to be about 45°.

623] Exp. 11. *Saturated solution in tube* 14 *compared with salt in* 149 *of water in tube* 15.

Tube 15	14	Electrometer at $3\frac{1}{2}$. Th. = 45. [R. = 474000.]
1·6	41·8	sensib. greater.
2·6		sensib. less.

Sat. sol. conducts 20·5 times better than salt in 149.

* [See figure in facsimile of MS. on opposite page.]

† [The reading here is doubtful; see facsimile of MS. on opposite page. Cavendish says, Arts. 601, 602, 616, that he took the shock with *metals* in each hand, but the word here cannot be read "metal." The word occurs also in Arts. 585 and 637.]

Examination whether comparative resistance of salt in 2999 & salt in 150 000 was the same when tried in above mentiond manner or when passed through 2 wires in glass of water as in fig.

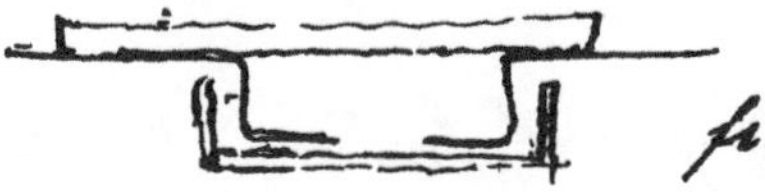

Jan. 6 The tubes 12 & 20 filled with salt in about 105 of water salt in 150 000 of water in glass 2 jars el. to $1\frac{3}{4}$ & commun. to rest

If dist. wires in tube 12 was $\begin{cases}33,5\\18,5\end{cases}$ the shock was sensib. {less / greater} than that through the wires in glass

— — — — — — —

The same tried as before only with the jars el. to 2 & the shock received with shock metter

If dist wires in tube was {31,5 / 20,3} shock was plainly {less / greater} than through wires in glass

---

The glass filled with salt in 2999 & the shock compared with that through tube 20 with same sol. salt in 105

The jars el. to 2 & received with shock metter

If dist. wires in tube 20 was {15 / 1,} shock was {greater / less} than through wires in glass

N.B. Great irregularity was found in trying the last experiment the cause of which I am unacquainted with

Therefore salt in 2999 conducts 70/3,5 times better than salt in 150 000

To face p. 326.

p. 14

~~Some sal Sylvii, Sal Amm., quadr. nitre & Glaub. salt & Calc. S. S. were dissolved in in mixed with such a proportion of water that the~~

2.3 of sal Amm. 3.2 of sal Sylvii 3.17 of quadr nitre 2.21. of glas calcined glaub. salt & 14.10 of calc. S.S A were dissolved in water the solut. of each being 3.10.12 that is such that the quant. acid in each should be equiv: to that in a solut. salt in 29 of water

Jan. 13
Sal Sylvii in tube 15 comp. with salt in 29 in tube 22 el. = 2 th. 55

tube 15	22	
5,	18,2	seemed gr. but doubt.
4,5		plainly greater
7		seemed less but doubt
7,7		seemed less
8,2		plainly less

$\frac{7,4+5}{2} = 6,2$ supposed right

7 :

Jan. 14 sal. amm. tried same way

7,2	.2	seemed less but doubt
7,9		plainly less
5,3		sensibly greater
5,8		seemed gr. but doubt

To face p. 327

Exp. 12. Jan. 8. *Sat. sol. in tube 22 compared with salt in 149 in tube 23.*

Tube 23	Tube 22	Electrom. $1\frac{1}{4}$. Th. = 42. [R. = 146000.]
1·6	41	plainly greater.
1·8		seeming. greater, but doubtful.
2·4		plainly less.
2·2		rather less.
2·0		not sensib. different.
1·8		seemed greater.

$\frac{1\cdot8+2\cdot1}{2}$ in tube 23 supposed = 41 in tube 22.

Therefore sat. sol. conducts 19·6 times better than salt in 149.

Exp. 13. *Salt in 149 in tube 5 compared with salt in 2999 in tube 17.*

Tube 17	5	Electrometer = 3. Th. = 44. [R. = 652000.]
1·8	40·5	sensib. greater.
2·8		sensib. less.

Salt in 149 conducts 17·3 times better than salt in 2999.

By new measure of tubes.

Exp. 14. *Same solutions in tubes 18 and 19.*

Tube 19	18	Electrometer = $1\frac{3}{4}$. [R. = 308000.]
2·2	42·8	sensib. greater.
2·4		not sensib.
3·0		sensib. less.
2·8		seemed less, but doubtful.

$\frac{2\cdot2+2\cdot9}{2}$ in tube 19 supposed equal to 42·8 in tube 18.

Salt in 149 conducts 16·7 times better than salt in 2999.

Exp. 15. Jan. 9. *Sat. sol. in tube 22 compared with salt in 29 in tube 23.*

Tube 23	22	Electrometer = $1\frac{1}{4}$. Th. = 42. [R. = 128000.]
9·7	35·8	sensib. less.
9·4		seemed less, but doubtful.
6·6		seemed greater.
6·9		not sensib. gr.
6·3		sensib. greater.

6·5 + 9·7 supp. = 35·8 in tube 23.

Sat. sol. conducts 4·38 times better than salt in 29.

624] *Comparison of water purged of air by boiling and plain water.*

Jan. 12. Salt in 2999 in tube 12. Salt in 150,000 in tube 20.

Tube 20	12	Electrometer = $4\frac{1}{2}$. Th. = 50.
1·2		plainly greater.
1·4		seemed greater, but doubtful.
1·8		seemed less, but doubtful.
2·0		plainly less.

The water was then boiled over lamp in the same vial in which it had been kept some time, and then cooled in water and compared in the same manner.

2·0	sensib. less.
1·8	seemed less, but doubtful.
1·2	seemed greater, but doubtful.
1·0	plainly greater.

Therefore if anything water conducted better before boiling than after, but the difference might very likely proceed from the error of the experiment.

In order to see whether the water had absorbed much air by being exposed to the air in the trial, some of the boiled water was exposed to the air as much as that which was tried in the tube was supposed to have been, and boiled over again in a vial. It did not begin to discharge air till it was heated to 190°, and then discharged but little. Some more of the boiled water which had not been poured out of the vial seemed to discharge as much air. But some distilled water which had not been boiled began to discharge air almost as soon as heated, and discharged a great deal before it began to boil*.

625] *Comparison of water impregnated with fixed air and plain water.*

Some distilled water was impregnated with fixed air produced by oil of vitriol and marble, and compared with salt in 2999 in same tubes and manner as in former experiments.

		Electrometer = $4\frac{1}{2}$. Th. = 55.
1·6		seemed greater, but doubtful.
1·4		plainly greater.
2·4		not sensibly less.
2·6		sensibly less, scarce doubtful.
2·8		plainly less.

* [See Art. 692.]

The same water deprived of its fixed air by boiling, and tried as before.

Tube 20	12	
1·2		plainly less.
1·0		sensib. less.
·9		seemed less, but doubtful.
·6		plainly greater.
·7		not sensib. greater*.

626] 2·3 of Sal. Amm., 3·2 of Sal. Sylvii, 3·17 of Quadr. nitre, 2·21 of calcined Glauber's salt and 14·10 of calc. S. S. A.† were dissolved in water, the solution of each being 3.10.12, that is, such that the quant. acid in each should be equiv. to that in a solut. salt in 29 of water.

Jan. 13. *Sal. Sylvii in tube* 15 *compared with salt in* 29 *in tube* 22.

Tube 15	22	Electrometer = 2. Th. = 55. [R. = 253000.]
5	18·2	seemed greater, but doubtful.
4·5		plainly greater.
7		seemed less, but doubtful.
7·7		seemed less.
8·2		plainly less.

$\frac{7\cdot4 + 5}{2} = 6\cdot2$ supposed right.

Jan. 14. *Sal. amm. tried same way.*

7·2		seemed less, but doubtful.
7·9		plainly less.
5·3		sensibly greater.
5·8		seemed greater, but doubtful.

Calc. S S. tried same way.

4·0		plainly greater.
4·5		scarce sensib. gr.
6·0		plainly less.
5·5		sensib. less.

Salt in 29.

4·5		plainly greater.
5·0		scarce sensib. greater.
7·0		plainly less.
6·5		scarce sensib. less.

* [See Art. 693.]

† [See facsimile on the opposite page. The results are given in Art. 694. See Note 34 and Preface.]

Glauber's salt.

Tube 15	22	
4·2		plainly less.
3·8		no sens. diff.
4·2		scarce sens. less.
4·4		do.
4·6		just sens. less.
3·4		sensibly greater.
3·6		scarce sensib. less.

Quadrang. nitre.

4·0		plainly greater.
4·3		seemed greater, but doubtful.
6·0		seemed less, rather doubtful.
6·2		plainly less.

627] 2·0 of oil of vitriol, 2, 5·10 of spirit of salt 2, and 5·19 of f. alk. D, were diluted with water, the solution being 3 . 10 . 12. Consequently the quantity of acid in 2 first were equivalent to that in salt in 59 of water, and the alk. in last was equivalent to that in salt in 29; compared in the same manner as the former.

Jan. 15. *F. alk.* Th. = 55°.

4·5		scarce sensib. greater.
4·0		seemed greater, but doubtful.
3·7		plainly greater.
5·7		plainly less.
5·4		seemed less, but doubtful.

Diluted oil of vitriol.

4·0		sensib. greater.
4·3		not sensib.
5·0		sensib. less.
4·7		not sensib.

Diluted spirit of salt.

11·8		seemed less, rather doubtful.
8·0		seemed greater, rather doubtful.

Another diluted spirit of salt was made of same strength as the former. Being tried with wires at 9·9 inch. distance no sensible difference was perceived, which agrees with former.

Another diluted oil of vitriol was made and tried, Jan. 16.

5·2		sensib. greater.
7·7		sensib. less.

EXPERIMENTS IN JANUARY, 1781*.

628] Some basket salt† was dried before fire, and a saturated solution made with it which contained $\frac{1}{3{\cdot}78}$ of salt‡, and also other solutions of different strengths, all being made with distilled water.

Sat. sol. in tube 14, *salt in* 69 *in tube* 15. [R. = 399000.]

15	14	Electrometer $3\frac{1}{4}$. Th. = 53.	
3·6	39·1	sensib. greater.	
5·8		plainly less.	4·5 supposed right.
5·4		sensib. less.	

Sat. sol. conducts 8·63 times better than salt in 69.

Same solutions in same tubes.

5·4		plainly less.	
5·1		seemed rather less.	4·3 supposed right.
3·6		scarce sensib. greater.	
3·3		sensib. greater.	

Sat. sol. conducts 9·03 times better than salt in 69.

Sat. sol. in tube 14, *salt in* 29 *in tube* 15. El. = $3\frac{1}{4}$.

7·3		sensib. greater.	9·1 supposed right.
11·0		sensib. less.	

Sat. sol. conducts 4·1 times better than salt in 29.

The same solutions in same tubes.

10·7		seemed rather less.
7·5		sensib. greater.
7·8		seemed rather greater.
9·7		supposed right.

Sat. sol. conducts 3·85 times better than salt in 29.

The same solutions in same tubes. El. = $1\frac{1}{2}$. [R. = 136000.]

2·6	13·3	seemed rather greater.	
2·6		D° scarce sensib.	
3·7		plainly less.	2·95 supposed right.
3·5		sensib. less.	
3·3		seemed rather less.	

Sat. sol. conducts 3·95 times better than salt in 29.

* [The results of these experiments are collected in Art. 695. See Note 33.]

† ["Salt made up in form of sugar loaves, in small wicker baskets, which is thence called *loaf salt* or *basket salt*." *Rees' Cyclopædia*.]

‡ [26·45 per cent. Saturated solution at 18°C is 26·4 per cent. by Kohlrausch.]

Sat. sol. in tube 14. *Salt in* 11 *in tube* 15. El. $1\frac{3}{8}$.

15	14	[R. = 136000.]	
4·7	13·3	sensib. greater.	
8		plainly less.	5·95 supposed true.
7·6		do.	
7·2		sensib. less.	

Sat. sol. conducts 1·92 times better than salt in 11.

The same again.

7		seemed rather less.	
4·9		sensib. greater.	6·05 supposed right.
5·1		seemed rather greater.	

Sat. sol. conducts 1·88 times better than salt in 11.

The same again.

5·1		seemed rather greater, but doubtful.	
4·9		sensib. greater.	
7		sensib. less.	5·95 supposed true.

Sat. sol. conducts 1·92 times better than salt in 11.

629] *Salt in* 142 *put in tubes* 5 *and* 15 *in order to find what power of velocity the resistance is proportional to.*

15	5	Electrometer = 3. [R. = 579000.]	
3·3	41·9	plainly less.	
3·1		do.	
2·9		scarce sensib. less.	2·75 supposed right.
2		plainly greater.	
2·2		D°.	
2·4		sensib. greater.	
2·6		seemed rather greater.	
3·1		sensib. less.	

Therefore log. vel. in 15 by do. in 5 = 1·2122,

log. length in 5 by do. in 15 = 1·1829.

Therefore resistance is as $\frac{1 \cdot 1829}{1 \cdot 2122} = \cdot 976$ power of velocity.

The same repeated.

15	5		
3·2		sensib. less.	
3		hardly sensib.	
2·6		seemed sensib. greater.	
2·5		sensib. greater.	2·85 supposed true.
2·7		hardly sensib.	

Log. vel. in 15 by D° in 5 = 1·2122,

Log. length in 5 by D° in 15 = 1·2122.

Therefore resistance is directly as velocity*.

630] *Salt in* 69 *in tube* 22. *Salt in* 999 *in* 23.

23	22	Electrometer = $3\frac{1}{2}$. [R. = 1335000.]	
4·9	41·5	plainly less.	
4·6		D°.	
4·3		scarce sensib. less.	
3·6		sensib. greater.	
3·7		scarce sensib. greater.	4 supposed true.
3·4		seemed greater.	
3·2		plainly greater.	
3·4		sensib. greater.	

Salt in 69 conducts 9·91 times better than salt in 999.

The same repeated.

3·4		seemed greater.	
3·6		scarce sensib. greater.	
3·2		sensib. greater.	3·85 supposed true.
4·3		scarce sensib. less.	
4·5		sensib. less.	

Salt in 69 conducts 10·3 times better than salt in 999.

The same liquors in tubes 5 *and* 17.

17	5	Electrometer = $1\frac{1}{2}$. [R. = 288000.]	
4·2	42·2	plainly less.	
4		sensib. less.	
3·8		hardly sensib.	
3·1		sensib. greater.	3·55 supposed right.
3·3		hardly sensib.	

Salt in 69 conducts 11·31 times better than salt in 999.

* [This is the first experimental proof of what is now known as Ohm's Law.]

The same repeated.

17	5	
3·3		hardly sensib. greater.
3·1		sensib. greater.
4		not sensib. less.
4·2		seemed rather less.
4·3		D°.
4·4		sensib. less.

Salt in 69 conducts 10·75 times better than salt in 999.

Salt in 999 *in tube* 12; *distilled water in tube* 20. El. = $2\frac{1}{2}$.
[R. = 494000.]

20	12	
·3	43·3	seemed rather less.

2 or 3 hours after it seemed rather greater at ·5.

Next morning was plainly greater at ·7.

The water being changed for fresh, seemed rather less at ·3

The distilled water changed for salt in 20,000.

2·1		sensib. less.	
2		not sensib. less.	
1·7		sensib. greater.	
1·8		seemed rather greater.	1·95 supposed right.
1·9		not sensib. greater.	
2·1		sensib. less.	

Salt in 999 conducts 20 times better than salt in 20,000.

Salt in 20,000 conducts about 7 times better than distilled water; therefore if distilled water contains $\frac{1}{120000}$ of salt their conducting powers will be as the quantity of salt in them.

The same repeated. Electrometer = 2.

2·1		not sensib. less.	
2·2		seemed rather less.	
2·3		D°.	
2·4		D°.	
2·5		plainly less.	
2·3		seemed rather less.	2 supposed right.
1·7		seemed rather greater.	
1·6		plainly greater.	

Therefore salt in 999 conducts 19·5 times better than salt in 20,000.

The waters changed for fresh. El. = 2½.

20	12		
2		scarce sensib. different.	
1·9		D°.	
1·8		sensib. greater.	
2·1		not sensib. less.	2·05 supposed right.
2·3		sensib. less.	

Salt in 999 conducts 19 times better than salt in 20,000.

The same repeated. El. = 2.

2·3	42·9	sensib. less.	
2·1		scarce sensib. less.	
1·8		seemed rather greater.	1·95 supposed right.
1·7		D°.	
1·6		sensib. greater.	

Salt in 999 conducts 19·8 times better than salt in 20,000.

Salt in 69 *in tube* 22. *Salt in* 142 *in tube* 23.

23	22	El. = 2. [R. = 407000.]	
7·3	12·65	semed rather less.	
7·5		sensibly less.	
6·3		not sensib. greater.	6·75 supposed right.
6		sensib. greater.	

Salt in 69 conducts 1·74 times better than salt in 142.

The same repeated.

6		seemed rather greater.	
5·8		doubtful.	
5·6		plainly greater.	
5·8		not sensib. greater.	6·45 supposed right.
7·3		plainly less.	
7		seemed rather less.	
6·8		scarce sensibly less.	

Salt in 69 conducts 1·84 times better than salt in 142.

Salt in 999 *in tube* 12, *distilled water in tube* 20. El. = 2½.
[R. = 494000.]

·83		sensib. less.	N.B. The tubes had been measured
·7		sensib. greater.	between the last trial and this.

The distilled water was then changed for fresh.

·3		sensib. less.

Another bottle was filled with distilled water and tube 20 filled up again with that.

·3		seemed rather less.

631] *The tube* 20 *filled with the same distilled water mixed with* $\frac{1}{200}$ *of spirits of wine.*

·3		seemed not sensibly less.

Same mixture mixed with $\frac{1}{20}$ *more of spirits of wine,* id est, *sp^t wine in* 18 *of distilled water.*

·3		scarce sensib. less.

Equal bulks of sp^t wine and distilled water.

·3		seemed scarce sensib. less.

Pure spirits of wine.

·3		seemed of 2 rather greater.

Therefore there is not much difference between the resisting power of the above distilled water and spirits of wine and mixtures of the 2: but of the 2, spirits of wine resists least.

[Calibration of Tubes.]

632] Tube 14.

Dist. mid. col. from str. end.	Length col.	
3·6	4·08	col. = 2·45.
9·2	4·03	40·6 inc. = 28·5 gr.
13·6	3·9	4·26 = 3 gr.
18·1	3·7	12·2 inches of tube next to bend contain
25·5	3·54	$\frac{1}{8}$ part of ☿ of that in col. 42·5 long.
31·1	3·46	
36·8	3·09	
42	2·8	

Tube 15.

− 3·6	5·32	col. = 3·25 gr.
+ 1·7	5·16	
+ 7·9	4·95	
+ 9·9	4·47	
+ 11·6	4·67	

After it was measured 7 inches were cut off from straight end, and the numbers in the first col. are the distances from the shortened end. A new bend was also made 12·8 from new end, and the part where the tube is equal to tube 14 is at 10·3 from D°.

633] Jan. 1781. The following tubes were measured over again by introducing a col. ☿ and measuring its length in 3 different places, the beginning of the 1st col. being at ½ inch from bend, and the beginning of the second at the end of the first. Another column was afterwards introduced whose length was pretty nearly equal to the sum of the 3 former, and weighed.

N° of tube.	1st col.	2nd col.	3rd col.		
14	12·2	14·3	16	41·8 inc.	= 29·3 gr.
15	3·47	3·71	3·86	11.55	= 7·7
22	13·03	13·65	14·6	41·8	= 100
23	3·6	3·3	3·09	10	= 24·5
5	12·65	14·	13·5	42·1	= 489
17	3·25	3·18	3·08	9·9	= 116

634] The two following tubes were measured by stopping up the end near bend and weighing them with different quantities of ☿ in them, and measuring the distance of the top of the column from straight end, whence it was found that in

	14·8	long beginning	1448
N° 12 a col.	28·6	at ½ inch	= 2777
	42·8	from bend	4033
	3·17	long beginning	270
N° 20 a col.	6·27	at ½ inch from	543
	9·97	bend weighed	881

635] In the following result the column whose length is given in the 2nd column is supposed to begin at ½ inch from the bend.

By the resistance of each is meant $\frac{1}{\text{grains ☿ in each inch}}$.

[The resistance of a column of mercury one inch long weighing one grain is ·13 Ohms, and the resistance of saturated solution of salt at t° Centigrade is to that of mercury as 10^8 is to $2015 + 45{\cdot}1\ (t - 18)$. Hence the resistance as given by Cavendish must be multiplied by $6907 + 82{\cdot}2\ (59 - T)$ to convert it into Ohms when the tube contains saturated solution at T° Fahrenheit]

RESULT.

No.	Length column.	Resist.	Log. do.	Resist. for each inch.	Log. do.
14	12·2	14·99	1·1758	1·229	·0894
	26·5	35·58	1·5512	1·343	·1280
	42·5	61·36	1·7879	1·444	·1595
15	3·47	4·907	·6908	1·414	·1505
	7·18	10·518	1·0219	1·465	·1658
	11·04	16·591	1·2199	1·503	·1769
22	13·03	5·157	·7124	·3958	9·5975
	26·68	10·817	1·0341	·4054	9·6079
	41·28	17·294	1·2379	·4190	9·6222
23	3·6	1·588	·2009	·4412	9·6446
	6·9	2·923	·4658	·4237	9·6270
	9·99	4·093	·6120	·4096	9·6124
5	12·65	1·030	·0126	·08138	8·9105
	26·65	2·291	·3600	·08596	8·9343
	40·15	3·463	·5395	·08626	8·9358
17	3·25	·2844	9·4539	·08750	8·9420
	6·43	·5566	9·7455	·08656	8·9373
	9·51	·8121	9·9096	·08539	8·9314
12	14·8	·1513	9·1798	·01022	8·0095
	28·6	·2946	9·4692	·01030	8·0128
	42·8	·4552	9·6582	·01064	8·0268
20	3·17	·03722	8·5708	·01174	8·0697
	6·27	·07243	8·8599	·01155	8·0626
	9·97	·11293	9·0528	·01133	8·0541

COMPARISON OF RESISTANCE OF COPPER WIRE WITH THAT OF SAT. SOL.

636] The wire was wound on reel on bars of glass about $\frac{3}{4}$ inch broad, the distance of one round of wire from the next on same bar being ·6.

The mean circumference of reel $= 46{\cdot}7 \times 2\sqrt{2}$*.

There were 8 rows of glass bars, and 28 rounds of wire on each row, and on one row there was $\frac{1}{4}$ round over. Therefore whole length of wire $= 93{\cdot}4 \times \sqrt{2} \times 8 \times 28 + \frac{1}{4} = 29623$ inches.

* [The reel was probably square, with glass bars at the corners, the length of the diagonal being 46·7 inches.]

This weighed 2967 grains, consequently there are 9·984 inches to 1 grain.

N.B. There were many knots in the wire.

637] The resistance of this wire was attempted to be compared with that of sat. sol. in tube 17 by shock melter* as in former experiments, but without success. It was therefore compared by the sound of the explosion by discharging the jars by a wire without its passing through my body; but in this there was considerable difficulty, as the light of the spark passed through the wire was very different from that passed through the water, the first being reddish and the latter white. The sound also was of a different kind, the latter being sharper.

Distance of wires in tube 17.

Distance		
		El. = 4.
·68	not sensib. diff.	
·6	scarce sensib. stronger.	
·55	doubtful.	
·5	seemed rather greater.	
·45	sensib. greater.	
·9	seemed sensib. less.	
1	do.	
1·2	sensib. less.	
·9	scarce sensib. less.	El. = 5½.
1·	not sensib. less.	
1·2	seemed rather less.	
·6	plainly greater.	
·7	not sensib. greater.	
·4	not sensib. gr.	El. = 3.
·3	seemed rather gr.	
·2	plainly gr.	
·8	seemed rather less.	
1·	D°.	
1·2	plainly less.	
1·2	plainly less.	El. = 3.
1	D°.	
·8	scarce sensib. less.	
·4	sensib. gr.	
·5	scarce sensib.	
·5	sensib. gr.	El. = 4.
·6	seemed rather gr.	
·7	not sensib. diff.	
1	seemed rather less.	

* [See Arts. 585 and 622.]

638. Two Leyden vials were made of barometer tubes filled with ☿ and coated on outside with tinfoil. The quantity of electricity in them was found to be very nearly the same, but that in N° 1 rather the greatest.

The charge of each of these tubes is about 714 inches, and that of the large jars about 6100, and that of the three jars 1, 2 and 4 together is also 6100*.

The shock of these tubes was received through my body in the same manner as in trying the large jars, either by making the shock pass through the copper wire or through the sat. sol. or receiving it in the simple manner without passing through either: the experiment being tried as usual by charging both tubes from the same conductor and receiving the charges of one one way and the other the other.

639. It was found by repeated trials that the shock received through the copper wire was plainly greater than the simple shock. When received through the sat. sol. with wires

in contact not sensib. less than simple.			
at	·1	dist. seemed rather less, but doubtful.	
	·5	D°.	
	1	D°.	scarce doubtful.
	2		not doubtful.
	4	D°.	
	6·6		considered less.

The tubes charged to $1\frac{1}{2}$ by old electrometer.

640] It was also found by the small jars 1, 2 or 4 that the shock received through the wire was stronger than the simple shock.

The shock through the wire was also much greater than the simple shock when the covering which was put over the wire to defend it from accidents was taken away.

It was also plainly greater when the shock passed through only 3 rows of the wire instead of the 8.

If the shock was received through 166 inches of the same wire not stretched upon glass, without any knots in it, it seemed not at all greater, but if anything less than the simple shock. It was the same if received through a piece of wire of about the same length with 37 knots in it.

641] Some more of the same wire was stretched by silk into 32 × 12 rows, each 78·7 inches long; consequently the whole length was 30220 inches. It weighed 3272 grains, *id est*, 9·24 inches to a grain.

* [Probably globular inches. The numbers do not agree with those in Art. 583.]

The shock of the above-mentioned tubes was sensibly greater when received through this last wire than when received simply, but was considerably less than when received through the first wire.

They were then compared by sound with the same tube charged to $5\frac{1}{2}$, when the sound of the shock passed through the new wire was sharper, and the other fuller.

The sound of the shock passed through the new wire seemed full as brisk, and the light as white as of that passed through ·55 of sat. sol., but not near so strong as when the wires in sat. sol. were in contact, the sound and light, however, seemed nearly of the same kind. When distance in tube was 1·1 the sound was evidently less loud than that with the wire.

When the shock was allowed to pass through both wires, the sound, I thought, seemed much of the same kind as when passed through new wire singly.

The shock passed through both wires felt plainly greater than the simple shock, and the difference seemed greater than that between the new wire simply and the simple shock.

In the foregoing the shock passed at the same time through both wires, but it was then tried so that it should first pass through old and from thence through new wire.

The shock felt then evidently stronger than the simple shock or that through new wire alone, but I could not tell whether it was greater or less than that through the old wire alone.

642] A piece of the same wire was wound about 150 times round one of the slips of glass, and was laid flat on another of these slips which lay flat on a table.

The shock of these tubes seemed rather greater when received through this wire than when received simply, but the difference was not considerable, but it seemed evidently less than the shock received through the new wire.

643] The wire was taken from off the reel with the slips of glass, and all except a small part of it stretched round the garden in 14 rounds. The shock of the above-mentioned tubes received through this wire felt plainly greater than that passed through the wire stretched by the silk threads, and much greater than the plain shock.

The shock passed through the sat. sol., wires in contact, seemed about equal to the plain shock.

The spark passed through garden wire seemed rather redder than that through the silk wire, but the difference was not remarkable.

The spark passed through garden wire seemed about as strong as that through about ·8 of an inch of saturated solution, but sensibly redder.

644] The reel was altered, and some copper wire silvered stretched upon it.

The mean circumference of reel $= 44{\cdot}05 \times 2\sqrt{2}$.

There were 12 rows of glass bars and 42 rounds of wire on each row, therefore whole length of wire $= 88{\cdot}1 \times \sqrt{2} \times 12 \times 42 = 62790$ inches. This weighed 5747 grains. Consequently there are 10·93 inches to 1 grain.

The shock received through this wire felt vastly stronger than the simple shock; the shock of tube 2 received through the wire with electrometer at $1\frac{1}{4}$ seeming little less strong than the simple shock with the same tube and the electrometer at $1\frac{3}{4}$, but considerably stronger than with electrometer at $1\frac{1}{2}$.

645] The above-mentioned wire compared with sat. sol. by sound.

1·46. Seemed more brisk. The light of salt water white, the other very red.
1·7 D°
2·5 D°
3·5 D°
5·5 D° } El. $= 5\frac{1}{2}$.
8·7 I believe nearly the same.

8·3 seemed much weaker. El. = 3.
6 seemed rather greater.
7·5 doubtful.
8·5 seemed rather weaker.
6 seemed rather stronger.

7·2 doubtful. El. = 3.
8·5 seemed rather weaker.
6 doubtful.
5 D°.
4 seemed stronger.
8·5 seemed rather weaker.
8·5 doubtful.
9·5 I believe rather less, certainly a sharper sound, but I believe rather less loud.
7 seemed greater.

$\frac{1\cdot65}{2} = \cdot82$	
$\frac{1\cdot8}{2} = \cdot9$	
$\frac{1\cdot1}{2} = \cdot55$	$\frac{3\cdot72}{5} = \cdot74$
$\frac{1\cdot3}{2} = \cdot65$	
$\frac{1\cdot6}{2} = \cdot8$	
$\frac{14\cdot5}{2} = 7\cdot25$	
$\frac{12\cdot5}{2} = 6\cdot25$	$\frac{21\cdot75}{3} = 7\cdot25$
$\frac{16\cdot5}{2} = 8\cdot25$	

646] ·74 of sat. sol. in tube 17 is equivalent to 29623 inches of copper wire, 9·984 inches of which = 1 grain.

7·25 of sat. sol. in do. = 62790 of copper wire, 10·93 of which = 1 grain*.

* [Length of wire.	Resistance of pure copper calculated from Matthiessen annealed.	hard drawn.	Resistance of saturated solution in tube 17 calculated from Kohlrausch.
29623	424	433	413
62790	984	1004	4046]

RESULT.

647] 1773, p. 92 [Art. 557]. The connecting wire to the two plates of 9·3 inches contains 1·4 inc. el. The connecting wire to the rosin plates of p. 86 [Art. 554], should contain rather more in proportion to its length than this, *id est*, rather more than ·28.

By p. 93 [Art. 557], the 4 rosin plates seemed to contain about $\frac{1}{2}$ inc. el. less when placed close together than at dist. Let us therefore suppose that the charge of 2 rosin plates placed close together with connecting wire between them exceeds twice the charge of 1 plate by ·28 inc. el., and that the charge of 4 plates exceeds 4 times the charge of 1 by $2\frac{1}{2}$ times that quantity, or ·7 inc. el. Let us suppose, too, that the charge of the 2 double plates A & B with connecting wire exceeds twice the charge of 1 by ·28.

648] 8 square inches of elect. = 9 circular inches.

	L.*	
$\frac{\text{glob. inc. el.}}{\text{circ. inc. el.}} = 1{\cdot}54$	$= {\cdot}1880.$	Res. p. 5 [Art. 654].
$\frac{\text{circ. inc. el.}}{\text{glob. inc. el.}} = {\cdot}649$	$= 9{\cdot}8120.$	

N.B. By inc. el. is meant circular inches of electricity.

649] Mar. 13. P. 85 [Art. 553].

	[Side of square equivalent to trial plate when the balls separate negatively,	positively.]	Difference.	Mean.
Circle $18\frac{1}{2}$	17·66	13·34	4·32	15·50
Double B	17·89	13·34	4·55	15·61
Double A	17·89	13·34	4·55	15·61
Circle 36	33·65	26·56	7·09	30·10
2 doub.	31·38	24·08	7·30	27·73
D	29·61	22·53	7·08	26·07

* [These logarithms are correct only to three places of decimals, they should be 0·1875 and 9·8122. See Note 35.]

Mon. Mar. 15. P. 85.

	[Side of square equivalent to trial plate when the balls separate negatively,	positively.]	Difference.	Mean.
Circ. 36	33·65	27·18	6·47	30·41
Circ. 30	28·42 28·10			
Plate air	30·87	24·39	6·48	27·63
2 doub.	31·13	24·39	6·74	27·76
D	29·86	22·84	7·02	26·35
Doub. B	18·22	13·82	4·40	16·02
Doub. A	18·22	13·82	4·40	16·02
Circle 18½	18·11	13·58	4·53	15·84
Mar. 19. P. 86.				
Circle 9	9·28	6·48	2·80	7·88
Rosin 1	10·59	7·13	3·46	8·86
2	10·47	6·91	3·56	8·69
3	10·59	7·24	3·35	8·91
4	10·35	7·02	3·33	8·68
1 + 2 Rosin	18·56	14·51	4·05	16·53
3 + 4	18·67	14·62	4·05	16·64
Circle 18½	18·00	13·82	4·18	15·91
Circle 36	33·90	27·49	6·51	30·74
4 rosin	32·14	25·94	6·20	29·04
Mar. 23. P. 90 [Art. 554].				
Circle 9·3	9·28	6·48	2·80	7·88
Rosin 1	10·22	7·13	3·09	8·67
2	10·09	7·02	3·07	8·55
3	10·22	7·13	3·09	8·67
4	10·22	7·13	3·09	8·67
Rosin 1 + 2	18·56	14·06	4·50	16·31
3 + 4	18·56	14·06	4·50	16·31
Circle 18½	17·66	13·34	4·32	15·50
Circle 36	34·40	26·56	7·84	30·48
4 rosin	33·40	26·25	7·15	29·82
Mar. 24. P. 91.				
Circle 36	33·65	27·49	6·16	30·57
Plate air 1	30·75	25·01	5·74	27·88
4 rosin	31·76	25·32	6·44	28·54
rosin 1 + 2	18·34	14·51	3·83	16·42
3 + 4	18·78	14·51	4·27	16·64
Circle 18	18·11	13·82	4·29	15·96
Circle 9·3	9·56	6·48	3·08	8·02
Rosin 1	10·83	7·35	3·48	9·09
2	10·83	7·46	3·37	9·14
3	10·83	7·46	3·37	9·14
4	10·83	7·46	3·37	9·14

650] [Results of Art. 649.] By Mar. 13. [Art. 553.]

Double plate = circ. $18\frac{1}{2}$ + ·11 sq. inc., or ·12 inc. el.

Circ. 36 = 2 doub. + 2·67 inc. el. without allowance for communicating wire, &c., or 2·95 with.

D = 2 doub. − 1·60 with allowance.

$$\left.\begin{matrix}\text{Circ. } 36\\ \text{D}\end{matrix}\right\} = \text{Circ. } 18\tfrac{1}{2} \times 2 \begin{cases} +3{\cdot}19 \\ -1{\cdot}36 \end{cases}.$$

By Mar. 15 [Art. 553].

Doub. pl. = $18\frac{1}{2}$ + ·20.

$$\left.\begin{matrix}\text{D}\\ \text{pl. air}\\ \text{circ. } 36\end{matrix}\right\} = \text{circ. } 18\tfrac{1}{2} \times 2 \begin{cases} -\ \ {\cdot}91 \\ +\ \ {\cdot}53 \\ +3{\cdot}66 \end{cases} = 2 \text{ doub.} \begin{cases} -1{\cdot}51 \\ +\ \ {\cdot}13 \\ +3{\cdot}26 \end{cases} \text{with allowance.}$$

651] All the following are with allowance.

Mar. 19 [Art. 554]. 1 ros. = circ. 9·3 + 1·01.
circ. $18\frac{1}{2}$ = 2 ros. − ·47 = circ. 9·3 × 2 + 1·55.
circ. 36 = 4 ros. + 2·61 = circ. $18\frac{1}{2}$ × 2 + 3·55.

Mar. 23 [Art. 554]. 1 ros. = circ. 9·3 + ·85.
circ. $18\frac{1}{2}$ = 2 ros. − ·63 = circ. 9·3 × 2 + 1·07.
circ. 36 = 4 ros. + 1·44 = circ. $18\frac{1}{2}$ × 2 + 2·76.
1·32

Mar. 24 [Art. 554]. 1 ros. = circ. 9·3 + 1·25.
circ. $18\frac{1}{2}$ = 2 ros. − ·36 = circ. 9·3 × 2 + 2·14.
circ. 36 = 4 ros. + 2·98 = circ. $18\frac{1}{2}$ × 2 + 3·70.
Plate air 1 = circ. 36 − 3·03 = circ. $18\frac{1}{2}$ × 2 + ·67.

By mean of all

circ. 36 = circ. $18\frac{1}{2}$ × 2 + 3·37.
circ. $18\frac{1}{2}$ = circ. 9·3 × 2 + 1·59.

Therefore charge of circle of

$$\left.\begin{matrix}37 \text{ inc.}\\ 18\tfrac{1}{2} \text{ inc.}\end{matrix}\right. \text{ exceeds } 2^{ce} \text{ charge of circ. of } \begin{matrix}18\tfrac{1}{2} \text{ by } 4{\cdot}47\\ 9\tfrac{1}{4} \text{ by } 1{\cdot}69\end{matrix},$$

37 inc. exceeds 4 times charge of $9\frac{1}{4}$ by 7·85.

652] *If charge circle is greater than it would be if placed at a great distance from any other body in ratio of $a : a - 36$,

charge circ. of $18\frac{1}{2}$ should exceed in ratio of $a : a - 18\frac{1}{2}$ and so on.

Therefore, if we suppose $a = 167$,

$$\text{charge circ.} \begin{cases} 36 \\ 18\tfrac{1}{2} \\ 9{\cdot}3 \end{cases} \text{should exceed its true charge by} \begin{cases} 9{\cdot}89 \\ 2{\cdot}31 \\ \ \ {\cdot}55 \end{cases},$$

* [See Art. 338, and Note 24.]

and charge circ. 36 should exceed

$$\left.\begin{matrix}2^{ce}\text{ charge of }18\tfrac{1}{2}\text{ by }4{\cdot}27\\ 4\text{ times charge of }9{\cdot}3\text{ by }6{\cdot}49\end{matrix}\right\},\text{ which is }\left\{\begin{matrix}{\cdot}90\text{ greater}\\ {\cdot}06\text{ less}\end{matrix}\right.$$

than by experiment, and charge circ. $18\frac{1}{2}$ should exceed 2^{ce} charge of 9·3 by 1·11, which is ·58 less than by experiment.

We will therefore suppose that the charge of circ. $18\frac{1}{2}$ or of globe 12·1, as found by experiment, exceeds the true charge in the ratio of 9 to 8, as it should do if $a = 167$.

653] 1771. P. 15 [Art. 456].

$$\begin{matrix}\text{doub. B contains }\frac{3}{16}\\ \text{doub. A contains }\frac{1}{8}\end{matrix}\text{ sq. inc. or }\left\{\begin{matrix}{\cdot}21\\ {\cdot}14\end{matrix}\right.\text{ circ. inc. less than circ. }18\tfrac{1}{2}.$$

1772. P. 12 [Art. 478].

$$\begin{matrix}\text{doub. B}\\ \text{doub. A}\end{matrix}\text{ contains }\begin{matrix}{\cdot}11\\ {\cdot}23\end{matrix}\text{ sq. inc. or }\begin{matrix}{\cdot}12\\ {\cdot}26\end{matrix}\text{ circ. inc. more than circ. }18\tfrac{1}{2}.$$

1773. P. 85 [Arts. 553 & 650]. Each doub. plate contains ·16 circ. inc. more than circ. $18\frac{1}{2}$.

654] P. 15. 1771 [Art. 456]. Globe cont. $\frac{1{\cdot}25}{4}$ sq. inc. or ·35 circ. inc. more than circ. 18·5.

P. 12. 1772 [Art 478]. Globe contains same as circ.

By mean, globe of 12·1 = circ. of 18·67, or
globe of 12 inc. = circ. of 18·5.

Therefore 1 circ. inc. = ·65 glob. inc.
or 1 sq. inc. = ·73 glob. inc.

DEF. *The charge of globe* 1 *inc. diam. placed at great dist. from any other body is called* 1 *glob. inc.*

The circ. 18·5 = 13·5 glob. inc.*

The doub. plate A or B is supp. = 13·6 glob. inc.

655] P. 18, 1772 [Art. 483],
D, E, F & G cont. ·68 inc. el. less than 2 doub.

P. 19, 1773 [Art. 509],
D & F cont. 1 inc. less than do.

P. 59, 1773 [Art. 533],
D, E & F cont. $1\frac{1}{4}$ inc. less than do.

* [See Note 35.]

P. 85, 1773 [Art. 553],

———	D	cont. 1·6 less than 2 doub.
———	D	cont. 1·31 less than do.
———	D	cont. 1·36 less than 2^{ce} circ. 18½.
	D	cont. ·91 less than do.

D is supposed to cont. 1·3 circ. inc. or ·85 glob. inc. less than 2 doub., *id est*, 26·3 glob. inc.

656] 1773, P. 28 [Art. 515],

M cont. 1 inc. el. more than D + E + F.

P. 29 [Art. 515],

M cont. same as ....

P. 54 [Art. 528],

M cont. 2½	} more than D + E + F.	N. 16½.
K & L 1		

1773, P. 57 [Art 530],

M cont. 2	more than D + E + F.	N. 14½.
K & L 1		

P. 57 [Art. 530],

M cont. 3½	more than D + E + F.	N. 16.
K & L 2½		

It is supp. that M cont. 2·7 / K & L — 1·5 more than D + E + F, *id est*, 80·7 / 79·9 glob inc. el.

657] 1773, P. 55 [Art. 529],

A cont. 30·3	inc. el. less than K + L + M.	N. 15.
B or C 32		

P. 57 [Art. 530],

each cont. 33·7 less than do. N. 14½.

P. 58 [Art. 531],

each cont. 38·6 less. N. 16.

It is supp. that A, B, and C each cont. 34·8 inc. less than K + L + M, *id est*, 29·1 less than 9D, *id est*, 217·8 glob. inc.

658] By exper. of 1772,

F or G	cont.	2	inc. more than D.
E	—	1·6	—
M	cont.	3·86	less than E + F + G.
K & L		12·02	less.
F		10·72	less.
Therefore E + F + G	cont.	5·6	more than 3D.
M		1·7	more.
K or L		6·4	less.
F		5·1	less.
A, B & C each contain		15·2	less than F + K + L.
or		33·1	less than 9D.

1773, P. 56 [Art. 530],

H cont. 10· inc. more than A + B + C.

1772, P. 29 [Art. 493],

H cont. the same as A + B + C.

H is supposed to contain 654 glob. inc.*

659] *Instantaneous spreading of el.*† Measures P. 19 [Art. 593].

The area of the old coatings of	A	= 33·9	and circumf. =	20·6
	C	= 33·2		20·4
	H	= 36·3		21·4
Area of slit coatings of	A	= 31·8	and circumf.	73·5
	C	= 30·4		76·5
	H	= 33·3		80·1
	crown	= 24·7		69·6
Area of oblong coatings of	A	= 34·1	& circumf.	23·4
	C	= 33·3		23·2
	H	= 36·4		24·1
	crown	= 29·0		21·6

660] P. 15, 1773, 504,

White Cyl. cont. 7 inc. el. less than H.

P. 13. 502, 5

By mean it cont. 6 less than H.

P. 62 [Art. 536],

H with slit coat. cont. 77·5 more than white cyl.

crown with oblong coat. 33·7 .... N. 12.

P. 63 [Art. 536],

H with D° cont. 99·1 more than wh. cyl. N. 11.

crown 43·8 ....

* [See Art. 318.] † [See Art. 319.]

P. 65 [Art. 537],

H with D°	70·8 more than wh. cyl.	N. 14.
crown with slits	34	

P. 66 [Art. 537],

crown D° cont.	20 more than wh. cyl.	N. 12½.

P. 71 [Art. 541],

H D° cont.	74·1 more than wh. cyl.	N. 21.
crown	67·3	

P. 81 [Art. 550],

H with obl. cont.	20·2 more than wh. cyl., st. el.* at	2 + 3	N. 15.
H	34	3 + 1	
crown	57·3		
H	9 less than wh. cyl.	1 + 3	
crown	14·6 more than		

P. 82 [Art. 550],

H	18·5 more than wh. cyl.	N. 14½.

A and C with circ. coatings are supposed to contain same as B.

P. 62 [Art. 536],

A with slit. coat. cont.	13·5	more than B	us. el.†	
C with slit. coat. cont.	13·5			
A	15·2		1 + 3	N. 12½.
C	11·8			
A	33·7		3 + 1 very irreg.	
C	33·7			

P. 63 [Art. 536],

A	18·5	us. el.	N. 11.
C	15·2		
A	13·5	1 + 3	
C	10·1		
A	18·5	3 + 1 very irreg.	
C	18·5		

P. 65 [Art. 537],

A with obl.	3·4	more than B	us. el.	N. 14.
C with obl.	5·1			
A	5·1	———	1 + 3	
C	1·7			
A	0		3 + 1	
C	1·7	less than B		

* [Straw electrometer. See Art. 560, note.]
† [Usual degree of electrification. See Art. 329 and note 10.]

P. 66 [Art. 537],

A	3·4	more than B	us. el.	N. 12½.
C	5·1			

661] By mean H with slits cont. 78 inc. el. } more than wh. cyl.
with oblong 19 —— }

Crown with slits contains 27 inc. el. more than wh. cyl.
oblong 39 ————————

N.B. This is meant in dry weather & with usual deg. el.

The crown with slits exceeded wh. cyl. by 42·7 more with electrom. at 3 + 1 than at 1 + 3, and H with oblong exceeded wh. cyl. by 43 more with electrom. at 3 + 1 than at 1 + 3, but it must be observed that this was only one day's observ.

With usual deg. el. A/C exceeded B by 16 / 14·3

with electrom. at 1 + 3 by 14·3 / 11

& at 3 + 1 by 26·1 / 26·1

A/C with oblong exceeded B with us. el. by 3·4 / 5·1

with electrom. at 1 + 3 by 5·1 / 1·7

and at 3 + 1 by 0 / −1·7

662] Hence we have the following results:—

	L. inc. el. in each sq. inc. circ. or obl. coa.	Inc. el. in slit coat more than in oblong.	Sq. inc. coating answering to D°.	Sq. inc. of slit coat equiv. to obl.	Sq. inc. in obl.	Diff.	Excess circum. slit coat. above oblong.	Spreading of elect.
A with us. el.	9·959	12·6	1·27	30·53	34·1	3·57	50·1	·072
el. at 1+3		9·2	·93	30·87		3·23		·065
el. at 3+1		26·1	2·63	29·17		4·93		·098
C with us. el.	0·050	9·2	·91	29·49	33·3	3·81	53·3	·071
el. at 1+3		9·3	·92	29·48		3·82		·072
3+1		27·8	2·75	27·65		5·65		·106
H with us. el.	4·437	59	2·12	31·18	36·4	5·22	56·0	·094
Crown do.	5·552	−12	−·33	25·03	29·0	3·97	48·0	·083

663]

	Inc. el. in oblong coating −D° in circular.	Sq. inc. of coating equiv. to D°.	Sq. inc. of circular coating equiv. to oblong.	Sq. inc. in oblong.	Diff.	Excess circumf. oblong above circular.	Sq. inc. equiv. to excess of spreading of elect. in oblong above that in circular.
A	3·4	·34	34·24	34·1	·14	2·8	·20
C	5·1	·51	33·71	33·3	·41	2·8	·20
H	1·3	·46	36·76	36·4	·36	2·7	·25

It is plain that the numbers in the 8th or last col. ought to be equal to those in the 6th, as is nearly the case.

664] *Whether charge of coated glass bears the same proportion to that of another body whether el. is strong or weak* *.

P. 61 [Art. 535], E on neg. side tried against sliding tin plates on pos.

Charge of E was $\frac{1}{60}$ part less with straw el. at 3 + 1 than at 1 + 3, the diff. between neg. and pos. el. was much too small to be certain of.

P. 66 [Art. 538], a ball blown at end of therm. tube tried in same manner. Charge just the same whether electrom. at 1 + 3 or 3 + 1.

P. 68 [Art. 538], charge D° $\frac{1}{50}$ less with el. at 3 + 1 than at 1 + 3.
——— D° $\frac{1}{37}$ ———

P. 82 & 84 [Arts. 551 & 55], tried with machine for finding quant. el. in common plates. No perceptible diff. between charge of E whether tried with el. at 1 + 3 or 3 + 1.

665] By P. 9 [Art. 661], it should seem that el. spread. ·034 inc. more on surface with greater degree of el. than with smaller, and therefore as the diam. coating of E or D is 2·16.

So that it should seem as if the charge of a coated plate in which the spreading of the el. was prevented would be at least $\frac{1}{16}$ less with the stronger degree el. than with the weaker.

666] By exper. of P. 69 [Art. 539], it appeared that the charge of tin cyl. was to that of D + E when electrified very weakly as 1·28 to 1, and by P. 70 [Art. 539] as 1·24 to 1. By mean as 1·26 to 1†.

By mean of P. 76 [Art. 545], the charge of the same cyl. was to that of D + E when electrified in the usual degree as 1·33 to 1.

By mean of P. 77 [Art. 546], it came out as 1·37 to 1, but this last can not be depended on, as wire for making communication with ground was forgot to be fixed‡.

* [Arts. 356, 451, 463, 535, 539, 551.] † [See Arts. 358, 539, 545.]

‡ The comp. charge of the cyl. is 48·4 glob. inc. The real charge, supposing that the wire contains 3·6 glob. inc. less when joined to cyl. than to D + E = 73·6, and therefore its real charge exceeds the computed in the ratio of 1·52 to 1. [See Note 25.]

667] It should seem that the charge of D and E is increased $\frac{10}{78}$ by spreading of el. when elect. in usual degree, therefore if we suppose that the spreading is insensible when electrified in very small degree, the charge of a glass plate is less in proportion to that of another body when electrified with usual degree el. than when elect. with a very small one in ratio of 1·26 to 1·51, or of 5 to 6.

668] *On plate air**.

[By Art. 517],

P. 32 pl. air 1 cont. 1 inc. el. more than D }
33 $\frac{3}{4}$ } by mean $\frac{7}{8}$ more than D.

The same plate air contained 2 inc. el. less when resting intirely on machine than when resting by 1 corner.

[By Art. 517], Pl. air 2 cont. 1 inc. el. less than D + E.
P. 32, pl. air 3 10·5 inc. el. less than D + E + F.
P. 33, pl. air 4 1 inc. el. less than D + E + F.
P. 36, pl. air 5 $\frac{1}{4}$ more than D.
P. 37, Do.

By res. P. 5 [Art. 653], D, E, and F cont. 26·3 glob. inc.

Therefore pl. air 1 contains 27 glob. inc.
2 52
3 72·1
4 78·3
5 26·5

669] [Table of plates of air given in Art. 343.]

670]

Plate air.	Log. diam. by thickness.				
1	1·1017	·7919	7452	6928	6332
2	1·4375	·9426	8689	7799	6679
3	1·6013	1·0163	9235	8054	6427
4	1·6525	·9747	8566	6939	4307
5	1·3895	·9458	8809	8045	7117

The 4 last columns are the log. of $\frac{\text{diam.}}{\text{thickness}} \times$ excess $\frac{\text{real charge}}{\text{computed charge}}$ above N, the value of N in 3rd col. being 1, in 2nd 1·05, in 3rd 1·1, and in 4th 1·15.

The numbers in the 3rd column seem most uniform, and therefore it seems likely that [if] the $\frac{\text{diam.}}{\text{thickness}}$ was very great, $\frac{\text{real charge}}{\text{comp. charge}}$ would equal 1·1†.

* [See Art. 340.]

† [See Art. 347.]

671] If we suppose the el. to spread ·07 inc. on surf. thick plates and ·09 on surf. thin ones, the result of Nairne's plates is as follows*.

	Diam.	D° corrected.	Thick-ness.	Computed charge.	Real charge.	Real charge by com-puted.	Real charge by diam.
D	2·155	2·295	·2057	3·20	26·3	8·22	
E	2·16	2·3	·2065	3·20	26·3	8·22	11·4
F	2·175	2·315	·2115	3·17	26·3	8·30	
K	2·265	2·445	·07712	9·69	79·9	8·29	
L	2·335	2·515	·08205	9·63	79·9	8·29	
M	2·195	2·375	·07187	9·81	80·7	8·23	
A	6·57	6·71	·2112	26·6	217·8	8·18	32·6
B	6·6	6·74	·2132	26·6	217·8	8·18	
C	6·5	6·64	·2065	26·7	217·8	8·16	
H	6·8	6·98	·07556	80·6	654	8·11	93·7

672] *Computations of other flat plates of glass, &c.*

						Mean charge.
[Art. 507] P. 18	thick white	=D				
[Art. 508] P. 19		D°.			26·3	
P.	thin white	=D+E−·5	·33		52·3	
P. 19	N	=D+E−1·8	1·2		51·9	
[Art. 509] P. 20	P	=M−15	9·7		71	71·9
[Art. 515] P. 28		=D+E+F−9·5	6·1		72·8	
[Art. 509] P. 20	Q	=M−9	5·7		74·8	76·5
[Art. 515] P. 28		=D+E+F−1·1	·7		78·2	
[Art. 509] P. 20	O	=M−9	5·7		74·8	75
[Art. 515] P. 28		=D+E+F−5·8	3·8		75·1	
[Art. 509] P. 20	white plate	=M−7·7	5		75·7	
[Art. 515] P. 28		=D+E+F−7·7	5		73·9	
[Art. 509] P. 20	old G	=M−7·3	4·8		75·9	
[Art. 515] P. 28		=D+E+F−5·8	3·8		75·1	
[Art. 510] P. 21	crown A	=A−13	8·5	211·3		
[Art. 533] P. 59		−6·7	4·4			
[Art. 510] P. 21	crown C	=A−13	8·5	208·7		
[Art. 533] P. 59		−15	9·8			
[Art. 527] P. 53	small ground crown	=D+E+F−4½				
[Art. 528] P. 54		−3½	2·4		76·5	
[Art. 531] P. 57		−3				
[Art. 527] P. 53	large ground crown	=C−13				
[Art. 528] P. 54		−2	2·7		215·1	
[Art. 531] P. 57		0				
[Art. 533] P. 59		−1·7				

* [In Art. 324 the "Real charges" of this table are multiplied by ·122 for easy comparison with the computed charges.]

						Mean charge.
[Art. 507]	P. 18	exper. rosin 1	= doub. B − 1	·06	13·5	
		rosin 2	E − 2			
[Art. 519]	P. 36		− 1·7	1·1	25·2	
[Art. 509]	P. 20	rosin 3	= M − 17	11·1	69	
[Art. 515]	P. 28		= D + E + F − 16	10·4		
[Art. 518]	P. 35	rosin 4	= D + E + F		78·9	
[Art. 519]	P. 36		= D°.			
[Art. 527]	P. 53	rosin 5	= doub. B − 1	·65	13	
[Art. 528]	P. 54		− 1			
[Art. 518]	P. 35	1st made ros.	= D + 2		·6	26·9
[Art. 519]	P. 36		0			
[Art. 518]	P. 35	deph. bees wax 1	= D − $2\frac{1}{2}$		1·8	24·5
[Art. 519]	P. 36		− 3			
[Art. 518]	P. 35	deph. bees wax 2	= E + F − 4		2·1	50·5
			− 2·5			
[Art. 527]	P. 53	deph. bees wax 3	= E + F − 7·5			
[Art. 528]	P. 54		− 11	$\frac{30}{3}$	6·5	46·1
[Art. 533]	P. 59		− 11			
[Art. 527]	P. 53	plain bees wax	= E + F − 3·5			
[Art. 528]	P. 54		0	$\frac{6}{3}$	1·3	51·3
[Art. 533]	P. 59		− 2·5			
[Art. 518]	P. 35	lac	= D + E + F + 1·5	$\frac{3·7}{2}$	1·1	80
[Art. 519]	P. 36		+ 2·2			

673] [Table given in Art. 370.]

The diam. was corrected on supposition that elect. spreads ·07 if the thickness of glass = ·21, and ·09 if thickness = ·08, and so in proportion in other thicknesses.

674] [Table given in Art. 371.]

The correction of the diameter is the same as would be used according to the preceding rule to a glass plate of 2ce the thickness, only the correction used is never less than $\frac{1}{10}$ inch.

675] *On the glass cylinders.*

503, P. 14,	gr. cyl. 2.	= H + 45 inc. el.	inc. el. H + D + 14·3	= 690 glob. inc.
504, P. 15,		H + 55·6		
503, P. 14,	white jar	= H + 74 = H + M − 34	H + M − 27	= 717
504, 15,		H + M − 20		
504, P. 15,	gr. cyl. 1	= H + M + 30		= 754
502, P. 13,	gr. cyl. 4	= C + K + J × $\frac{23}{24}$		= 353
545, P. 75,	therm. tube 1	= D + E + F + 2 inc. el.		= 80·2
546, P. 77,	therm. tube 2	= D + E + F + 2·8		= 80·7

676] [Table given in Art. 383.]

The white jar and cyl. and the 3 green cyl. are corrected for the spreading of the electricity in the same manner as the flat plates, but the 2 therm. tubes are not.

677] *On the compound plates**.

P. 60, Art. 534.

The 3 plates A, B and C placed over each other with bits of lead between contained 8·9 inc. el. less than K or L, therefore its charge =74 inc. The 3rd part of the charge of A, B, or C is 72·6 inc.

The coatings taken from the 3 plates A, B, & C, the plates placed close together and the outside surfaces coated with circles 6·6 in diam.

544, P. 75, it contained 7·5 inc. less than D + E + F.
546, P. 77, 6 less.
By mean it contains 6·7 less, therefore charge = 74·5.

The thickness of the 3 plates together is ·6309. The computed charge of a plate of that thickness with a coating 6·6 in diam. supposing the el. to spread ·07 inc. is 9·00, and the real charge of such a plate according to the mean ratio of the real and computed charges of D, E, and F is 74·2.

678] A plate of exper. rosin about 8 inc. square was pressed out, thickness irregular, but at a medium about ·122. It was coated with circles 6·61 in diam.†

Art. 548, P. 79, its charge = $K + D + E \times 1 + \frac{1}{66}$,
in afternoon $\times 1 + \frac{1}{44}$.

By mean it = $K + D + E \times 1 + \frac{1}{55} = 135$.

The real charge of this plate is to its computed, supposing the el. to spread ·07 inc., as 2·89 to 1.

The charge of this plate is the same as that of a glass one ·345 thick, supposing ratio of real and computed charge the same as in A or B.

679] The coatings being taken from this plate it was included between the plates B and H, and the outside surfaces coated with circles of 6·6 in diam.

552, P. 83, it cont. 6·9 inc. el. less than K,
6·5 ———
5·2 less than D + E + F,

by mean it contains 75·5 glob. inc.

The charge of plate glass of the same sort as Nairne's ·634 thick (*id est*, equal to the sum of the thicknesses of the two glass plates and a glass plate equiv. to the rosin) = 73·3, supposing the el. to spread the same on this plate as on the rosin.

* [Arts. 379, 534.] † [Arts. 381, 552.]

680] [Same as Art. 368.]

681] By res. P. 5 [Art. 654] a globe of 12 inc. contains as much el. as a circle of 18·5*, therefore by Prop. XXIX. $p = \frac{11}{13}$, therefore

	Charge of both plates when distant 18	26	36	Sing. pl.
if $p = \frac{11}{13}$	1	1·046	1·078	1·172
0	1	1·062	1·108	1·249
inf[inite]	1	1·035	1·059	1·126
or if $p = \frac{11}{13}$	·853	·892	·920	1
0	·801	·851	·887	1
inf.	·888	·919	·940	1

By 1st exp. 1772 [Art. 473] the proportions were

thus	·811	·859	·899	1†
By 2nd exp. [Art. 475]	·798	·840	·894	1

682] The charges of the following bodies are supposed to bear the following proportions to each other‡.

globe	12·1 diam.		= 1
circle	18·5 diam.		= ·992
square of	15·5		·958
oblong	17·9	× 13·4	·964
cyl.	35·9 by	2·53	1·028
	54·2	·73	·978
	72	·185	·966

Charge by theory of {sh[ort] cyl. / long cyl. / wire} is between {·887 & 1·469 / ·896 & 1·573 / ·894 & 1·619} that of globe being one, and if charge cyl. is supposed to be to that of globe whose diam. = length cyl. :: $\frac{9}{8}$: N. L. $\frac{2 \text{ length}}{\text{diam.}}$, their charge = ·998 / 1·008 / 1·006.

This ratio approaches about 5 times nearer to the first proportion than the 2nd§.

The area of the oblong is the same as that of the square, and their charges are very nearly the same.

The charge of a square is to that of a circle whose diam. = side square as 1·153 to 1 ||.

* [Note 35.]

† [Note 21.]

‡ [Exp. VII., Art. 281. The numbers here are different. See Art. 478.]

§ [Note 12.]

|| [This ratio is given in Art. 283 as 1·53 by a mistake of the Editor in copying, see Note 22.]

683] In exper. P. 11, 1772 [Art. 477], the large wire should contain about $\frac{1}{93}$ less el. than if its diam. was double the small ones. Allowing for this, the charges of the large wire at 36, 24 & 18 inc. dist. should be between the two following proportions:

1	·942	·915	·891,
1	·901	·868	·844,

but I believe ought to approach about 5 times nearer to the former. The observed proportions are*

1	·903	·860	·850.

* [Note 13.]

RESULTS
[OF EXPERIMENTS ON RESISTANCE OF SOLUTIONS].

684] Resistance of Pump-water is $4\frac{1}{8}$, Salt in 1000 of rain water 9, Sea water 100 } times less than that of rain water*.

685] A shock is diminished very nearly the same, but if anything rather more, by passing through 9 tubes, 37 inches of which hold 3373 grains of ☿, than through one tube, 37 inches of which hold 3480 grains of ☿ †.

686] A shock is as much diminished in passing through 6·8 inches of a tube, 37 inches of which hold 567 grains, as through $44\frac{1}{4}$ of one 37 inches of which hold 3480. So that resistance should seem as 1·03 power of velocity ‡.

687] If resistance is as $\begin{cases}1{\cdot}03\\1{\cdot}08\end{cases}$ power of velocity, the resistance of iron wire is $\begin{cases}437000\\607000\end{cases}$ times less than that of saturated solution of sea salt ||.

688] Resistance of sat. sol. S. S in 99 of distilled water is 39 times greater than that of the sat. sol.

Resistance of distilled water is 18 times greater than that of sat. sol. in 99 of distilled water §.

689¶] *Experiments in* 1776 *and* 1777.

No. of Exp.		Conducts times better than		Tubes.	Electro-meter.	
1	Sat. sol.	8·6	salt in 69	14 & 15	4	
2		8·94	——	22 & 23	$1\frac{1}{4}$	
3		9·61	——	14 & 15	3	New solutions.
6		10·05		5 & 17	1	D° jars 1 & 2, new sol. battery.
7		10·31		5 & 17		
10		9·02		22 & 23	1	
9		7·79	salt in 29 diluted with $1\frac{1}{3}$ of water supp. $2\frac{1}{3}$.	22 & 23	1	

* [Arts. 398, 524.] † [Arts. 574, 575, &c.] ‡ [Arts. 575, 629.]
|| [Arts. 398, 576, and note 32.] § [Art. 577.]
¶ [Arts. 617—623.]

No. of Exp.		Conducts times better than		Tubes.	Electro-meter.
8	sat. sol.	3·51	salt in 29	22 & 23	1
15	——	4·38	——	22 & 23	1¼
11	sat. sol.	20·5	salt in 149	14 & 15	3½
12	——	19·6	——	22 & 23	1¼
4	salt in 69	9·57	salt in 999	22 & 23	1¾
5	salt in 999	9·92	salt in 20,000	12 & 20	1¾
13	salt in 149	17·3	salt in 2999	5 & 17	3
14	——	16·7	——	18 & 19	1¾

N. B. It is not said what water the solutions were made with. By the comparison of salt in 999 with salt in 20,000, it should seem either that they were not made with distilled water, or that some mistake was made in the experiment.

690] In Jan. 1777, salt in 2999 conducted about 70 or 90 times better than some water distilled in the preceding summer, or about 25 or 50 times better than the distilled water used in the year 1776 *.

Salt in 2999 conducted about 25 times better than salt in 150,000.

691] Salt in 69 conducts 1·97 times better in heat of 105° than in that of 58°½ †.

The proportion of the resistance of sat. sol. and salt in 999 to each other seems not much altered by varying heat from 50° to 95° ‡.

692] Salt in 150,000 seemed to conduct rather better than the same water deprived of air by boiling in the same vial in which it was kept, and cooled quick in water to prevent its absorbing much air. But the difference was not more than might arise from error of experiment §.

693] Distilled water impregnated with fixed air from oil of vitriol and marble conducted 2½ times better than the same water deprived of its air by boiling ||.

694] Conducting power of other saline solutions compared with that of salt in 29 of water ¶.

Sal. Sylvii	1·08
Sal. amm.	1·13
Calc. S. S.	·852
Glaub. salt	·696
Quadran. Nitre	·887
F. alk.	·819
Spt. salt	1·72
Oil vitr.	·783
D° another parcel	1·12

* [Art. 621.] † [Art. 619.] ‡ [Art. 620.]
§ [See Art. 624.] || [Art. 625.] ¶ [Art. 626 and Note 34.]

N. B. The solutions of the neutral salts were all of such strength that the acid in them was equiv. to that in salt in 29.

The f. alk. also was equiv. to that in salt in 29, but the acids were equiv. to that in salt in 59.

695] *Experiments in Jan.*, 1781*.

					Tubes.	Electro-meter.	
Sat. Sol.	conducts	8·63	times better than	salt in 69	14 & 15	$3\frac{1}{4}$	8·8
——		9·03		——	——	—	
Sat. Sol.		4·1		salt in 29	——	—	3·97
——		3·85		——	——	—	
——		3·95		——	——	$1\frac{3}{8}$	
Sat. Sol.		1·92		salt in 11	——	$1\frac{3}{8}$	1·91
——		1·88		——	——	—	
——		1·92		——	——	—	
Salt in 69		1·74		salt in 142	22 & 23	2	1·79
——		1·84		——	——	—	
Salt in 69		9·91		salt in 999	——	$3\frac{1}{2}$	10·57
——		10·3		——		—	
——		11·31		——	5 & 17	$1\frac{1}{2}$	
——		10·75		——		—	
Salt in 999		20		salt in 20,000	12 & 20	$2\frac{1}{2}$	19·6
——		19·5		——		2	
——		19		——		$2\frac{1}{2}$	
——		19·8		——		2	

Salt in 20,000 conducts about 7 times better than distilled water.

696] Therefore the resistance of water with different quantities of salt in [it] are as follows †:

Quantity salt.	Resistance.	Log. do.	Resist. × quant. salt.	Log. do.
1 by 3·78	1			
12	1·91	·2810	·602	9·7793
30	3·97	·5988	·500	9·6992
70	8·8	·9445	·475	9·6769
143	15·75	1·1973	·416	9·6195
1000	93·02	1·9686	·352	9·5461
20000	1823	3·2608	·345	9·5373

* [Art. 628.]

† [See Note 33.]

NOTES

BY THE EDITOR.

NOTE 1, ARTS. 5 AND 67.

On the theory of the Electric Fluid.

The theory of One Electric Fluid is here stated very completely by Cavendish*. The fluid, as imagined by him, is not a purely hypothetical substance, which has no properties except those which are attributed to it for the purpose of explaining phenomena. He calls it an elastic fluid, and supposes that its particles and those of other matter have certain properties of mutual repulsion or of attraction, just as he supposes that the particles of air are indued with a property of mutual repulsion, but according to a different law. See Art. 97 and Note 6. But in addition to these properties, which are all that are necessary for the theory, he supposes that the electric fluid possesses the general properties of other kinds of matter. In Art. 5 he speaks of the weight of the electric fluid, and of one grain of electric fluid, which implies that a certain quantity of the electric fluid would be dynamically equivalent to one grain, that is to say, in the language of Boscovich and modern writers, it would be equal in *mass* to one grain.

We must not suppose that the word weight is here used in the modern sense of the force with which a body is attracted by the earth, for in the case of the electric fluid this force depends entirely on the electrical condition of the earth, and would act upward if the earth were overcharged and downward if the earth were undercharged.

Cavendish also supposes that there is a limit to the quantity of the electric fluid which can be collected in a given space. He speaks (Art. 20) of the electric fluid being pressed close together so that its particles shall touch each other. This implies that when the centres of the particles approach to within a certain distance, the repulsion, which up to that point varied as the nth power of the distance, now varies much more rapidly, so that for an exceedingly small diminution of distance the mutual repulsion increases to such a degree that no force which we can bring to bear on the particles is able to overcome it.

We may consider this departure from the simplicity of the law of

* For an earlier form of Cavendish's theory of electricity, see "Thoughts concerning electricity" (Arts. 195—216), and Note 18.

force as introduced in order to extend the property of "impenetrability" to the particles of the electric fluid. It leads to the conclusion that there is a certain maximum density beyond which the fluid cannot be accumulated, and that therefore the stratum of the electric fluid collected at the surface of electrified bodies has a finite thickness.

No experimental evidence, however, has as yet been obtained of any limit to the quantity of electricity which can be collected within a given volume, or any measure of the thickness of the electric stratum on the surface of conductors, so that if we wish to maintain the doctrine of a maximum density, we must suppose this density to be exceedingly great compared with the density of the electric fluid in saturated bodies.

A difficulty of far greater magnitude arises in the case of undercharged bodies. It is a consequence of the theory that there is a stratum near the surface of an undercharged body which is entirely deprived of electricity, the rest of the body being saturated. Hence the electric phenomena of an undercharged body depend entirely upon the matter forming this stratum. Now, though on account of our ignorance of the electric fluid we are at liberty to suppose a very large quantity of it to be collected within a small space, we cannot make any such supposition with respect to ordinary matter, the density of which is known.

In the first place, it is manifestly impossible to deprive any body of a greater quantity of the electric fluid than it contains. It is found, indeed, that there is a limit to the negative charge which can be given to a body, but this limit depends not on the quantity of matter in the body but on the area of its surface, and on the dielectric medium which surrounds it. Thus it appears from the experiments of Sir W. Thomson and those of Mr Macfarlane, that in air at the ordinary pressure and temperature a charge of more than 5 units of electricity, either positive or negative, can exist on the surface of an electrified body without producing a discharge. In other media the maximum charge is different. In paraffin oil, and in turpentine, for instance, it is much greater than in air*. In air of a few millimetres pressure it is much less, but in the most perfect vacuum hitherto made, the charge which may be accumulated before discharge occurs is probably very great indeed.

Now this charge, or undercharge, whatever be its magnitude, can be accumulated on the surface of the thinnest gold leaf as well as on the most massive conductors. Suppose that there is a deficiency of five units of electricity for each square centimetre of the surface on both sides of a sheet of gold leaf whose thickness is the hundred thousandth part of a centimetre. We have no reason to believe the gold leaf to be entirely deprived of electricity, but even if it were, we must admit that every cubic centimetre of gold requires more than a million units of electricity to saturate it.

* By Messrs Macfarlane and Playfair's experiments the maximum electromotive intensity is 364 for paraffin oil and 338 for turpentine. For air it is 73, between disks one centimetre apart. (*Trans. R. S.* Ed. 1878.) They have since found that the electric strength of the vapour of a certain liquid paraffin at 50 mm. pressure is 1·7 times that of air at the same pressure, and that the electric strength of a solid paraffin which melts at 22°·7 C. is 2·5 when liquid and 5 when solid, that of air being 1.

But we have by no means reached the limit of our experimental evidence. For Cavendish shows in Art. 49 that if in any portion of a bent canal the repulsion of overcharged bodies is so great as to drive all the fluid out of that portion, then the canal will no longer allow the fluid to run freely from one end to the other, any more than a siphon will equalize the pressure of water in two vessels, when the water does not rise to the bend of the siphon.

Hence if we could make the canal narrow enough, and the electric repulsion of bodies near the bend of the canal strong enough, we might have two conductors connected by a conducting canal but not reduced to the same potential, and this might be tested by afterwards connecting them by means of a conductor which does not pass close to any overcharged body, for this conductor will immediately reduce the two bodies to the same potential.

Such an experiment, if successful, would determine at once which kind of electricity ought to be reckoned positive, for, as Cavendish remarks in Art. 50, the presence of an undercharged body near the bend of the canal would not prevent the flow of electricity.

But even if the electric fluid were not all driven out of the canal, but only out of a stratum near the surface, the effective conducting channel would thereby be narrowed, and the resistance of the canal to an electric current increased.

Now we may construct the canal of a strip of the thinnest gold leaf, and we may measure its electric resistance to within one part in ten thousand, so that if the presence of an overcharged body near the gold leaf were to drive the electric fluid out of a stratum of it amounting to the ten thousandth part of its thickness, the alteration might be detected. Hence we must admit either that the one-fluid theory is wrong, or that every cubic centimetre of gold contains more than ten thousand million units of electricity.

The statement which Cavendish gives of the action between portions of the electric fluid and between the electric fluid and ordinary matter is nearly, but not quite, as general as it can be made.

Since the mode in which the force varies with the distance is the same in all cases, we may suppose the distance unity. Two equal portions of the electric fluid which at this distance repel each other with a force unity are defined to be each one unit of electricity.

Let the attraction between a unit of the electric fluid and a gramme of matter be a. Since we may suppose this force different for different kinds of matter, we shall distinguish the attraction due to different kinds of matter by different suffixes, as a_1 and a_2. Let the repulsion between two grammes of matter entirely deprived of electricity be r_{12}, these two portions of matter being of the kinds corresponding to the suffixes 1 and 2.

Now consider a body containing M grammes of matter and F units of the electric fluid. The repulsion between this body and a unit of the electric fluid at distance unity is

$$F - Ma. \qquad (1)$$

If this expression is zero, the body will neither repel nor attract the electric fluid. In this case the body is said to be saturated with the electric fluid, and the condition of saturation is that every gramme of matter contains a units of the electric fluid. From what we have already said, it is plain that a must be a number reckoned by thousands of millions at least. The definition of saturation as given by Cavendish is somewhat different from this, although on his own hypothesis it leads to identical results. He makes the condition of saturation to be (in Art. 6) "that the attraction of the electric fluid in any small part of the body on a given particle of *matter* shall be equal to the repulsion of the matter in the same small part on the same particle." Hence this condition is expressed by the equation

$$Fa = Mr. \tag{2}$$

But as the essential property of a saturated body is that it does not disturb the distribution of electricity in neighbouring conductors, we must consider the true definition of saturation to be that there is no action on the *electric fluid.*

Now consider two bodies of different kinds of matter M_1 and M_2, and let each of them be saturated.

The quantity of electric fluid in the first will be

$$F_1 = M_1 a_1, \tag{3}$$

and that in the second $$F_2 = M_2 a_2. \tag{4}$$

The repulsion between the two bodies will be

$$F_1 F_2 - F_1 M_2 a_2 - F_2 M_1 a_1 + M\, M_2 r_{12}, \tag{5}$$

or, substituting the values of F_1 and F_2, and changing the signs, it will be an *attraction* equal to $$M_1 M_2 (a_1 a_2 - r_{12}). \tag{6}$$

Now we know that the action between two saturated bodies is an attraction equal to $$M_1 M_2 k, \tag{7}$$

where k is the constant of gravitation.

Hence we must make $$a_1 a_2 - r_{12} = k \tag{8}$$

for every two kinds of matter, k being the same for all kinds of matter.

According to Baily's repetition of Cavendish's experiment for determining the mean density of the earth*,

$$k = 6{\cdot}506 \times 10^{-8} \frac{(\text{centimetre})^3}{\text{gramme . second}}. \tag{9}$$

This number is exceedingly small compared to the product $a_1 a_2$,

* Baily's adopted mean for the earth's density is 5·6604, which, with the values of the earth's dimensions and of the intensity of gravity at the earth's surface *used by Baily himself*, gives the above value of k as the direct result of his experiments.

which is of the order 10^{20} at least. Hence r_{12}, the repulsion between two grammes of matter entirely deprived of electricity, is of the same order as a_1a_2.

If we consider the attraction of gravitation as something quite independent of the attractions and repulsions observed in electrical phenomena, we may suppose $a_1a_2 - r_{12} = 0$, (10)
so that two saturated bodies neither attract nor repel each other.

Now we have adopted as the condition of saturation, that neither body acts on the electric fluid in the other. But since neither body acts on the other as a whole, each has no action on the matter in the other, so that our definition of saturation coincides with that given by Cavendish.

Lastly, let the two bodies not be saturated with electricity, but contain quantities $F_1 + E_1$ and $F_2 + E_2$ respectively, where $F_1 = a_1M_1$, and $F_2 = a_2M_2$, and E_1 and E_2 may be either positive or negative, provided that $F + E$ must in no case be negative.

The repulsion between the bodies is

$$(F_1 + E_1)(F_2 + E_2) - (F_1 + E_1) M_2a_2 - (F_2 + E_2) M_1a_1 + M_1M_2r_{12}, \quad (11)$$

and this by means of equations (3) (4) and (10) is reduced to

$$E_1 E_2.$$

Theory of Two Fluids.

In the theory of Two Electric Fluids, let V denote the quantity of the Vitreous fluid and R that of the Resinous.

Let the repulsion between two units of the same fluid be b, and let the attraction between two units of different fluids be c.

Let the attraction between a unit of either fluid and a gramme of matter be a, and let the repulsion between two grammes of matter be r.

If a body contains V_1 units of vitreous, R_1 units of resinous electricity, and M_1 grammes of matter, its repulsion on a unit of vitreous electricity will be $V_1b - R_1c - M_1a_1$,
and the repulsion on a unit of resinous electricity

$$-V_1c + R_1b - M_1a_1.$$

The definition of saturation is that there shall be no action on either kind of electricity. Hence, equating each of these expressions to zero, we find as the conditions of saturation

$$V_1 = R_1 = M_1 \frac{a_1}{b - c}.$$

The total repulsion between the two bodies is

$$(V_1V_2+R_1R_2)b-(V_1R_2+V_2R_1)c-(V_1+R_1)M_2a_2-(V_2+R_2)M_1a_1+M_1M_2r_{12}.$$

If we now put
$$V_1=M_1\frac{a_1}{b-c}+\tfrac{1}{2}S_1+\tfrac{1}{2}E_1,$$

$$R_1=M_1\frac{a_1}{b-c}+\tfrac{1}{2}S_1-\tfrac{1}{2}E_1,$$

$$V_2=M_2\frac{a_2}{b-c}+\tfrac{1}{2}S_2+\tfrac{1}{2}E_2,$$

$$R_2=M_2\frac{a_2}{b-c}+\tfrac{1}{2}S_2-\tfrac{1}{2}E_2,$$

the total repulsion becomes

$$M_1M_2\left(r_{12}-\frac{2a_1a_2}{b-c}\right)+E_1E_2\frac{b+c}{2}+S_1S_2\frac{b-c}{2}-S_1M_2a_2-S_2M_1a_1.$$

The first term of this expression, with its sign reversed, represents the attraction of gravitation, and the second term represents the observed electric action, but the other terms represent forces of a kind which have not hitherto been observed, and we must modify the theory so as to account for their non-existence.

One way of doing so is to suppose $b=c$ and $a_1=a_2=0$. The result of this hypothesis is to reduce the condition of saturation to that of the equality of the two fluids in the body, leaving the amount of each quite undetermined. It also fails to account for the observed action between the bodies themselves, since there is no action between them and the electric fluids.

The other way is to suppose that $S_1=S_2=0$, or that the sum of the quantities of the two fluids in a body always remains the same as when the body is saturated. This hypothesis is suggested by Priestley in his account of the two-fluid theory, but it is not a dynamical hypothesis, because it does not give a physical reason why the sum of these two quantities should be incapable of alteration, however their difference is varied.

The only dynamical hypothesis which appears to meet the case is to suppose that the vitreous and resinous fluids are both incompressible, and that the whole of space not occupied by matter is occupied by one or other of them. In a state of saturation they are mixed in equal proportions.

The two-fluid theory is thus considerably more difficult to reconcile with the facts than the one-fluid theory.

NOTE 2, ARTS. 27 AND 282.

The problem of the distribution, in a sphere or ellipsoid, of a fluid, the particles of which repel each other with a force varying inversely as the nth power of the distance, has been solved by Green*. Green's method is an extremely powerful one, and allows him to take account of the effect of any given system of external forces in altering the distribution.

If, however, we do not require to consider the effect of external forces, the following method enables us to solve the problem in an elementary manner. It consists in dividing the body into pairs of corresponding elements, and finding the condition that the repulsions of corresponding elements on a given particle shall be equal and opposite.

(1) *Specification of Corresponding Points on a line.*

Let A_1A_2 be a finite straight line, let P be a given point in the line, and let Q_1 and Q_2 be corresponding points in the segments A_1P and PA_2 respectively, the condition of correspondence being

$$\frac{1}{Q_1P} - \frac{1}{A_1P} = \frac{1}{PQ_2} - \frac{1}{PA_2}. \tag{1}$$

It is easy to see that when Q_1 coincides with A_1, Q_2 coincides with A_2, and that as Q_1 moves from A_1 to P, Q_2 moves in the opposite direction from A_2 to P, so that when Q_1 coincides with P, Q_2 also coincides with P.

Let Q_1' and Q_2' be another pair of corresponding points, then

$$\frac{1}{Q_1'P} - \frac{1}{A_1P} = \frac{1}{PQ_2'} - \frac{1}{PA_2}. \tag{2}$$

Subtracting (1) from (2)

$$\frac{1}{Q_1'P} - \frac{1}{Q_1P} = \frac{1}{PQ_2'} - \frac{1}{PQ_2}, \tag{3}$$

or

$$\frac{Q_1Q_1'}{Q_1'P \,.\, Q_1P} = \frac{Q_2'Q_2}{PQ_2' \,.\, PQ_2}. \tag{4}$$

If the points Q_1 and Q_1' are made to approach each other and ultimately

* "Mathematical Investigations concerning the laws of the equilibrium of fluids analogous to the electric fluid, with other similar researches," *Transactions of the Cambridge Philosophical Society*, 1833. Read Nov. 12, 1832. See Mr Ferrers' Edition of Green's Papers, p. 119.

to coincide, Q_1Q_1' ultimately becomes the fluxion of Q, which we may write $Q_1^{\bullet}$, and we have

$$\frac{Q_1^{\bullet}}{Q_1P^2} = \frac{Q_2^{\bullet}}{PQ_2^{\,2}}, \tag{5}$$

or corresponding elements of the two segments are in the ratio of the squares of their distances from P.

Let us now suppose that A_1PA_2 is a double cone of an exceedingly small aperture, having its vertex at P; let us also suppose that the density of the redundant fluid at Q_1 is ρ_1, and at Q_2 is ρ_2; then since the areas of the sections of the cone at Q_1 and Q_2 are as the squares of the distances from P, and since the lengths of corresponding elements are also, by (5), as the squares of their distances from P, the quantities of fluid in the two corresponding elements at Q_1 and Q_2 are as $\rho_1Q_1P^4$ to $\rho_2PQ_2^{\,4}$. If the repulsion is inversely as the nth power of the distance, the condition of equilibrium of a particle of the fluid at P under the action of the fluid in the two corresponding elements at Q_1 and Q_2 is

$$\rho_1Q_1P^{4-n} = \rho_2PQ_2^{\,4-n}. \tag{6}$$

We have now to show how this condition may be satisfied by one and the same distribution of the fluid when P is any point within an ellipsoid or a sphere. We must therefore express ρ so that its value is independent of the position of P.

Transposing equation (1) we find—

$$\frac{1}{Q_1P} + \frac{1}{PA_2} = \frac{1}{PQ_2} + \frac{1}{A_1P}. \tag{7}$$

Multiplying the corresponding members of equations (1) and (7) and omitting the common factor $A_1P\,.\,PA_2$,

$$\frac{A_1Q_1\,.\,Q_1A_2}{Q_1P^2} = \frac{A_1Q_2\,.\,Q_2A_2}{PQ_2^{\,2}}, \tag{8}$$

we may therefore write, instead of equation 6,

$$\rho_1(A_1Q_1\,.\,Q_1A_2)^{\frac{4-n}{2}} = \rho_2(A_1Q_2\,.\,Q_2A_2)^{\frac{4-n}{2}}. \tag{9}$$

Let us now suppose that A_1A_2 is a chord of the ellipsoid, whose equation is

$$\frac{x^2}{a^2} + \frac{y^2}{b^2} + \frac{z^2}{c^2} = 1. \tag{10}$$

If we write

$$1 - \frac{x^2}{a^2} - \frac{y^2}{b^2} - \frac{z^2}{c^2} = p^2, \tag{11}$$

then the product of the segments of the chord at Q_1 is to the product of the segments at Q_2 as the values of p^2 at these points respectively, or

$$A_1Q_1\,.\,Q_1A_2 : A_1Q_2\,.\,Q_2A_2 :: p_1^2 : p_2^2. \tag{12}$$

We may therefore write, instead of equation (9),

$$\rho_1 p_1^{4-n} = \rho_2 p_2^{4-n}. \tag{13}$$

If, therefore, throughout the ellipsoid,

$$\rho = Cp^{n-4}, \tag{14}$$

where C is constant, every particle of the fluid within the ellipsoid will be in equilibrium.

We have in the next place to determine whether a distribution of this kind is physically possible.

Let E be the quantity of redundant fluid in the ellipsoid,

$$E = C\int_0^1 p^{n-4}\, 4\pi abc\, p\,(1-p^2)^{\frac{1}{2}}\,dp$$

$$= 4\pi abc\, C\int_0^1 p^{n-3}(1-p^2)^{\frac{1}{2}}\,dp \tag{15}$$

$$= 2\pi abc\, C\,\frac{\Gamma\left(\frac{3}{2}\right)\Gamma\left(\frac{n-2}{2}\right)}{\Gamma\left(\frac{n+1}{2}\right)}. \tag{16}$$

Let ρ_0 be the density of the redundant fluid if it had been uniformly spread through the volume of the ellipsoid, then

$$E = \frac{4\pi}{3}\, abc\,\rho_0, \tag{17}$$

and if ρ is the actual density of the redundant fluid,

$$\rho = \rho_0 \cdot \frac{2}{3}\,\frac{\Gamma\left(\frac{n+1}{2}\right)}{\Gamma\left(\frac{3}{2}\right)\Gamma\left(\frac{n-2}{2}\right)}\, p^{n-4}. \tag{18}$$

When n is not less than 2, there is no difficulty about the interpretation of this result.

The density of the redundant fluid is everywhere positive.

When $n = 4$ it is everywhere uniform and equal to ρ_0.

When n is greater than 4 the density is greatest at the centre and is zero at the surface, that is to say, in the language of Cavendish, the matter at the surface is saturated.

When n is between 2 and 4 the density of the redundant fluid at the centre is positive and it increases towards the surface. At the surface itself the density becomes infinite, but the quantity collected on the surface is insensible compared with the whole redundant fluid.

When n is equal to 2, $\Gamma\left(\frac{n-2}{2}\right)$ becomes infinite, and the value of ρ is zero for all points within the ellipsoid, so that the whole charge is collected on the surface, and the interior parts are exactly saturated, and this we find to be consistent with equilibrium.

When n is less than 2 the integral in equation (15) becomes infinite. Hence if we assume a value for C in the interior parts of the ellipsoid, we cannot extend the same law of distribution to the surface without introducing an infinite quantity of redundant fluid. We might therefore conclude that if the quantity of redundant fluid is given, we must make $C=0$, and suppose the redundant fluid to be all collected at the surface, and the interior to be exactly saturated. But, on trying this distribution, we find that it is not consistent with equilibrium. For when n is less than 2, the effect of a shell of fluid on a particle within it is a force directed from the centre. If, therefore, a sphere of saturated matter is surrounded by a shell of electric fluid, the fluid in the sphere will be drawn towards the shell, and this process will go on till the different parts of the interior of the sphere are rendered undercharged to such a degree that each particle of fluid in the sphere is as much attracted to the centre by the matter of the sphere as it is repelled from it by the fluid in the sphere and the shell together. This is the same conclusion as that stated by Cavendish.

Green solves the problem, on the hypothesis of two fluids, in the following manner.

Suppose that the sphere, when saturated, contains a finite quantity, E, of the positive fluid, and an equal quantity of the negative fluid, and let a quantity, Q, of one of them, say the positive, be introduced into the sphere.

Let the whole of the positive fluid be spread uniformly over the surface of the sphere whose radius is a, so that if P' is the surface-density,

$$4\pi a^2 P' = E + Q.$$

Green then considers the equilibrium of fluid in an inner and concentric sphere of radius b, acted on by the fluid in the surface whose radius is a, and shows that if the density of the fluid is

$$\rho = \frac{2}{\pi} P'a \sin\frac{n-2}{2}\pi\,(a^2-b^2)^{\frac{2-n}{2}}(a^2-r^2)^{-1}(b^2-r^2)^{\frac{n-2}{2}},$$

there will be equilibrium of the fluid within the inner sphere.

The value of ρ is evidently negative if n is less than 2.

Green then determines, from this value of the density, the whole quantity of fluid within the sphere whose radius is b, and then by equating this to $-E$, the whole quantity of negative fluid, he determines the radius, b, of the inner sphere, so that it shall just contain the whole of the negative fluid.

The whole of the positive fluid is thus condensed on the outer surface, the whole of the negative fluid distributed within the inner sphere, and the shell between the two spherical surfaces is entirely deprived of both fluids.

At the outer surface, the force on the positive fluid is from the centre, but the fluid there cannot move, because it is prevented by the insulating medium which surrounds the sphere.

In the shell between the two spherical surfaces the force on the positive fluid would be from the centre. Hence if any positive fluid enters this shell, it will be driven to the outer surface, and if any negative fluid enters, it will be driven to the inner surface.

But all the positive fluid is already at the outer surface, and all the negative fluid is already in the inner sphere, where, as Green has shown, it is in equilibrium, and thus the fluids are in equilibrium throughout the sphere.

It may be remarked that this solution, according to which a certain portion of matter becomes entirely deprived of both fluids, is inconsistent with the ordinary statements of the theory of two fluids, which usually assert that bodies, under all circumstances, contain immense quantities of both fluids.

In the two-fluid theory, by depriving matter of both fluids, we get an inactive substance which gives us no trouble, but in the one-fluid theory, matter deprived of fluid exerts a strong attraction on the fluid, the consideration of which would considerably complicate the mathematical problem.

Infinite plate with plane parallel surfaces.

The distribution of the fluid in an infinite plate with plane parallel surfaces is given in the general solution which we have obtained for a body bounded by a quadric surface, namely, $\rho = Cp^{n-4}$.

In the case of the plate we must suppose it bounded by the planes $x = +a$, and $x = -a$, and then p is defined by the equation

$$x^2 = a^2(1 - p^2).$$

If σ is the quantity of fluid in a portion of the plate whose area is unity,

$$\sigma = \int_{-a}^{+a} \rho dx = Ca \frac{\Gamma\left(\frac{n-2}{2}\right)\Gamma\left(\frac{1}{2}\right)}{\Gamma\frac{n-1}{2}}.$$

Thin disk.

The distribution in an infinitely thin disk may be deduced from that in an ellipsoid by making one of the axes infinitely small. It is better however to proceed by the method which we have already employed, only that instead of supposing the line A_1PA_2 (Fig. p. 368) to be a double cone, we suppose it to be a double sector cut from the disk. The breadth of this sector is proportional to the distance from P, so that the condition of equilibrium of the repulsions of two corresponding elements whose surface-densities are σ_1 and σ_2 is

$$\sigma_1 Q_1 P_1^{3-n} = \sigma_2 Q_2 P_2^{3-n},$$

whence we find, as before, that if the equation of the edge of the disk is

$$\frac{x^2}{a^2} + \frac{y^2}{b^2} = 1,$$

and if
$$1 - \frac{x^2}{a^2} - \frac{y^2}{b^2} = p^2,$$

then the surface-density at any point is

$$\sigma = Cp^{n-3}.$$

The quantity of fluid in the disk is found by integrating over the surface of the disk, and is

$$Q = \frac{2\pi ab\, C}{n-1}.$$

Hence if σ_0 is the mean surface-density, the surface-density at any point is given by the equation

$$\sigma = \frac{n-1}{2} \sigma_0 p^{n-3}.$$

Thin rod.

The distribution on an infinitely thin rod is found by considering A_1PA_2 a rod of uniform section, which leads to the equation

$$\lambda_1 Q_1 P^{2-n} = \lambda_2 P Q_2^{2-n},$$

where λ is the linear density, and if the length of the rod is $2a$, and if x is the distance from the middle, and $x^2 = a^2(1-p^2)$, the distribution of the linear density is given by

$$\lambda = Cp^{n-2}.$$

The charge of the whole rod is

$$Ca \frac{\Gamma\left(\frac{n}{2}\right)\Gamma\left(\frac{1}{2}\right)}{\Gamma\left(\frac{n+1}{2}\right)} = 2,$$

so that if λ_0 denotes the mean linear density,

$$\lambda = \lambda_0 \frac{2\Gamma\left(\frac{n+1}{2}\right)}{\Gamma\left(\frac{n}{2}\right)\Gamma(\frac{1}{2})} p^{n-2},$$

when $n = 2$, $\lambda = \lambda_0$, or the density is uniform.

Since the fluid is in equilibrium in all these cases, the potential is uniform throughout the body. We may therefore determine the value of the potential at any point within the body by finding its value at any selected point, as for instance at the centre. If de be an element of the fluid, and r its distance from the given point, the corresponding element of the potential due to the force whose value is er^{-n} is $\frac{1}{n-1}er^{1-n}$.

We thus find for the potential of the sphere

$$V = C 2\pi a^{4-n} \frac{1}{n-1} \Gamma\frac{n-2}{2} \Gamma\frac{4-n}{2}$$

$$= Qa^{1-n} \frac{1}{n-1} \frac{\Gamma\frac{n+1}{2}\Gamma\frac{4-n}{2}}{\Gamma\frac{3}{2}}.$$

When n becomes equal to 4, V becomes infinite.

When n is equal to 2, $V = Qa^{-1}$.

For the plate bounded by parallel planes, V is infinite, except for values of n between 3 and 4, for which

$$V = \frac{2\pi\sigma_0}{(n-1)(n-3)} a^{3-n} \frac{\Gamma\left(\frac{n-1}{2}\right)\Gamma\left(\frac{4-n}{2}\right)}{\Gamma(\frac{1}{2})},$$

where σ_0 is the quantity of fluid in unit of area of the plate.

For a circular disk

$$V = Qa^{1-n}\tfrac{1}{2}\,.\,\Gamma\left(\frac{n-1}{2}\right).\,\Gamma\left(\frac{3-n}{2}\right),$$

in which n must be between 1 and 3.

When $n = 2$, $V = \frac{\pi}{2}Qa^{-1}$.

For an infinitely narrow rod

$$V = Qa^{1-n} \frac{\Gamma\left(\frac{n+1}{2}\right)\Gamma\left(\frac{2-n}{2}\right)}{(n-1)}.$$

NOTE 3, ART. 69.

On canals of incompressible fluid.

It appears from several passages (Arts. 40, 236, 273, 276, 278, 294, 348) that Cavendish considered that the weakest point in his theory was the assumption that the condition of electric equilibrium between two conductors connected by a fine wire is the same as if, instead of the wire, there were a canal of incompressible fluid defined as in Art. 69.

It is true that the properties of the electric fluid, as defined by Cavendish in Art. 3, are very different from those of an incompressible fluid. But it is easy to show that the results deduced by Cavendish from the hypothesis of a canal of incompressible fluid are applicable to the actual case in which the bodies are connected by a fine wire.

In what follows, when we speak of the electrified body or bodies, the canal or the wire is understood not to be included unless it is specially mentioned.

Cavendish supposes the canal to be everywhere exactly saturated with the electric fluid, and that the only external force acting on the fluid in the canal is that due to the electrification of the other bodies.

Since this resultant force is not in general zero at all points of the canal, the fluid in the canal cannot be in equilibrium unless it is prevented from moving by some other force. Now the condition of incompressibility excludes any such displacement of the fluid as would alter the quantity of fluid in a given volume, and the stress by which such a displacement is resisted is called isotropic (or hydrostatic) pressure. In a hypothetical case like this it is best, for the sake of continuity, to suppose that negative as well as positive values of the pressure are admissible.

In the electrified bodies themselves the properties of the fluid are those defined in Art. 3. The fluid is therefore incapable of sustaining pressure except when its particles are close packed together, and as it cannot sustain a negative pressure, the pressure must be zero in the electrified bodies, and therefore also in the canal at the points where it meets these bodies.

The condition of equilibrium of the fluid in the canal is

$$\rho \frac{dV}{ds} + \frac{dp}{ds} = 0,$$

where V denotes the potential of the electric forces due to the electrified bodies, ρ the density, and p the pressure of the fluid in the canal, and s the length of the canal reckoned from a fixed origin to the point under consideration.

Since by the hypothesis of incompressibility, ρ is constant,

$$\rho V + p = C,$$

where C is a constant; and if we distinguish by suffixes the symbols belonging to the two ends of the canal where it meets the bodies A_1 and A_2,

$$\rho V_1 + p_1 = \rho V_2 + p_2.$$

But we have seen that $p_1 = p_2 = 0$. Hence dividing by ρ we find for the condition of equilibrium

$$V_1 = V_2,$$

or the electric potential of the two bodies must be equal.

We arrive at precisely the same condition if we suppose the bodies connected by a fine wire which is made of a conducting substance.

Let V as before be the potential at any given point due to the electrified bodies, and let V_1 be its value in A_1, and V_2 its value in A_2, and let V' be the potential due to the electrification of the wire at the given point, then the condition of equilibrium of the electricity in the wire is that $V + V'$ must be constant for all points within the substance of the wire. Hence at the two ends of the wire

$$V_1 + V_1' = V_2 + V_2'.$$

Hence the actual potential due to the bodies and the wire together is the same in A_1 and A_2.

The only difference, then, between the actual case of the wire and the hypothetical case of the canal is that the surface of the wire is charged with electricity in such a way as to make its potential everywhere constant, whereas the canal is exactly saturated, and the effect of variation of potential is counteracted by variation of pressure.

Hence the canal produces no effect in altering the electrical state of the other bodies, whereas the wire acts like any other body charged with electricity.

The charge of the wire, however, may be diminished without limit by diminishing its diameter. It is approximately inversely proportional to the logarithm of the ratio of a certain length to the diameter of the wire. Hence by making the wire fine enough, the disturbance of the distribution of electricity on the bodies may be made as small as we please.

From the Preface to Green's "Essay on the Application of Mathematical Analysis to the Theories of Electricity and Magnetism."

"CAVENDISH, who having confined himself to such simple methods as may readily be understood by any one possessed of an elementary knowledge of geometry and fluxions, has rendered his paper accessible to a great number of readers; and although, from subsequent remarks, he appears dissatisfied with an hypothesis which enabled him to draw some important conclusions, it will readily be perceived, on an attentive perusal of his paper, that a trifling alteration will suffice to render the whole perfectly legitimate.

In order to make this quite clear, let us select one of Cavendish's propositions, the twentieth for instance [Art. 71], and examine with some attention the method there employed. The object of this proposition is to show, that when two similar conducting bodies communicate by means of a long slender canal, and are charged with electricity, the respective quantities of redundant fluid contained in them will be proportional to the $n-1$ power of their corresponding diameters; supposing the electric repulsion to vary inversely as the n power of the distance.

This is proved by considering the canal as cylindrical, and filled with incompressible fluid of uniform density: then the quantities of electricity in the interior of the two bodies are determined by a very simple geometrical construction, so that the total action exerted on the whole canal by one of them shall exactly balance that arising from the other; and from some remarks in the 27th proposition [Arts. 94, 95] it appears the results thus obtained agree very well with experiments in which real canals are employed, whether they are straight or crooked, provided, as has since been shown by Coulomb, n is equal to two. The author, however, confesses he is by no means able to demonstrate this, although, as we shall see immediately, it may very easily be deduced from the propositions contained in this paper.

For this purpose let us conceive an incompressible fluid of uniform density, whose particles do not act on each other, but which are subject to the same actions from all the electricity in their vicinity, as real electric fluid of like density would be; then supposing an infinitely thin canal of this hypothetical fluid, whose perpendicular sections are all equal and similar, to pass from a point a on the surface of one of the bodies through a portion of its mass, along the interior of the real canal, and through a part of the other body, so as to reach a point A on its surface, and then proceed from A to a in a right line, forming thus a closed circuit, it is evident from the principles of hydrostatics, and may be proved from our author's 23rd proposition [Art. 84], that the whole of the hypothetical canal will be in equilibrium, and as every particle of the portion contained within the system is necessarily so, the rectilinear portion aA must therefore be in equilibrium.

This simple consideration serves to complete Cavendish's demonstration, whatever may be the form or thickness of the real canal, provided the quantity of electricity in it is very small compared with that contained in the bodies.

An analogous application of it will render the demonstration of the 22nd proposition [Art. 74] complete, when the two coatings of the glass plate communicate with their respective conducting bodies by fine metallic wires of any form."

NOTE 4, ART. 83.

On the charges of two equal parallel disks, the distance between them being small compared with the radius.

The theory of two parallel disks, charged in any way, may be deduced from the consideration of two principal cases.

The first case is when the potentials of the two disks are equal. If the distance between the disks is very small compared with their diameter, we may consider the whole system as a single disk, the charge of which is approximately the same as if it were infinitely thin. Hence if V be the potential, and if we write A for the capacity of the first disk, and B for the coefficient of induction between the two disks, the charge of the first disk is

$$Q_1 = AV_1 - BV_2,$$

and that of the second is

$$Q_2 = AV_2 - BV_1.$$

If we make

$$V_1 = V_2 = V,$$

$$Q_1 + Q_2 = 2(A - B)V.$$

Hence, by note 2,

$$A - B = \frac{a^{n-1}}{\Gamma\left(\frac{n-1}{2}\right)\Gamma\left(\frac{3-n}{2}\right)}.$$

The second case is when the charges of the disks are equal and opposite. The surface-density in this case is approximately uniform except near the edges of the disks. I have not attempted to ascertain the amount of accumulation near the edge except when $n = 2$. If we suppose the density uniform, then for a charge of the first disk equal to πa^2, its potential, when b the distance between the disks is small compared with a the radius, will be approximately

$$V_1 = \frac{2\pi}{(n-1)(3-n)} b^{3-n}.$$

Hence, since $V_2 = -V_1$

$$A + B = \tfrac{1}{2}(n-1)(3-n)a^2 b^{n-3},$$

and we find

$$A = \tfrac{1}{4}(n-1)(3-n)a^2 b^{n-3} + \frac{a^{n-1}}{2\Gamma\left(\frac{n-1}{2}\right)\Gamma\left(\frac{3-n}{2}\right)},$$

$$B = \tfrac{1}{4}(n-1)(3-n)a^2 b^{n-3} - \frac{a^{n-1}}{2\Gamma\left(\frac{n-1}{2}\right)\Gamma\left(\frac{3-n}{2}\right)}.$$

When $n=2$,

$$A=\tfrac{1}{4}\frac{a^2}{b}+\frac{a}{2\pi},$$

$$B=\tfrac{1}{4}\frac{a^2}{b}-\frac{a}{2\pi}.$$

In this case, however, we can carry the approximation further, for it is shown in Note 20 that

$$A-B=\frac{1}{\pi}\left(a+\frac{1}{2\pi}b\log\frac{a}{b}\right).$$

It is shown in "Electricity and Magnetism," Art. 202, that when two disks are charged to equal and opposite potentials, the density near the edge of each disk is greater than at a distance from it, and the whole charge is the same as if a strip of breadth $\frac{b}{2\pi}$ had been added all round the disk.

Hence
$$A+B=\frac{1}{2b}\left(a+\frac{b}{2\pi}\right)^2$$
$$=\frac{a^2}{2b}+\frac{a}{2\pi}+\frac{b}{8\pi^2},$$

and
$$A=\frac{a^2}{4b}+\frac{3}{4\pi}a+\frac{1}{4\pi^2}b\left(\log\frac{a}{b}+\tfrac{1}{4}\right),$$
$$B=\frac{a^2}{4b}-\frac{1}{4\pi}a-\frac{1}{4\pi^2}b\left(\log\frac{a}{b}-\tfrac{1}{4}\right).$$

Note 5, Art. 90.

This proposition seems intended to justify those experimental methods in which the potential of the earth is assumed as the zero of potential.

Cavendish, by introducing the idea of degrees of electrification, as distinguished from the magnitudes of overcharge and undercharge, very nearly attained to the position of those who are in possession of the idea of potential. But the very form of the phrases "positively or negatively electrified," which Cavendish uses, confers an importance on the limiting condition of "no electrification," which we hardly think of attributing to "zero potential." For we know that all electrical phenomena depend on differences of potential, and that the particular potential which we assume for our zero may be chosen arbitrarily, because it does not involve any physical consequences.

It is true that the mathematicians define the zero of potential as the potential at an infinite distance from the finite system which includes the electric charges. This, however, is not a definition of which the experimentalist can avail himself, so he takes the potential of the earth as a zero accessible to all terrestrial electricians, and each electrician "makes his own earth."

The earth-connexion used by Cavendish is described in Art. 258. But when the whole apparatus of an electrical experiment is contained in a moderate space, such as a room, it is convenient to make an artificial "earth" by connecting by metal wires the case of the electrometer with all those parts of the apparatus which are intended to be at the same potential, and calling this potential zero.

It appears by observation, that in fine weather the electric potential at a point in the air increases with the distance from the earth's surface up to the greatest heights reached by observers, and in all parts of the earth. It is only when there are considerable disturbances in the atmosphere that the potential ever diminishes as the height increases. Hence the potential of the earth is probably always less than that of the highest strata of the atmosphere.

If the earth and its atmosphere together contain just as much electricity as will saturate them, and if there is no free electricity in the regions beyond, then the potential of the outer stratum of the atmosphere will be the same as that at an infinite distance, that is, it will be the zero of the mathematical theory, and the potential of the earth will be negative.

NOTE 6, ART. 97, p. 43.

On the Molecular Constitution of Air.

The theory of Sir Isaac Newton here referred to is given in the *Principia*, Lib. II., Prop. XXIII.

Newton supposes a constant quantity of air enclosed in a cubical vessel which is made to vary so as to become a cube of greater or smaller dimensions. Then since by Boyle's law the product of the pressure of the air on unit of surface into the volume of the cube is constant; and since the volume of the cube is the product of the area of a face into the edge perpendicular to it, it follows that the product of the total pressure on a face of the cube into the edge of the cube is constant, or the total pressure on a face is inversely as the edge of the cube.

Now if an imaginary plane be drawn through the cube parallel to one of its faces, the mutual pressure between the portions of air on opposite sides of this plane is equal to the pressure on a face of the cube. But the number of particles is the same, and their configuration is geometrically similar whether the cube is large or small. Hence the distance between any two given molecules must vary as the edge of the cube, and the force between the two molecules must vary as the total force between the sets of molecules separated by the imaginary plane, and therefore the product of the repulsion between two given molecules into the distance between them must be constant, in other words the repulsion varies inversely as the distance.

In this demonstration the repulsion considered is that between two *given* molecules, and it is shown that this must vary inversely as the distance between them in order to account for Boyle's law of the elasticity of air.

If, however, we suppose the same law of repulsion to hold for every pair of molecules, Newton shows in his Scholium that it would require a greater pressure to produce the same density in a larger mass of air.

We must therefore suppose that the repulsion exists, not between every pair of molecules, but only between each molecule and a certain definite number of other molecules, which we may suppose to be defined as those nearest to the given molecules. Newton gives as an example of such a kind of action the attraction of a magnet, the field of which is contracted when a plate of iron is interposed, so that the attractive power appears to be bounded by the nearest body attracted.

If the repulsion were confined to those molecules which are within a certain *distance* of each other, then, as Cavendish points out, the pressure arising from this repulsion would vary nearly as the square of the density, provided a large number of molecules are within this distance. Hence this hypothesis will not explain the fact that the pressure varies as the density.

On the other hand, if the repulsion were limited to particular pairs of particles, then since the particles are free to move, these pairs of particles would move away from each other till only those particles were near each other between which the repulsive force is supposed not to exist.

It would appear therefore that the hypothesis stated by Newton and adopted by Cavendish is the only admissible one, namely, that the repulsive force is inversely as the distance, but is exerted only between the nearest molecules.

Newton's own conclusion to his investigation of the properties of air on the statical molecular hypothesis is as follows:—"An vero Fluida Elastica ex particulis se mutuo fugantibus constent, Quæstio Physica est. Nos proprietatem Fluidorum ex ejusmodi particulis constantium mathematice demonstravimus, ut Philosophis ansam præbeamus Quæstionem illam tractandi."

The theory that the molecules of elastic fluids are in motion satisfies the conditions of the question as pointed out by Newton in a much more natural manner than any modification of the statical hypothesis.

According to the kinetic theory of gases, each molecule is in motion, and this motion is during the greater part of its course undisturbed by the action of other molecules, and is therefore uniform and in a straight line. When however it comes very near another molecule, the two molecules act on each other for a very short time, the courses of both are changed and they go on in the new courses till they encounter other molecules.

It would appear from the observed properties of gases that the mutual action between two molecules is insensible at all sensible distances. As the molecules approach, the action is at first attractive, but soon changes to a repulsive force of far greater magnitude, so that the general character of the encounter depends mainly on the repulsive force.

On this theory, the elasticity of the gas may still be said in a certain sense to arise from the repulsive force between its molecules, only instead of this repulsive force being in constant action, it is called into play only during the encounters between two molecules. The intensity of the impulse is not the same for all encounters, but as it does not depend on the interval between the encounters, we may consider its mean value as constant. The average value of the force between two molecules is in this case the value of the impulse divided by the time between two encounters. Hence we may say that the force is inversely as the distance between the molecules, and that it acts between those molecules only which encounter each other.

For an earlier investigation by Cavendish of the properties of an elastic fluid, see Note 18.

Note 7, Art. 101.

Here Cavendish endeavours to fix a precise meaning to the terms "positively and negatively electrified," terms which he found current among electricians, but not well defined. The meaning which he here fixes to them, and which he afterwards makes much use of, is equivalent to the meaning of the modern term potential, as used by practical electricians. The idea of potential as used by mathematicians is expressed by Cavendish in his theory of canals of incompressible fluid.

In the "Thoughts concerning Electricity," and in the unpublished papers, degrees of electrification are spoken of. These degrees of electrification are measured in the experimental researches by means of electrometers of different kinds, and since he has compared the indications of his electrometers with the degrees of electrification required to make a spark pass between the balls of Lane's discharging electrometer, we may express all these measurements in modern units, though Cavendish's original electrometers no longer exist.

I have not been able to trace the idea of electric potential in the work of Œpinus, so that Cavendish seems the first to have made use of it. The relation between the charge of a body and the degree of its electrification is the main object of Cavendish's experimental researches, and the results of his work were expressed in the material form of a collection of coated plates, each of which had a capacity equal to that of a sphere of known diameter.

The leading idea in the great experimental work of Coulomb seems to be the measurement of the charges of the different bodies of a system and of parts of these bodies. Perhaps the most valuable of Coulomb's many contributions to experimental physics was the measurement of the surface-density of the distribution of electricity on a conductor on different parts of its surface by means of the proof plane. The numerical results obtained by Coulomb led directly to the great mathematical work of Poisson. I have not been able, however, to trace, even in those parts of Coulomb's papers where it would greatly facilitate the exposition, any idea of potential as a quantity which has the same value for all parts of a system of conductors communicating with each other.

NOTE 8, p. 51.

Cases of Attraction and Repulsion.

The statements of Cavendish may be illustrated by the case of two spheres A and B, whose radii are a and b, and the distance between their centres c.

If the charge of A is 1, and that of B is 0, the attraction is

$$2\frac{b^3}{c^5} + 3\frac{b^5}{c^7} + 4\frac{b^7}{c^9} + 5\frac{b^9 + 4b^6a^3}{c^{11}} + \&c.$$

an expression which shows that it depends chiefly on the value of b, the radius of the sphere without charge.

If the sphere B, instead of being without charge, is at potential zero, that is, if it is not insulated, the attraction is

$$\frac{b}{c^3} + 2\frac{b^3}{c^5} + 3\frac{a^3b^2 + b^5}{c^7} + \&c.$$

This expression exceeds the former by

$$\frac{b}{c^3} + 3\frac{a^3b^2}{c^7} + \&c.$$

The number of times that the attraction of an uninsulated sphere exceeds that of a sphere without charge is therefore approximately

$$2\frac{c^2}{b^2},$$

which is greater as the sphere is smaller. This agrees with what Cavendish says in Art. 108.

With respect to two bodies at the same potential, Cavendish remarks in Art. 113, that it may be said that one of them may be rendered undercharged in the part nearest to the other, and he shows that even in this case, the two bodies must repel each other. But it may be shown that each of the bodies must be overcharged in every part of its surface. For in the first place no part can be undercharged, for the lines of force which terminate in an undercharged surface must have come from an overcharged surface at which the potential is higher than at the surface. But there is no body in the field at a higher potential than the two bodies considered. Hence no part of their surface can be undercharged.

Nor can any finite part of the surface be free from charge, for it may be shown that if a finite portion of the surface of a conductor is free from charge, every point which can be reached by continuous motion from that part of the surface without passing through an electrified surface must be at the same potential. Hence no finite portion of a surface can be free from charge, unless the whole surface is free from charge.

Note 9, Art. 124.

The rate at which electricity passes from a conductor to the surrounding air or from the surrounding air to a conductor was believed to be much greater by Cavendish and his contemporaries than is consistent with modern experiments. Judging from the statements of the electricians of each generation, it would seem as if this rate had been diminishing steadily during the last hundred years in exact correspondence with the improvements which have been made in the construction of solid insulating supports for electrified conductors.

Whenever the intensity of the electromotive force at the surface of a conductor is sufficiently great, the air no doubt becomes charged.* This is the case at a sharp point connected with the conductor even when the potential is low, but when the curvature of the surface is continuous and gentle, the conductor must be raised to a high potential before any discharge to air begins to take place.

Thus in Thomson's portable electrometer, in which there are two disks placed parallel to each other at different potentials, the percentage loss of electricity from day to day is very small, and seems to depend principally on the solid insulators, for when the disks are placed very near each other, less loss is observed than when they are further apart, though the intensity of the force urging the electricity through the intervening stratum of air is greater the nearer the disks are to each other.

* M. R. Nahrwold (Wiedemann's *Annalen* v. (1878) p. 440) finds that the discharge from a sharp point communicates a charge to dusty air which can be detected in the air for some time afterwards. This does not occur in air free from dust. But the discharge from an incandescent platinum wire communicates a lasting charge even to air free from dust.

On the surface density of electricity near the vertex of a cone.

Green has given in a note to his Essay, section (12), the following results of an investigation which, so far as I am aware, he never published.

"Since this was written, I have obtained formulæ serving to express, generally, the law of the distribution of the electric fluid near the apex O of a cone, which forms part of a conducting surface of revolution having the same axis. From these formulæ it results that, when the apex of the cone is directed inwards, the density of the fluid at any point p, near to it, is proportional to r^{n-1}; r being the distance Op, and the exponent n very nearly such as would satisfy the simple equation

$$(4n+2)\,\beta = 3\pi,$$

where 2β is the angle at the summit of the cone.

If 2β exceeds π, this summit is directed outwards, and when the excess is not very considerable, n will be given as above: but 2β still increasing, until it becomes $2\pi - 2\gamma$, the angle 2γ at the summit of the cone, which is now directed outwards, being very small, n will be given by

$$2n \log \frac{2}{\gamma} = 1,$$

and in case the conducting body is a sphere whose radius is b, on which P represents the mean density of the electric fluid; ρ, the value of the density near the apex O, will be determined by the formula

$$\rho = \frac{2Pbn}{(a+b)\,\gamma} \left(\frac{r}{a}\right)^{n-1},$$

a being the length of the cone.

Professor F. G. Mehler* of Elbing has investigated the distribution of electricity on a cone under the influence of a charged point on the axis, and the inverse problem of the distribution on a spindle formed by the revolution of the segment of a circle about its chord.

He finds that when the segment is a very small portion of the circle, so that the conical points of the spindle are very acute, the surface-density at any point is inversely proportional to the product of the distances of that point from the two conical points.

* Ueber eine mit den Kugel- und Cylinderfunctionen verwandte Function, und ihre Anwendung in der Theorie der Electricitätsvertheilung. (Elbing, 1870.)

NOTE 10, p. 63.

Sir W. Thomson* has determined in absolute measure the electromotive force required to produce a spark in air between two electrodes in the form of disks, one of which was plane, and the other slightly convex, placed at different distances from each other. Mr Macfarlane† has recently made a more extensive series of experiments on the disruptive discharge of electricity. He finds that in air at the ordinary pressure and temperature the electromotive force required to produce a spark between disks, 10 cm. diameter, and from 1 to 0·025 cm. apart, is expressed by the empirical equation

$$V = 66{\cdot}940\,(s^2 + {\cdot}20503s)^{\frac{1}{2}},$$

where s is the distance between the disks.

If we suppose that in the space between the disks the potential varies uniformly, as it does between two infinite planes, then the resultant electromotive intensity is $R = \frac{V}{s}$.

If, on the other hand, we suppose that the variation of the potential near the surface of the disks is affected by unknown causes, we would get a better estimate of the intensity by taking

$$R = \frac{dV}{ds}.$$

Both $\frac{V}{s}$ and $\frac{dV}{ds}$ diminish as the distance increases, approximating to the limit 66·940.

This corresponds to a surface-density of 5·327 units of electricity per square centimetre, and to a tension of 178·3 dynes per square centimetre. As the ordinary pressure of the atmosphere is about a million dynes per square centimetre, the pressure with which the electricity tends to break through the air is only about $\frac{1}{5600}$ of the pressure of the atmosphere.

If the electrodes are convex surfaces, whose radii of curvature, a and b, are large compared with the least distance c between the surfaces, then if

$$\frac{1}{s} = \frac{1}{c} + \frac{2}{3a} - \frac{1}{3b},$$

the greatest electric force at the surface whose radius is a will be equal to that at either of two parallel plane surfaces at the same potentials whose distance is s.

* *Proc. R. S.*, 1860, or *Papers on Electrostatics*, chap. XIX.

† *Trans. R. S. Edin.*, Vol. XXVIII., Part II. (1878), p. 633.

Hence the electromotive force required to produce a spark between convex surfaces, as in Lane's electrometer, is less than if the surfaces had been plane and at the same distance.

When the air-space is large, the path of the sparks, and therefore the electromotive force required to produce them, is exceedingly irregular. The accompanying figure is from a photograph of a succession of sparks taken between the same electrodes from four Leyden jars charged by Holtz's machine.

A portion of the path near the positive electrode is nearly straight, there is then a sharp turn, which, in all the sparks represented, is in the same direction. Beyond this the course of the spark is very irregular, although its general direction is deflected towards the same side as the first sharp turn.

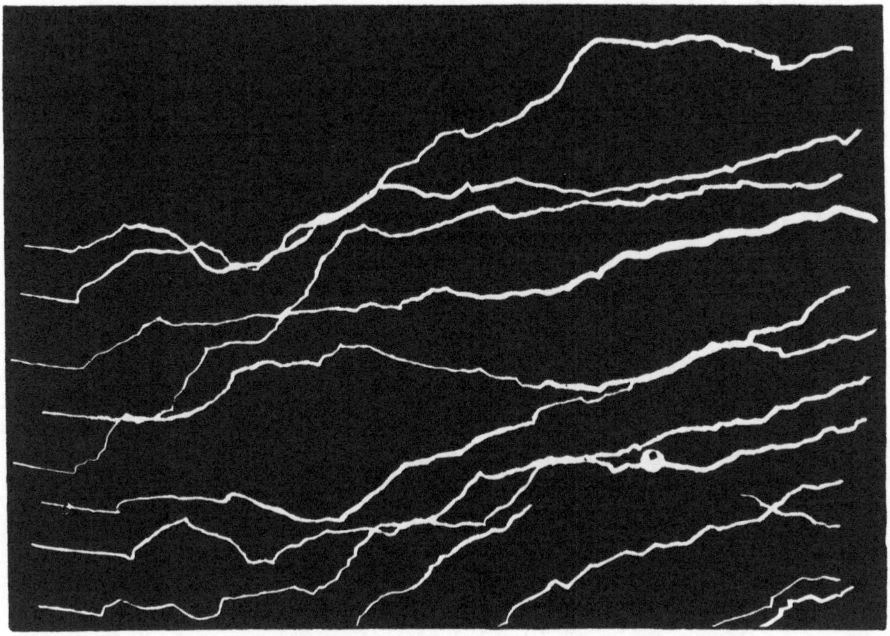

NOTE 11, Art. 141.

Theory of two circular disks on the same axis, their radii being small compared with the distance between them.

A circular disk may be considered as an ellipsoid, two of whose axes are equal, while the third is zero, and we may apply the method of ellipsoidal co-ordinates to the calculation of the potential*. In the

* See Ferrers' *Spherical Harmonics*, p. 136.

case before us everything is symmetrical about the axis, so that we have to consider only the zonal harmonics, and of these only those of even order, unless we wish to distinguish between the surface density on opposite sides of the same element of the disk, for this depends on the harmonics of odd orders.

Let a be the radius of the first disk, b that of the second, and c the distance between them.

We shall use ellipsoidal co-ordinates confocal with the first disk. Let r_1 and r_2 be the greatest and least distances respectively of a given point from the edge of the disk, and let

$$a^2 - \tfrac{1}{4}(r_1 - r_2)^2 = a^2\mu^2, \tag{1}$$

$$\tfrac{1}{4}(r_1 + r_2)^2 - a^2 = a^2\nu^2, \tag{2}$$

then if z is the distance of the point from the plane of the disk, and r its distance from the axis,

$$z = a\mu\nu, \tag{3}$$

$$r^2 = a^2(1-\mu^2)(\nu^2 - 1). \tag{4}$$

If the surface-density of the electricity on the disk is a function of the distance from the axis, it may be expressed in the form

$$\sigma = \sigma_0 + \sigma_2 + \&c., \tag{5}$$

where

$$\sigma_n = \frac{1}{2\pi a^2\mu} A_{2n} P_{2n}(\mu), \tag{6}$$

and P_{2n} is the zonal harmonic of order $2n$. Only even orders are admissible, for since every element of the disk corresponds to two values of μ, numerically equal but of opposite signs, a term involving an harmonic of odd order would give the surface-density everywhere zero.

The potential arising from this distribution at any point whose ellipsoidal co-ordinates are $\omega = a\mu$ and $\eta = a\nu$

is

$$V = V_0 + V_2 + \&c. + V_n, \tag{7}$$

where

$$V_{2n} = A_{2n}\frac{1}{a}\frac{2n\,!}{2^{2n}n\,!\,n\,!}P_{2n}(\mu)\,Q'_{2n}(\nu). \tag{8}$$

In this expression $Q'_{2n}(\nu)$ denotes a series, the terms of which are numerically equally to those of $Q_{2n}(\nu)$, the zonal harmonic of the second kind, but with the second and all even terms negative. If we put i for $\sqrt{-1}$, we may write

$$Q'_{2n}(\nu) = (-)^n i\, Q_n(i\nu) \tag{9}$$

$$= \frac{1\,.\,2\,.\,3 \;\; 2n}{1\,.\,3\,.\,5 \;\; 4n+1}\nu^{-(2n+1)} - \frac{3\,.\,4\,.\,5 \ldots (2n+2)}{3\,.\,5\,.\,7 \;\; (4n+3)}\nu^{-(2n+3)} + \&c. \tag{10}$$

This expression is an infinite series, the terms of which increase without limit when ν is diminished without limit.

It may, however, be expressed in the finite form*

$$Q'_{2n}(\nu) = P'_{2n}(\nu) \tan^{-1}\left(\frac{1}{\nu}\right) - Z_{2n}(\nu), \tag{11}$$

where

$$P'_{2n}\nu = (-)^n P_{2n}(i\nu), \tag{12}$$

that is to say $P'_{2n}(\nu)$ is a zonal harmonic of the first kind with all its terms positive, and $Z_{2n}(\nu)$ is a rational and integral function of ν of $2n-1$ degrees, which is such as to cancel all the terms of $P'_{2n}(\nu) \tan^{-1}\left(\frac{1}{\nu}\right)$ which do not vanish when ν becomes infinite.

The expression (11) is applicable to small as well as great values of ν. Thus we find when ν is 0, as it is at the surface of the disk,

$$Q'_{2n}(0) = \frac{2n\,!}{2^{2n} n\,!\, n\,!} \frac{\pi}{2}. \tag{13}$$

The potential at any point of the disk is therefore the sum of a series of terms, the general form of which is

$$V_{2n} = A_{2n} \frac{\pi}{2a} \frac{(2n\,!)^2}{2^{4n}(n\,!)^4} P_{2n}(\mu). \tag{14}$$

On the axis, $\mu = 1$ and $a\nu = z$, and the potential is the sum of a series of terms, the general form of which is

$$U_{2n} = A_{2n} \frac{1}{a} \frac{2n\,!}{2^{2n} n\,!\, n\,!} Q'_{2n}(\nu). \tag{15}$$

Since we have to determine the value of the potential arising from the first disk at a point in the second disk for which $z = c$ at a distance r from the axis, and if we write

$$r^2 = b^2(1-p^2), \tag{16}$$

where b is the radius of the second disk, and p is a quantity corresponding to μ in the first disk, then the most convenient expression for the potential due to the first disk at a point (p) in the second, is

$$V = U - \frac{1}{2^2} \frac{b^2}{a^2} \frac{d^2U}{d\nu^2}(1-p^2) + \frac{1}{2^2 \, . \, 4^2} \frac{b^4}{a^4} \frac{d^4U}{d\nu^4}(1-p^2)^2 - \&c., \tag{17}$$

where U denotes the value of the potential at the axis, and where, after the differentiations, νa is to be made equal to c.

To investigate the mutual action of the two disks, let us assume that the surface-density on the second disk is the sum of a number of terms of which the general form is

$$\frac{1}{2\pi b^2} B_{2n} \frac{1}{p} P_{2n}(p). \tag{18}$$

* See Heine, *Handbuch der Kugelfunctionen*, § 28, 20.

The potential at the surface of the second disk arising from this distribution will be the sum of a series of terms of the form

$$\frac{\pi}{2}\frac{1}{b}\frac{(2n\,!)^2}{2^{4n}(n\,!)^4}B_{2n}P_{2n}(p). \tag{19}$$

The potential arising from the presence of the first disk is given in equation (17).

Having thus expressed the most general symmetrical distribution of electricity on the two disks and the potentials thence arising, we are able to calculate the potential energy of the system in terms of the squares and products of the two sets of coefficients A and B.

If W denotes the potential energy,

$$W=\tfrac{1}{2}\iint\sigma V ds, \tag{20}$$

when the integration is to be extended over every element of surface ds.

Confining our attention to the second disk, the part of W thence arising is

$$\pi b^2\int_0^1\sigma Vp dp, \tag{21}$$

and the part arising from the term in the density whose coefficient is B_{2n} is

$$\tfrac{1}{2}B_{2n}\int_0^1 VP_{2n}(p)\,dp. \tag{22}$$

The part of the value of V which arises from the electricity on the second disk itself is the sum of a series of terms of the form (19). The surface-integral of the product of any two of these of different orders is zero, so that in finding the potential energy of the disk on itself we have to deal only with terms of the form

$$B^2_{2n}\frac{\pi}{4}\frac{1}{b}\frac{(2n\,!)^2}{2^{4n}(n\,!)^4}\frac{1}{4n+1}. \tag{23}$$

The energy arising from the mutual action of the disks consists of terms whose coefficients are products of A's and B's, and in calculating these we meet with the integral*

$$\int_0^1(1-p^2)^m P_{2n}(p)\,dp-(-1)^n\frac{2^{2m}.\,m+n\,!\,m\,!\,m\,!\,2n\,!}{2m+2n+1\,!\,m-n\,!\,n\,!\,n\,!}. \tag{24}$$

We have, therefore, for the harmonic of order zero.

Surface-density on the first disk, $\sigma_0=\dfrac{A}{2\pi a^2}\dfrac{1}{\mu}$, where A is the charge of the first disk.

* I am indebted for the general value of this integral to Mr W. D. Niven, of Trinity College.

Potential at the surface of the first disk $V_0 = \frac{\pi}{2}\,\frac{1}{a}\,A$.

Potential at the surface of the second disk, arising from this distribution of electricity on the first,

$$V_0 = A\frac{1}{c}\left[1 - \tfrac{1}{3}\frac{a^2}{c^2} + \tfrac{1}{5}\frac{a^4}{c^4} - \tfrac{1}{7}\frac{a^6}{c^6} + \&c.\right]$$

$$- A\,\frac{b^2(1-p^2)}{2^2c^3}\left[2 - 4\,\frac{a^2}{c^2} + 6\,\frac{a^4}{c^4} - \&c.\right]$$

$$+ A\,\frac{b^4(1-p^2)^2}{2^2.4^2.c^5}\left[2.3.4 - 4.5.6\,\frac{a^2}{c^2} + \&c.\right]$$

$$- \&c.$$

Order 2. $$\sigma_2 = A_2\,\frac{1}{2\pi a^2\mu}\left(\tfrac{3}{2}\mu^2 - \tfrac{1}{2}\right).$$

Potential at the surface of first disk,

$$V_2 = \frac{\pi}{2}\,\frac{1}{a}A_2\frac{1}{2^2}\left(\tfrac{3}{2}\mu^2 - \tfrac{1}{2}\right).$$

Potential at the surface of the second disk,

$$V_2 = A_2\tfrac{1}{2}\frac{a^2}{c^3}\left[\frac{2}{3.5} - \frac{4}{5.7}\,\frac{a^2}{c^2} + \frac{6}{7.9}\,\frac{a^4}{c^4} - \&c.\right]$$

$$- A_2\tfrac{1}{2}\,\frac{a^2b^2(1-p^2)}{2^2.c^5}\left[\frac{2.4}{5} - \frac{4.6}{7}\,\frac{a^2}{c^2} + \&c.\right]$$

$$+ A_2\tfrac{1}{2}\frac{a^2b^4(1-p^2)}{2^2.4^2.c^7}\left[2.4.6 - \&c.\right].$$

Order 4. $$\sigma_4 = A_4\,\frac{1}{2\pi a^2p}\left[\frac{35}{8}\mu^4 - \frac{30}{8}\mu^2 + \frac{3}{8}\right].$$

Potential at first disk,

$$V_4 = \frac{\pi}{2a}A_4\frac{1^2.3^2}{2^2.4^2}\left[\frac{35}{8}\mu^4 - \frac{30}{8}\mu^2 + \frac{3}{8}\right].$$

Potential at second disk,

$$V_4 = A_4\frac{1.3}{2.4}\,\frac{a^4}{c^5}\left[\frac{2.4}{5.7.9} - \frac{4.6}{7.9.11}\,\frac{a^2}{c^2} + \&c.\right]$$

$$- A\,\frac{1.3}{2.4}\,\frac{a^4b^2(1-p^2)}{2^2c^7}\left[\frac{2.4.6}{7.9} - \&c.\right],$$

and so on.

We have next to calculate the energy arising from this distribution on the first disk, together with a corresponding distribution on the second disk, the coefficients of the harmonics for the second disk being B, B_2, B_4, &c.

It will consist of three parts, the potential energy of the first disk on itself, of the first and second on each other, and of the second on itself.

The first part will involve only terms having for coefficients the squares of the coefficients A, for those involving products of harmonics of different orders will vanish on integration.

The third part will, for the same reason, involve only squares of the coefficients B.

The second part will involve all products of the form AB.

Performing the integrations, putting $a = cx$ and $b = cy$,

$$W = A^2 \frac{1}{a}\frac{\pi}{4} + A_2^2 \frac{1}{a}\frac{\pi}{4}\frac{1}{2^2 . 5} + A_4^2 \frac{1}{a}\frac{\pi}{4}\frac{1}{2^6}$$

$$+ B^2 \frac{1}{b}\frac{\pi}{4} + B_2^2 \frac{1}{b}\frac{\pi}{4}\frac{1}{2^2 . 5} + B_4^2 \frac{1}{b}\frac{\pi}{4}\frac{1}{2^6}$$

$$+ AB\frac{1}{c}\left[1 - \tfrac{1}{3}(x^2+y^2) + \tfrac{1}{5}(x^4+y^4) + \tfrac{2}{3}x^2y^2 - \tfrac{1}{7}(x^6+y^6) - x^2y^2(x^2+y^2)\right.$$
$$\left. + \tfrac{1}{9}(x^8+y^8) + \tfrac{4}{3}x^2y^2(x^4+y^4) + \tfrac{14}{5}x^4y^4 - \&c.\right]$$

$$+ AB_2 \frac{y^2}{c}\frac{1}{3 . 5}\left[1 - 2x^2 - \frac{2 . 3}{7}y^2 + 3x^4 + \frac{2 . 3 . 5}{7}x^2y^2 + \frac{5}{7}y^4\right.$$
$$\left. - 4x^6 - 3 . 4x^4y^2 - \frac{4 . 5}{3}x^2y^4 - \frac{4 . 5}{3 . 11}y^4 + \&c.\right]$$

$$+ AB_4 \frac{y^4}{c}\frac{1}{3.5.7}\left[1 - 5x^2 - \frac{3.5}{11}y^2 + 2.7x^4 + \frac{4.5.7}{11}x^2y^2 + \frac{2.3.5.7}{11.13}y^4 - \&c.\right]$$

$$+ A_2B \frac{x^2}{c}\frac{1}{3 . 5}\left[1 - \frac{2 . 3}{7}x^2 - 2y^2 + \frac{5}{7}x^4 + \frac{2 . 3 . 5}{7}x^2y^2 + 3y^4\right.$$
$$\left. - \frac{4 . 5}{3 . 11}x^6 - \frac{4 . 5}{3}x^4y^2 - 4 . 3x^2y^4 + 4y^6 - \&c.\right]$$

$$+ A_2B_2 \frac{x^2y^2}{c}\frac{2}{3 . 5^2}\left[1 - \frac{3 . 5}{7}(x^2+y^2) + \frac{2 . 5}{3}(x^4+y^4) + \frac{3.4.5}{7}x^2y^2 - \&c.\right]$$

$$+ A_2B_4 \frac{x^2y^4}{c}\frac{1}{3 . 5 . 7}\left[1 - 4x^2 - \frac{4 . 7}{11}y^2 + \&c.\right]$$

$$+ A_4B \frac{x^4}{c}\frac{1}{3.5.7}\left[1 - \frac{3.5}{11}x^2 - 5y^2 + \frac{2.3.5.7}{11.13}x^4 + \frac{4.5.7}{11}x^2y^2 + 2.7y^4 - \&c.\right]$$

$$+ A_4B_2 \frac{x^4y^2}{c}\frac{1}{3 . 5 . 7}\left[1 - \frac{4 . 7}{11}x^2 - 4y^2 + \&c.\right]$$

$$+ A_4B_4 \frac{x^4y^4}{c}\frac{1}{5 . 7}.$$

In this expression for the energy of the system the coefficients A_2, A_4, B_2, B_4 are treated as independent of A and B. To determine the nearest approach to equilibrium which can be obtained from a distribution defined by this limited number of harmonics, we must make W a minimum with respect to A_2, B_2, A_4 and B_4.

We thus find for the values of these coefficients

$$A_2 = -B\frac{2}{\pi}\frac{2^2}{3}x^3\left[1 - \frac{2.3}{7}x^2 - 2y^2 + \tfrac{5}{7}x^4 + \frac{2.3.5}{7}x^2y^2 + 3y^4 - \&c.\right]$$
$$+ A\frac{4}{\pi^2}\frac{2^5}{3^2.5}x^3y^5\left[1 - \&c.\right]$$

$$B_2 = -A\frac{2}{\pi}\frac{2^2}{3}y^3\left[1 - 2x^2 - \frac{2.3}{7}y^2 + 3x^4 + \frac{2.3.5}{7}x^2y^2 + \tfrac{5}{7}y^4 - \&c.\right]$$
$$+ B\frac{4}{\pi^2}\frac{2^5}{3^2.5}x^5y^3\left[1 - \&c.\right]$$

$$A_4 = -B\frac{2}{\pi}\frac{2^6}{3.5.7}x^5\left[1 - \frac{3.5}{11}x^2 - 5y^2 + \&c.\right] + A\frac{4}{\pi^2}\frac{2^8}{3^2.5.7}x^5y^5.$$

$$B_4 = -A\frac{2}{\pi}\frac{2^6}{3.5.7}y^5\left[1 - 5x^2 - \frac{3.5}{11}y^2 + \&c.\right] + B\frac{4}{\pi^2}\frac{2^8}{3^2.5.7}x^5y^5.$$

We are now able to express the energy in the form

$$W = \tfrac{1}{2}p_{11}A^2 + p_{12}AB + \tfrac{1}{2}p_{22}B^2,$$

where A and B are the charges, and p_{11}, p_{12}, and p_{22} are the coefficients of potential, the value of which we now find to be

$$p_{11} = \frac{\pi}{2a} - \frac{1}{\pi}\frac{2^3}{3^2.5}\frac{b^5}{c^6}\left[1 - 4\frac{a^2}{c^2} - \frac{12}{7}\frac{b^2}{c^2} + 10\frac{a^4}{c^4} + 12\frac{a^2b^2}{c^4} + \frac{2.3.13}{5.7}\frac{b^4}{c^4} - \&c.\right]$$

$$p_{12} = \frac{1}{c} - \tfrac{1}{3}\frac{a^2+b^2}{c^3} + \tfrac{1}{5}\frac{a^4+b^4}{c^5} + \tfrac{2}{3}\frac{a^2b^2}{c^5} - \tfrac{1}{7}\frac{a^6+b^6}{c^7} - \frac{a^2b^2(a^2+b^2)}{c^7} + \&c.$$

$$p_{22} = \frac{\pi}{2b} - \frac{1}{\pi}\frac{2^3}{3^2.5}\frac{a^5}{c^6}\left[1 - \frac{12}{7}\frac{a^2}{c^2} - 4\frac{b^2}{c^2} + \frac{2.3.13}{5.7}\frac{a^4}{c^4} + 12\frac{a^2b^2}{c^4} + 10\frac{b^4}{c^4} - \&c\right]$$

Note 12, Art. 151.

On the electrical capacity of a long narrow cylinder.

The problem of the distribution of electricity on a finite cylinder is still, so far as I know, in the state in which it was left by Cavendish. It is sometimes assumed that the electric properties of a long narrow cylinder may be represented, to a sufficient degree of approximation, by those of the ellipsoid inscribed in the cylinder. The electrical capacity of the cylinder must be greater than that of the ellipsoid, because the electric capacity of any figure is greater than that of any part of that figure.

It is easier to state the conditions of the problem than to obtain an exact solution.

Let $2l$ be the length of the cylinder, and let b be its radius.

Let the axis of the cylinder be the axis of x, and let the origin be taken at the middle point of the axis. Let y be the distance of any point from the axis.

Let λdx be the quantity of electricity on that part of the curved surface of the cylinder for which x is between x and $x + dx$; we may call λ the linear density of the electricity on the cylinder.

Let σ be the surface-density on the flat ends.

Let ψ be the potential at a point on the axis for which $x = \xi$.

$$\psi = \int_{-l}^{+l} \lambda \left[(\xi - x)^2 + b^2\right]^{-\frac{1}{2}} dx + \int_0^b 2\pi\sigma y \left[(l - \xi)^2 + y^2\right]^{-\frac{1}{2}} dy$$
$$+ \int_0^b 2\pi\sigma y \left[(l + \xi)^2 + y^2\right]^{-\frac{1}{2}} dy, \qquad (1)$$

the first integral representing the part of the potential due to the curved surface, and the other two the parts due to the positive and the negative flat ends respectively.

The condition of equilibrium of the electricity is that ψ must be constant for all points within the cylinder, and therefore for all points on the axis between the two ends.

If, by giving proper values to λ and σ, we can make the value of ψ constant for any finite length along the axis, then, by Art. 144 of "Electricity and Magnetism," ψ will be constant for all points within the surface of the cylinder.

It was shown in Note 2 that the distribution of electricity in equilibrium on a straight line without breadth is a uniform one. We may expect, therefore, that the distribution on a cylinder will approximate to uniformity as the radius of the cylinder diminishes.

If we suppose λ and σ to be each of them constant,

$$\psi = \lambda \log \frac{(f_1 + l - \xi)(f_2 + l + \xi)}{b_2} + 2\pi\sigma (f_1 + f_2 - 2l), \qquad (2)$$

where f_1 and f_2 are the distances of the point (ξ) on the axis from the edges of the curved surface at the + and − ends of the cylinder respectively.

Just within the positive flat end of the cylinder, where ξ is just less than l,

$$\frac{d\psi}{d\xi} = -\lambda \left(\frac{1}{b} - \frac{1}{f_2}\right) + 2\pi\sigma. \qquad (3)$$

If the electricity were in equilibrium, this would be zero, and if the cylinder is so long that we may neglect the reciprocal of f_2, we find

$$\lambda = 2\pi b\sigma, \tag{4}$$

or the surface-density on the end must be equal to the surface-density on the curved surface.

The whole charge is therefore

$$E = \lambda\,(2l + b). \tag{5}$$

The greatest value of the potential is at the middle of the axis, where $\xi = 0$. Calling it $\psi_{(0)}$ and putting $f = l$,

$$\psi_{(0)} = \lambda\left(2\log\frac{2l}{b} + \frac{b}{l}\right). \tag{6}$$

The potential at the end of the axis is

$$\psi_{(l)} = \lambda\left(\log\frac{4l}{b} + 1\right). \tag{7}$$

The potential at the curved edge is approximately

$$\psi_{(e)} = \lambda\left(\log\frac{4l}{b} + \frac{2}{\pi}\right). \tag{8}$$

This is the smallest value of the potential for any point of the cylinder.

The capacity of the cylinder cannot therefore be less than

$$\frac{E}{\psi_{(0)}} = \frac{2l + b}{2\log\frac{2l}{b} + \frac{b}{l}}, \tag{9}$$

nor greater than

$$\frac{E}{\psi_{(e)}} = \frac{2l + b}{\log\frac{4l}{b} + \frac{2}{\pi}}. \tag{10}$$

Cavendish does not take into account the flat ends of the cylinder, but in other respects these limits are the same as those between which he shows that the capacity must lie. The approximation, however, is by no means a close one, for when the cylinder is very narrow the upper limit is nearly double the lower. Indeed Cavendish, in Arts. 479, 682, has recourse to experiment to determine the best form of the logarithmic expression.

We may obtain a much closer approximation by the following method, which is applicable to many cases in which we cannot obtain a complete solution.

Let W be the potential energy of any arbitrary distribution of electricity on a body of any form

$$W = \tfrac{1}{2}\,\Sigma\,(e\psi), \tag{11}$$

where e is the charge of any element of the body, and ψ the potential at that element.

The charge is

$$E = \Sigma(e). \tag{12}$$

Let us now suppose the electricity to become moveable and to distribute itself so as to be in equilibrium. The potential will then be uniform. Let its value be ψ_0, and since the charge remains the same the potential energy of the electrification in the state of equilibrium will be

$$W_0 = \tfrac{1}{2}\psi_0 E. \tag{13}$$

If K_0 is the capacity of the conductor,

$$E = K_0\psi_0, \tag{14}$$

and

$$K_0 = \tfrac{1}{2}\frac{E^2}{W_0}. \tag{15}$$

Since W, the potential energy due to any arbitrary distribution of the charge, may be greater, but cannot be less than W_0, the energy of the same charge when in equilibrium, the capacity may be greater, but cannot be less, than

$$\tfrac{1}{2}\frac{E^2}{W} \text{ or } \frac{[\Sigma(e)]^2}{\Sigma(e\psi)}. \tag{16}$$

This inferior limit of the capacity is greater than that derived from the maximum value of the potential, and, as we shall see, sometimes gives a very close approximation to the true capacity.

In the case of the cylinder, if we suppose λ to be uniform, and neglect the electrification of the flat ends

$$E = 2\lambda l,\quad W = 2\lambda^2 l\left(\log\frac{4l}{b} - 1\right), \tag{17}$$

$$K_0 > \frac{l}{\log\frac{4l}{b} - 1}. \tag{18}$$

When the length of the cylinder is more than 100 times the diameter this value of the capacity is sufficiently exact for all practical purposes. The capacity of the inscribed ellipsoid is

$$\frac{l}{\log\frac{2l}{b}}.$$

To obtain a closer approximation let us suppose that the linear density λ is expressed in the form $\lambda_0 + \lambda_1 + \&c. + \lambda_i$, the general term being

$$\lambda_n = A_n P_n\left(\frac{x}{l}\right), \tag{19}$$

where P_n is the zonal harmonic of order n.

If we consider a line of length $2l$ on which there is a distribution of electricity according to this law, and if f_1 and f_2 are the distances of a given point from the ends of the line, and if we write

$$\alpha = \tfrac{1}{2}\frac{f_2 + f_1}{l}, \quad \beta = \tfrac{1}{2}\frac{f_2 - f_1}{l}, \tag{20}$$

then the potential, ψ_n, at the given point (α, β), due to the distribution λ_n, is

$$\psi_n = A_n Q_n(\alpha)\, P_n(\beta), \tag{21}$$

where P_n is the same zonal harmonic as in equation (19), and Q_n is the corresponding zonal harmonic of the second kind*, and is of the form

$$Q_n(\alpha) = P_n(\alpha) \log \frac{\alpha + 1}{\alpha - 1} + R_n \alpha, \tag{22}$$

where $R_n(\alpha)$ is a rational function of α of $n-1$ degrees, and is such that $Q_n(\alpha)$ vanishes when α is infinite. The values of the first four harmonics of the second kind are

$$\left.\begin{aligned}
Q_0(\alpha) &= \log \frac{\alpha+1}{\alpha-1}, \\
Q_1(\alpha) &= \alpha \log \frac{\alpha+1}{\alpha-1} - 2, \\
Q_2(\alpha) &= (\tfrac{3}{2}\alpha^2 - \tfrac{1}{2}) \log \frac{\alpha+1}{\alpha-1} - 3\alpha, \\
Q_3(\alpha) &= (\tfrac{5}{2}\alpha^3 - \tfrac{3}{2}\alpha) \log \frac{\alpha+1}{\alpha+1} - 5\alpha^2 + \tfrac{4}{3}, \\
Q_4(\alpha) &= (\tfrac{35}{8}\alpha^4 - \tfrac{15}{4}\alpha^2 + \tfrac{3}{8}) \log \frac{\alpha+1}{\alpha-1} - \tfrac{35}{4}\alpha^3 + \tfrac{55}{12}\alpha.
\end{aligned}\right\} \tag{23}$$

In applying these results to the determination of the potential at any point of the axis of the cylinder we must remember that a point on the axis is at the distance b from any one of the generating lines of the cylinder, and therefore the potential at any point on the axis is the same as if the whole charge had been collected on one generating line.

Hence at the point on the axis for which $x = \xi$, if we write

$$L = \log \frac{f_1 + l - \xi}{b} + \log \frac{f_2 + l + \xi}{b}, \tag{24}$$

the potential due to the distribution whose linear density is

$$\lambda_n = A_n P_n\left(\frac{x}{l}\right), \tag{25}$$

is

$$\psi_n = A_n P_n\left(\frac{\xi}{l}\right)\left[L - n + \frac{n(n-1)}{2\,.\,2} - \frac{n(n-1)(n-2)}{2\,.\,3\,.\,3} + \frac{n(n-1)(n-2)(n-3)}{2\,.\,3\,.\,4\,.\,4} - \&c.\right] \tag{26}$$

approximately, provided ξ is between $\pm l$.

* See Ferrers' *Spherical Harmonics*, chap. v.

Thus, if

$$\left.\begin{aligned}
\lambda_0 &= A_0, \\
\lambda_1 &= A_1 \frac{x}{l}, \\
\lambda_2 &= A_2 \left(\tfrac{3}{2}\frac{x^2}{l^2} - \tfrac{1}{2}\right), \\
\lambda_3 &= A_3 \left(\tfrac{5}{2}\frac{x^3}{l^3} - \tfrac{3}{2}\frac{x}{l}\right), \\
\lambda_4 &= A_4 \left(\tfrac{35}{8}\frac{x^4}{l^4} - \tfrac{15}{4}\frac{x^2}{l^2} + \tfrac{3}{8}\right),
\end{aligned}\right\} \tag{27}$$

then

$$\left.\begin{aligned}
\psi_0 &= A_0 L, \\
\psi_1 &= A_1 \frac{\xi}{l}(L-2), \\
\psi_2 &= A_2 \left(\tfrac{3}{2}\frac{\xi^2}{l^2} - \tfrac{1}{2}\right)(L-3), \\
\psi_3 &= A_3 \left(\tfrac{5}{2}\frac{\xi^3}{l^3} - \tfrac{3}{2}\frac{\xi}{l}\right)(L-\tfrac{11}{3}), \\
\psi_4 &= A_4 \left(\tfrac{35}{8}\frac{\xi^4}{l^4} - \tfrac{15}{4}\frac{\xi^2}{l^2} + \tfrac{3}{8}\right)(L-\tfrac{25}{6}).
\end{aligned}\right\} \tag{28}$$

These values of the potential are calculated for the axis of the cylinder. The potential at the curved surface may be found from that at the axis by remembering that within the cylinder $\nabla^2\psi = 0$. At a distance b from the axis the potential is therefore

$$\psi_b = \psi - \tfrac{1}{4}\frac{d^2\psi}{d\xi^2}b^2 + \tfrac{1}{64}\frac{d^4\psi}{d\xi^4}b^4 - \&c., \tag{29}$$

where the values of ψ and its derivatives are those at the axis.

For a uniform distribution

$$\frac{d^2\psi}{d\xi^2} = -A_0\left(\frac{l-\xi}{f_1^2} + \frac{l+\xi}{f_2^2}\right), \tag{30}$$

which is approximately $-\dfrac{2A_0}{l}$ when $\xi = 0$, and $-\dfrac{A_0}{2l}$, when $\xi = \pm l$. Hence, when the length of the cylinder is many times its diameter, the potential at the axis may be taken for that at the surface in approximations of the kind here made.

We have next to find the integral of the product of the density into the potential. We may consider the product of each pair of terms by itself. If we write $\mathfrak{L}$ for the value of L when $\xi = l$, or approximately

$$\mathfrak{L} = \log\frac{4l}{b}, \tag{31}$$

$$\left.\begin{aligned}
\int\lambda_0\psi_0 dx &= 4A_0^2 l(\mathfrak{L}-1),\\
\int\lambda_0\psi_2 dx &= \int\lambda_2\psi_0 dx = -\tfrac{2}{3}A_0A_2 l,\\
\int\lambda_2\psi_2 dx &= \tfrac{4}{5}A_2^2 l(\mathfrak{L}-\tfrac{101}{30}),\\
\int\lambda_0\psi_4 dx &= \int\lambda_4\psi_0 dx = -\tfrac{1}{5}A_0A_4 l,\\
\int\lambda_2\psi_4 dx &= \int\lambda_4\psi_2 dx = -\tfrac{2}{7}A_2A_4 l,\\
\int\lambda_4\psi_4 dx &= \tfrac{4}{9}A_4^2 l(\mathfrak{L}-\tfrac{6989}{1260}).
\end{aligned}\right\} \qquad (32)$$

The charge is

$$E = \int\lambda dx = 2A_0 l. \qquad (33)$$

Determining A_2 so as to make $\int(\lambda_0+\lambda_2)(\psi_0+\psi_2)dx$ a minimum, we find

$$A_2 = \tfrac{5}{6}A_0^2\frac{1}{L-\frac{101}{30}},$$

and we obtain a second approximation to K,

$$K > \frac{l}{\mathfrak{L}-1-\frac{5}{36}\dfrac{1}{\mathfrak{L}-\frac{101}{30}}}. \qquad (34)$$

This approximation is evidently of little use unless the length of the cylinder considerably exceeds 7·245 times its diameter, for this ratio makes the second term of the denominator infinite. It shows, however, that when the ratio of the length to the diameter is very great, the true capacity approximates to the value of K_0 given in (18).

We may proceed in the same way to determine A_2 and A_4 so that

$$\int(\lambda_0+\lambda_2+\lambda_4)(\psi_0+\psi_2+\psi_4)dx$$

shall be a minimum, and we thus find a third approximation to the value of the capacity, in which

$$A_2 = \tfrac{5}{6}A_0\frac{\mathfrak{L}-\frac{3373}{630}}{(\mathfrak{L}-\frac{101}{30})(\mathfrak{L}-\frac{6989}{1260})-\frac{45}{196}},\quad A_4 = \tfrac{9}{20}A_0\frac{\mathfrak{L}-\frac{457}{210}}{(\mathfrak{L}-\frac{101}{30})(\mathfrak{L}-\frac{6989}{1260})-\frac{45}{196}},$$

so that when $\mathfrak{L}$ is very large the distribution approximates to

$$\lambda = A_0\left[1+\frac{1}{\mathfrak{L}}\tfrac{7}{32}\left\{9\frac{x^4}{l^4}-2\frac{x^2}{l^2}-\tfrac{17}{15}\right\}\right].$$

The value of the inferior limit of the capacity, as given by this approximation is

$$K_4 > \frac{l}{\mathfrak{L}-1-\frac{5}{36}\dfrac{1}{\mathfrak{L}-\frac{101}{30}}-\frac{9}{400}\dfrac{(\mathfrak{L}-\frac{457}{210})^2}{(\mathfrak{L}-\frac{101}{30})\left[(\mathfrak{L}-\frac{101}{30})(\mathfrak{L}-\frac{6989}{1260})-\frac{45}{196}\right]}}$$

As $\mathfrak{L}$ increases, K approaches to the value found by the first approximation.

To indicate the degree of approximation, the value of $\mathfrak{L}$ and of the successive terms of the denominator are given below.

$\frac{l}{b}$.	$\mathfrak{L}$.	Denominator of (34) and (35). 1st term.	2nd term.	3rd term.
10	3·68888	2·68888	− 0.43151	
20	4·38203	3·38203	− 0·13680	
30	4·78749	3·78749	− 0·09775	
50	5·29832	4·29832	− 0·07191	
100	5·99146	4·99146	− 0·05291	− 0·13566
1000	8·29405	7·29405	− 0·02818	− 0·00892

The observed capacities of Cavendish's cylinders may be deduced from the numbers given in Art. 281 by taking the capacity of the globe of 12·1 inches diameter equal to 6·05, and their capacities as calculated by the formula of this note are given in the following table.

Length.	Diameter.	Capacity by formula.	As measured by Cavendish.
72	·185	5·668	5·669
54·2	·73	5·775	5·754
35·9	2·53	5·907	6·044

The agreement of the calculated and measured values is remarkable.

Note 13, Arts. 152, 280.

Two cylinders.

In the case of two equal and parallel cylinders at distance c, the linear densities being uniform and equal to λ_1 and λ_2, the part of the potential energy arising from their mutual action is

$$\tfrac{1}{2}\int \lambda_1 \psi_2 dx = \int \lambda_2 \psi_1 dx = \lambda_1 \lambda_2 \left(4l \log \frac{r + 2l}{c} - 2r\right),$$

where

$$r^2 = 4l^2 + c^2.$$

If the two cylinders are in electric communication with each other $\lambda_1 = \lambda_2$, and the capacity of the two cylinders together is approximately

$$\frac{2l}{\log \dfrac{4l}{b} - 1 + \log \dfrac{r + 2l}{c} - \dfrac{r - c}{2l}}.$$

If a cylinder is placed at a distance d from a conducting plane surface and parallel to it, then the electric image of the cylinder will be at a distance $c = 2d$, and its charge will be negative, so that the capacity of the cylinder will be increased. The capacity of the cylinder in presence of a conducting plane at distance $\frac{1}{2}c$, is

$$\frac{l}{\log \frac{4l}{b} - 1 - \log \frac{r + 2l}{c} + \frac{r - c}{2l}}.$$

Thus in Cavendish's experiment he used a brass wire 72 inches long and 0·185 in diameter. The capacity of this wire at a great distance from any other body would be 5·668 inches. Cavendish placed it horizontally 50 inches from the floor. The inductive action of the floor would increase its capacity to 5·994 inches; Cavendish, by comparison with his globe, makes it 5·844.

To compare with this he had two wires each 36 inches long and 0·1 inch diameter.

The capacity of one of these at a distance from any other body would be 2·8697 inches, or the two together would be 5·7394 inches.

The two wires were placed parallel and horizontal at 50 inches from the floor. Each wire was therefore influenced by the other wire, and also by the negative images of itself and the other wire.

The denominator of the fraction expressing the capacity is therefore

Distance.	Wire itself.		Other wire.		Own image.		Other image.		
18	6·2724	+	9·8256	−	0·1759	−	0·1754	=	6·7467
24	6·2724	+	0·6596	−	0·1759	−	0·1733	=	6·5828
36	6·2724	+	0·4672	−	0·1759	−	0·1678	=	6·3955

The numerator of the fraction which expresses the capacity of both wires together is 36, so that the capacity of the two is

		From Cavendish's results.
At 18 inches	5·334	4·967
24	5·469	5·026
36	5·629	5·277
Wire of 72 inches	5·994	5·844

Note 14, Art. 155.

Lemma XVI.

If we suppose the plate *AB* to be overcharged and the plate *DF* to be equally undercharged, the redundant fluid in any element of *AB* being numerically equal to the deficient fluid in the corresponding element of *DF*, then what Cavendish calls the repulsion on the column *CE* in opposite directions becomes in modern language the excess of the potential at *C* over that at *E*. Hence the object of the Lemma is to determine approximately the difference of the potentials of two curved plates when their equal and opposite charges are given, and to deduce their charges when the difference of their potentials is given.

NOTE 15, Art. 169.

On the Theory of Dielectrics.

Cavendish explains the fact discovered by him, that the charge of a coated glass plate is much greater than that of a plate of air of the same dimensions, by supposing that in certain portions of the glass the electric fluid is free to move, while in the rest of the glass it is fixed.

Probably for the sake of being able to apply his mathematical theorems, he takes the case in which the conducting parts of the glass are in the form of strata parallel to the surfaces of the glass. He is perfectly aware that this is not a true physical theory, for if such conducting strata existed in a plate of glass, they would make it a good conductor for an electric current parallel to its surfaces. As this is not the case, Cavendish is obliged to stipulate, as in this proposition, that the conducting strata conduct freely perpendicularly to their surfaces, but do not conduct in directions parallel to their surfaces.

The idea of some peculiar structure in plates of glass was not peculiar to Cavendish. Franklin had shewn that the surface of glass plates could be charged with a large quantity of electricity, and therefore supposed that the electric fluid was able to penetrate to a certain depth into the glass, though it was not able to get through to the other side, or to effect a junction with the negative charge on the other side of the plate.

The most obvious explanation of this was by supposing that there was a stratum of a certain thickness on each side of the plate into which electricity can penetrate, but that in the middle of the plate there was a stratum impervious to electricity. Franklin endeavoured to test this hypothesis by grinding away five-sixths of the thickness of the glass from the side of one of his vials, but he found that the remaining sixth was just as impervious to electricity as the rest of the glass*.

It was probably for reasons of this kind, as well as to ensure that his thin plates were of the same material as his thick ones, that Cavendish prepared his thin plate of crown glass by grinding equal portions off both sides of a thicker plate. [Art. 378.]

It appears, however, from the experiments, that the proportion of the thickness of the conducting to the non-conducting strata is the same for the thin plates as the thick ones, so that the operation of grinding must have removed non-conducting portions as well as conducting ones, and we cannot suppose the plate to consist of one non-conducting stratum with a conducting stratum on each side, but must suppose that the conducting portions of the glass are very small, but so numerous that they form a considerable part of the whole

* Franklin's Works, 2nd Edition, Vol. I. p. 301, Letter to Dr Lining, March 18, 1755.

volume of the glass. If we suppose the conducting portions to be of small dimensions in every direction, and to be completely separated from each other by non-conducting matter, we can explain the phenomena without introducing the possibility of conduction through finite portions of glass.

It was probably because Cavendish had made out the mathematical theory of stratified condensers, but did not see his way to a complete mathematical theory of insulating media, in which small conducting portions are disseminated, that he here expounds the theory of strata which conduct electricity perpendicularly to their surfaces but not parallel to them.

In forming a theory of the magnetization of iron, Poisson was led to the hypothesis that the magnetic fluids are free to move within certain small portions of the iron, which he calls magnetic molecules, but that they cannot pass from one molecule to another, and he calculates the result on the supposition that these molecules are spherical, and that their distances from each other are large compared with their radii.

When Faraday had afterwards rediscovered the properties of dielectrics, Mossotti, noticing the analogy between these properties and those of magnetic substances, constructed a mathematical theory of dielectrics, by taking Poisson's memoir and substituting electrical terms for magnetic, and Italian for French, throughout.

A theory of this kind is capable of accounting for the specific inductive capacity being greater than unity, without introducing conductivity through portions of the substance of sensible size.

Another phenomenon which we have to account for is that of the residual charge of condensers, and what Faraday called electric absorption. The only notice which Cavendish has left us of a phenomenon of this kind is that recorded in Arts. 522, 523, in which it appeared "that a Florence flask contained more electricity when it continued charged a good while than when charged and discharged immediately."

To illustrate this phenomenon, I gave in "Electricity and Magnetism," Art. 328, a theory of a dielectric composed of strata of different dielectric and conducting properties.

Professor Rowland has since shown* that phenomena of the same kind would be observed if the medium consisted of small portions of different kinds well mingled together, though the individual portions may be too small to be observed separately.

It follows from the property of electric absorption that in experiments to determine the specific inductive capacity of a substance, the result depends on the time during which the substance is electrified. Hence most of those who have attempted to determine the value of this quantity for glass have obtained results so inconsistent with

* *American Journal of Mathematics*, No. I. 1878, p. 53.

each other as to be of no use. It is absolutely necessary, in working with glass, to perform the experiment as quickly as possible.

Cavendish does not give the exact duration of one of his "trials," but each trial probably took less than two or three seconds. His results are therefore comparable with those recently obtained by Hopkinson*, who effected the different operations by hand.

The results obtained by Gordon†, who employed a break which gave 1200 interruptions per second, and those obtained by Schiller‡ by measuring the period of electric oscillations, which were at the rate of about 14000 per second, are much smaller than those obtained by Cavendish and by Hopkinson.

Hopkinson finds that the quotient of the specific inductive capacity divided by the specific gravity does not vary much in different kinds of flint glass. As Cavendish always gives the specific gravity, I have compared his results with those of Hopkinson for glass of corresponding specific gravity.

Electrostatic capacity of glass.

	Specific gravity.	Cavendish.	Hopkinson.	Wüllner.	Gordon.	Schiller.
Flint-glass	3·279	7·93				
Do., a thinner piece	3·284	7·65				
Light flint	3·2		6·85		3·013	2·96
Dense flint...............	3·66		7·4		3·054	3·66
Double extra-dense flint	4·5		10·1		3·164	
Very light flint.........	2·87		6·57			5·83
Plate-glass	2·8	8		6·10		6·43
Crown-glass	2·53	8·6			3·108	

Note 16, Art. 185.

Mutual Influence of two Condensers.

To find the effect on the capacity of a condenser arising from the presence of another condenser at a distance which is large compared with the dimensions of either condenser.

Let A and B be the electrodes of the first condenser, let L and N be the capacities of A and B respectively, and M their coefficient of mutual induction, then if the potential of A is 1 and that of B is 0, the charge of A will be L and that of B will be M, and if both A and B are at potential 1 the charge of the whole will be $L+2M+N$,

* *Proceedings of the Royal Society*, June 14, 1877; *Phil. Trans.*, 1878, Part I., p. 17.

† *Proc. R. S.* Dec. 12, 1878.

‡ *Pogg. Ann.* 152 (1874), p. 535.

and this cannot be greater than half the greatest diameter of the condenser.

Let a and b be the electrodes of the second condenser, let its coefficients be l, m, n, and let its distance from their first condenser be R.

Let us first take the condenser AB by itself, and let us suppose that the potentials of A and B are x and y respectively, then their charges will be $Lx + My$ and $Mx + Ny$ respectively.

At a distance R from the condenser the potential arising from these charges will be

$$\{Lx + M(x+y) + Ny\} R^{-1} = P,$$

and if the second condenser, whose capacity when its electrodes are in contact is $l + 2m + n$, is placed at a distance R from the first and connected to earth, its charge will be

$$-P(l + 2m + n) = Q.$$

This charge of the second condenser will produce a potential QR^{-1} at a distance R, and will therefore alter the potentials of A and B by this quantity, so that the potentials of A and B will be $x + QR^{-1}$ and $y + QR^{-1}$ respectively.

To find the capacity of A as altered by the presence of the second condenser, we must make the potential of $A = 1$ and that of $B = 0$, which gives

$$x - \{Lx + M(x+y) + Ny\}(l + 2m + n)R^{-2} = 1,$$

$$y - \{Lx + M(x+y) + Ny\}(l + 2m + n)R^{-2} = 0.$$

Hence $$x = y + 1,$$

and $$y = \{L + M + (L + 2M + N)\,y\}(l + 2m + n)R^{-2},$$

or $$y = \frac{(L+M)(l+2m+n)R^{-2}}{1-(L+2M+N)(l+2m+n)R^{-2}},$$

and the capacity of A is $Lx + My$ or $L + (L+M)\,y$, or

$$[AA] = L + \frac{(L+M)^2(l+2m+n)}{R^2-(L+2M+N)(l+2m+n)}.$$

The charge of B is $Mx + Ny$ or $M + (M+N)\,y$, or

$$[AB] = M + \frac{(L+M)(M+N)(l+2m+n)}{R^2-(L+2N+N)(l+2m+n)}.$$

The charges of a and b are $-(l+m)P$ and $-(m+n)P$ respectively, or

$$[Aa] = -\frac{R(L+M)(l+m)}{R^2-(L+2M+N)(l+2m+n)}$$

$$[Ab] = -\frac{R(L+M)(m+n)}{R^2-(L+2M+N)(l+2m+n)}$$

In these expressions we must remember that M is a negative quantity, that $L+M$ and $M+N$ can neither of them be negative, and that their sum $L+2M+N$ cannot be greater than the largest semidiameter of the condenser. Hence if R is large compared with the dimensions of the condensers, the second term of the values of $[AA]$ and $[AB]$ will be quite insensible, and even if the condensers are placed very near together these terms will be small compared with L, M, or N.

If a, instead of being part of a condenser, is a conductor at a considerable distance from any other conductor, we may put $m=n=0$, and if A is also a simple conductor, $M=N=0$, and we find

$$[AA]=L+\frac{L^2l}{R^2-Ll},$$

$$[Aa]=-\frac{RLl}{R^2-Ll},$$

by which the capacities and mutual induction of two simple conductors at a distance R can be calculated when we know their capacities when at a great distance from other conductors. See Note 24.

NOTE 17, ART. 194.

Theory of the Experiment with the Trial Plate.

Let A and B be the inner, a and b the outer coatings of the Leyden jars.

Let C be the body tried and D the trial plate, M the wire connecting A with C, and N the wire connecting b with D.

Let E be the electrometer with its connecting wires.

Let the coefficients of induction be expressed by pairs of symbols within square brackets, thus, let $[(A+C)(C+D)]$ denote the sum of the charges of A and C when C and D are both raised to potential 1 and all the other conductors are at potential 0.

First Operation.—The insides of the two jars are charged to potential P_0, the outsides and all other bodies being at potential 0.

The charge of A is $[A(A+B)]P_0$, and that of b is $[b(A+B)]P_0$.

Second Operation.—The outside coating of b is insulated, the charging wire is removed, and the inside of B is connected to earth. The charges of A and of b remain as before.

Third Operation.—A is connected to C by the wire M, and b is connected to D by the wire N.

The charge of A is communicated to A, C, and M, and the potential of this system is P_1, and the charge of b is communicated to b, D and N, and the potential of this system is P_2.

Hence we have the following equations to determine P_1 and P_2 in terms of P_0,

$$[(A+C+M)(A+C+M)]\,P_1+[(A+C+M)(b+D+N)]\,P_2 = [A\,(A+B)]\,P_0, \qquad (1)$$

$$[(A+C+M)(b+D+N)]\,P_1+[(b+D+N)(b+D+N)]\,P_2 = [b\,(A+B)]\,P_0. \qquad (2)$$

Fourth Operation.—The wires M and N are disconnected from C and D respectively, and the jars A and b are discharged and kept connected to earth.

The charges of C and D remain the same as before.

Fifth Operation.—The bodies C and D are connected with each other and with the electrometer E, and the final potential of the system CDE is observed by the electrometer to be P_3.

Equating the final charge of the system CDE to that of the system CD at the end of the fourth equation,

$$[(C+D+E)(C+D+E)]\,P_3=[(C+D)(A+C+M)]\,P_1 + [(C+D)(b+D+N)]\,P_2. \qquad (3)$$

Eliminating P_1 and P_2 from equations (1), (2) and (3),

$$P_3\,[(C+D+E)^2]\,\{[(A+C+M)^2]\,[(b+D+N)^2] - [(A+C+M)(b+D+N)]^2\}$$

$$=P_0\left\{\begin{array}{l}[A\,(A+B)]\,\{[(C+D)\,(A+C+M)]\,[(b+D+N)^2] \\ \qquad -[(C+D)(b+D+N)][(A+C+M)(b+D+N)]\} \\ +[b(A+B)]\,\{[(C+D)(b+D+N)][(A+C+M)^2] \\ \qquad -[(C+D)(A+C+M)]\,[(A+C+M)(b+D+N)]\}\end{array}\right\}. \qquad (4)$$

By means of his gauge electrometer, Art. 249, Cavendish made the value of P_0 the same in every trial, and altered the capacity of D, the trial plate, so that P_3 in one trial had a particular positive value, and in another an equal negative value. He then wrote down the difference of the two values of D as an indication to guide him in the choice of trial plates, and the sum of the two values, by means of which he compared the charges of different bodies.

He then substituted for C a body, C', of nearly equal capacity, and repeated the same operations, and finally deduced the ratio of C to C' from the equation

$$C : C' :: D_1+D_2 : D_1'+D_2'.$$

The capacities of the two jars were very much greater than any of the other capacities or coefficients of induction in the experiment, and of these $[b\,(B+b)]$ was less than half the greatest diameter of the second jar, and may therefore be neglected in respect of $[b^2]$ or $[Bb]$. We may therefore put $[Bb] = -[b^2]$, and in equation (4) neglect all terms except those containing the factors $[A^2][b^2]$ or $[A^2][Bb]$.

We thus reduce equation (4) to the form

$$P_3[(C+D+E)^2] = P_0\{[(C+D)(A+C+M)] - [(C+D)(b+D+N)]\}$$
$$= P_0\{[C^2] + [C(A+M)] - [C(b+N)] - [D^2] - [D(b+N)] + [D(A+M)]\}. \quad (5)$$

The bodies to be compared were either simple conductors, such as spheres, disks, squares and cylinders, and those trial plates which consisted of two conducting plates sliding on one another, or else coated plates or condensers.

Now the coefficient of induction between a coated plate and a simple conductor is much less than that between two simple conductors of the same capacity at the same distance, and the coefficient of induction between two coated plates is still smaller. See Note 16.

Hence if both the bodies tried are coated plates, the equation (5) is reduced to the form

$$P_3([C^2] + [D^2] + [E^2]) = P_0([C^2] - [D^2]), \quad (6)$$

so that the experiment is really a comparison of the capacities of the two bodies C and D.

But if either of them is a simple conductor, we must add to its capacity its coefficient of induction on the wire and jar with which it is connected, and subtract from it its coefficient of induction on the other wire and jar. These two coefficients of induction are both negative, but that belonging to its own wire and jar is probably greater than the other, so that the correction on the whole is negative.

Hence in Cavendish's trials the capacity deduced from the experiment will be less for a simple conductor than for a coated plate of equal real capacity.

This appears to be the reason why the capacities of the plates of air when expressed in "globular inches," that is, when compared with the capacity of the globe, are about a tenth part greater than their computed values. See Art. 347.

It would have been an improvement if Cavendish, instead of charging the inside of both jars positively and then discharging the outside of B, had charged the inside of A and the outside of B from the same conductor, and then connected the outside of both to earth, using the inside of B instead of the outside, to charge the trial plate negatively. In this way the excess of the negative electricity over the positive in B would have been much less than when the outside was negative.

With a heterostatic electrometer, such as those of Bohnenberger or Thomson, in which opposite deflections are produced by positive and negative electrification, the determination of the zero electrification may be made more accurately than any other, and with such an electrometer P_3 should be adjusted to zero. But the only electrometer which Cavendish possessed was the pith ball electrometer, in which the repulsion between the balls when at any given distance depends on the square of the electrification, and in which therefore the indications are very feeble for low degrees of electrification. Cavendish therefore first adjusted his trial plate so as to produce a given amount of separation of the balls by positive electrification, and then altered the trial plate so as to produce an equal separation by negative electrification. In each case he has recorded a number expressing the side of a square electrically equivalent to the trial plate, together with the difference and the mean of the two values.

He seems to have adopted the arithmetical mean as a measure of the charge of the body to be tried. It is easy to see, however, that the geometrical mean would be a more accurate value. For, if we denote the values of the final potential of the trial plate by accented letters in the second trial, we have

$$P_3'([C^2]+[D'^2]+[E^2]) = P_0([C^2]-[D'^2]). \qquad (7)$$

Since $P_3 + P_3' = 0$, we find by (6) and (7)

$$[C^2]([C^2]+[E^2]) = [D^2][D'^2] + \tfrac{1}{2}[E^2]([D^2]+[D'^2]).$$

If we neglect the capacity of the pith ball electrometer, which is much less than that of the bodies usually tried, this equation becomes

$$[C^2]^2 = [D^2][D'^2],$$

or the capacity of the body tried is the geometrical mean of the capacities of the trial-plate in its positive and negative adjustments.

NOTE 18, ART. 216.

On the "Thoughts Concerning Electricity," and on an early draft of the Propositions in Electricity.

The theory of electricity sketched in the "Thoughts" is evidently an earlier form of that developed in the published paper of 1771. We must therefore consider the "Thoughts" as the first recorded form of Cavendish's theory, and this for the following reasons.

(1) Nothing is said in the "Thoughts" of the forces exerted by ordinary matter on itself and on the electric fluid. The only agent considered is the electric fluid itself, the particles of which are supposed to repel each other. This fluid is supposed to exist in all bodies whether apparently electrified or not, but when the quantity of the fluid in any body is greater than a certain value, called the *natural* quantity for the body, the body is said to be overcharged, and when the quantity is less than the natural quantity the body is said to be undercharged.

The forces exerted by undercharged bodies are ascribed, not, as in the later theory, to the redundant matter in the body, but to the repulsion of the fluid in other parts of space.

The theory is therefore simpler than in its final form, but it tacitly assumes that the fluid could exist in stable equilibrium if spread with uniform density over all space, whereas it appears from the investigations of Cavendish himself that a fluid whose particles repel each other with a force inversely as any power of the distance less than the cube would be in unstable equilibrium if its density were uniform.

This objection does not apply to the later form of the theory, for in it the equilibrium of the electric fluid in a saturated body is rendered stable by the attraction exerted by the fixed particles of ordinary matter on those of the electric fluid.

(2) The hypotheses are reduced in the later theory to one, and the third and fourth hypotheses of the "Thoughts" are deduced from this.

(3) In the "Thoughts" Cavendish appears to be acquainted only with those phenomena of electricity which can be observed without quantitative experiments. Some of his remarks, especially those on the spark, he repeats in the paper of 1771, but in that paper (Art. 95) he refers to certain quantitative experiments, the particulars of which are now first published [Art. 265].

The "Thoughts," however, though Cavendish himself would have considered them entirely superseded by the paper of 1771, have a scientific interest of their own, as showing the path by which Cavendish arrived at his final theory.

He begins by getting rid of the electric atmospheres which were still clinging to electrified bodies, and he appears to have done this so completely that he does not think it worth while even to mention them in the paper of 1771.

He then introduces the phrase "degree of electrification" and gives a quantitative definition to it, so that this, the leading idea of his whole research, was fully developed at the early date of the "Thoughts."

Several expressions which Cavendish freely used in his own notes and journals, but which he avoided in his printed papers, occur in the "Thoughts."

Thus he speaks of the "compression" or pressure of the electric fluid.

Besides the "Thoughts," which may be considered as the original form of the introduction to the paper of 1771, there is a mathematical paper corresponding to the Propositions and Lemmata of the published paper, but following the earlier form of the theory, in which the forces exerted by ordinary matter are not considered, and referring directly to the "Hypotheses" of the "Thoughts."

The first part of this paper is carefully written out, but it gradually becomes more and more unfinished, and at last terminates abruptly, though, as this occurs at the end of a page, we may suppose that the end of the paper has been lost. I think it probable, however, that when

Cavendish had advanced so far, he was beginning to see his way to the form of the theory which he finally published, and that he did not care to finish the manuscript of the imperfect theory.

The general theory of fluids repelling according to any inverse power of the distance is given much more fully than in the paper of 1771, and the remarks on the constitution of air are very interesting.

I have therefore printed this paper, but in order to avoid interrupting the reader with a repetition of much of what he has already seen, I have placed it at the end of this Note.

CAVENDISH'S FIRST MATHEMATICAL THEORY *.

Let a fluid whose particles mutually repel each other be spread uniformly through infinite space. Let a be a particle of that fluid; draw the cone $ba\beta$ continued infinitely, and draw the section $b\beta$: if the repulsion of the particles is inversely as any higher power of the distance than the cube, the particle a will be repelled with infinitely more force from the particles between a and $b\beta$ than from all those situated beyond it, but if their repulsion is inversely as any less power than the cube, then the repulsion of the particles placed beyond $b\beta$ is infinitely greater than that of those between a and $b\beta$.

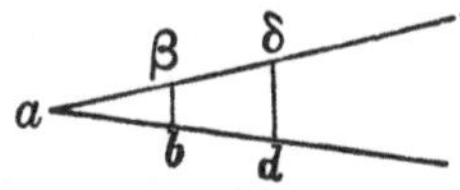

If the repulsion of the particles is inversely as the n power of the distance, n being greater than 3, it would constitute an elastic fluid of the same nature as air, except that its elasticity would be inversely as the $n+2$ power of the distance of the particles, or directly as the $\frac{n+2}{3}$ power of the density of the fluid.

But if n is equal to, or less than 3, it will form a fluid of a very different kind from air, as will appear from what follows.

COR. 1. Let a fluid of the above-mentioned kind be spread uniformly through infinite space except in the hollow globe BDE, and let the sides of the globe be so thin that the force with which a particle placed contiguous to the sides of the globe would be repelled by so much of the fluid as might be lodged within the space occupied by the sides of the globe should be trifling in respect of the repulsion of the whole quantity of fluid in the globe.

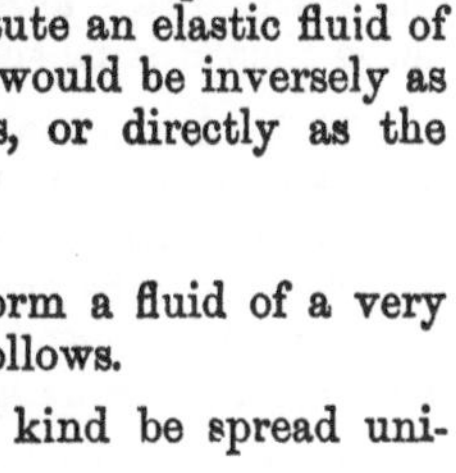

If the fluid within the globe was of the same density as without, the particles of the fluid adjacent to either the inside or outside surface of the globe would not press against those surfaces with any sensible force, as they would be repelled with the same force by the fluid on each side of them. But if the fluid within the globe is denser

* From MS. bundle 17.

than that without, then any particle adjacent to the inside surface of the globe will be pressed against by the repulsion of so much of the fluid within the globe as exceeds what would be contained in the same space if it was of the same density as without, and consequently will be greater if the globe be large than if it be small. Consequently the pressure against a given quantity (a square inch for example) of the inside surface of the globe will be greater if the globe is large than if it is small.

If the particles of the fluid repel each other with a force inversely as their distance, the pressure against a given quantity of the inside surface would be as the square of the diameter of the globe. So that it is plain that air cannot consist of particles repelling each other in the above-mentioned manner.

If the repulsion of the particles was inversely as some higher power of the distance than the cube, then any particle of the fluid would not be sensibly affected except by the repulsion of those particles which were almost close to it, so that the pressure of the fluid against a given quantity of the inside surface would be the same whatever was the size of the globe, but then the elasticity [would] be in a greater proportion than that of the $\frac{5}{3}$ power of the density.

If the repulsion of the particles is inversely as some less power than the cube of the distance, and the density of the fluid within the globe is less than it is without, then the particles on the outside of the globe will press against it, and the force will be greater if the globe is large than if it be small.

If the density of the fluid within the globe be greater than without, then the density will not be the same in all parts of the globe, but will be greater near the surface and less near the middle, for if you suppose the density to be everywhere the same, then any particle of the fluid, as *d*, would be pressed with more force towards *a*, the nearest part of the surface of the sphere, than it would [be] in the contrary direction.

If the repulsion of the particles is inversely as the square of the distance, I think the inside of the sphere would be uniformly coated with the fluid to a certain thickness, in which the density would be infinite, or the particles would be pressed close together, and in all the space within that, the density would be the same as on the outside of the sphere.

The pressure of a particle adjacent to the inside surface against it is equal to the repulsion of all the redundant matter in the sphere collected in the center, and the force with which a particle is pressed towards the surface of the sphere diminishes in arithmetical progression in going from the inside surface to that point at which its density begins to be the same as without, therefore the whole pressure against the inside of the sphere is equal to that of half the redundant matter in the sphere pressed by the repulsion of all the redundant matter collected in the center of the sphere.

Therefore, if the quantity of fluid in the sphere is such that its density, if uniform, would be $1+d$, and the radius of the sphere be called r, the whole pressure against the inside surface will be as $\frac{dr^3}{2} \times \frac{dr^3}{r^2}$, and the pressure against a given space of the inside surface will be as d^2r^2.

If this pressure be called P, d is as $\frac{\sqrt{P}}{r}$, and dr^3 is as $r^2\sqrt{P}$. Consequently, supposing the fluid to be pumped into different sized globes, the quantity of fluid pumped in will be as the square root [of the force] with which it is pumped, multiplied by the square of the diameter of the globe.

If the density within the sphere is less than without, then the density within the sphere will not be uniform, but will be greater towards the middle and less towards the outside, and if the repulsion of the particles is inversely as the square of the distance, there would be a sphere concentric to the hollow globe in which the density would be the same as on the outside of the globe, and all between that and the inside surface of the globe would be a vacuum.

From these corollaries it follows that if the electric fluid is of the nature here described, and is spread uniformly through bodies, except when they give signs of electricity, that then if two similar bodies of different sizes be equally electrified, the larger body will receive much less additional electricity in proportion to its bulk than the smaller one, and moreover when a body is electrified, the additional electricity will be lodged in greater quantity near the surface of the body than near the middle.

Let us now suppose the fluid within the globe *BDE* to be denser than without, and let us consider [in what manner] the fluid without will be affected thereby.

1st. There will be a certain space surrounding the globe, as $\beta\delta\epsilon$, which will be a perfect vacuum, for first let us suppose that the density without the globe is uniform, then any particle would be repelled with more force from the globe than in the contrary direction.

2ndly. Let us suppose that the space $\beta\delta\epsilon$, *BDE* is not a vacuum, but rarer than the rest of the fluid; still a particle placed close to the surface of the globe would be repelled from it with more force than in the contrary direction.

3rdly. Let us [suppose that] the density in the space between *BDE* and $\beta\delta\epsilon$ is greater than without, then according to some hypothesis of the law of repulsion a particle placed at *B* might be in equilibrium, but one placed at β could by no means be so.

So that there is no way by which the particles can be in equilibrium, unless there is a vacuum all round the globe to a certain distance. How the density of the fluid will be affected beyond this vacuum I cannot exactly tell, except in the following case:—

If the repulsion of the particles is inversely as the square of the distance, there will be a perfect vacuum between *BDE* and $\beta\delta\epsilon$, and beyond that the density will be perfectly uniform, $\beta\delta\epsilon$ being a sphere concentric to *BDE*, and of such a size, that if the matter in *BDE* was spread uniformly all over the sphere $\beta\delta\epsilon$, its density would be the same as beyond it.

For any quantity of matter spread uniformly over the globe $\beta\delta\epsilon$ or *BDE* affects a particle of matter placed without that sphere just in the same manner as if the whole fluid was collected in the center of the sphere, so that any particle of matter placed without the sphere $\beta\delta\epsilon$ will be in perfect equilibrio.

In like manner if the fluid within *BDE* is rarer than without, there will be a certain space surrounding the globe, as that between *BDE* and $\beta\delta\epsilon$, in which the density will be infinite, or in which the particles will be pressed close together, and if the repulsion of the particles is inversely as the square of the distance, the density of the fluid beyond that will be uniform: the diameter of $\beta\delta\epsilon$ being such that if all the matter within it was spread uniformly, its density would be the same as without.

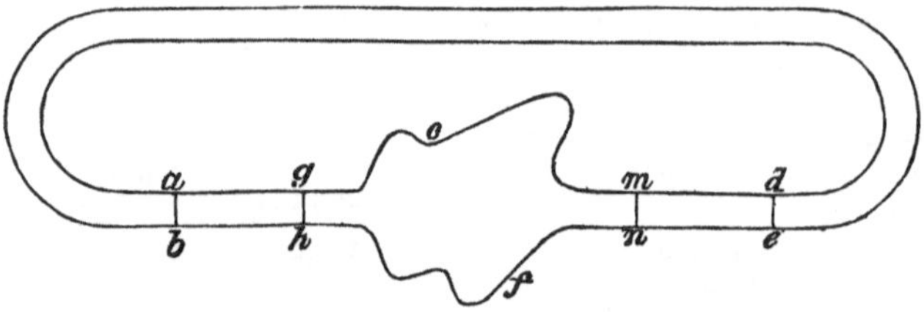

Let a fluid of the above-mentioned kind be spread uniformly through infinite space except in the canal *acdef* of any shape whatsoever, except that the ends *aghb* and *mden* are straight canals of an equal diameter, and of such a length that a particle placed at *a* or *d* shall not be sensibly affected by the repulsion of the matter in the part *gcmnfh*, and let there be a greater quantity of the fluid in this canal than in an equal space without.

Then the density of the fluid in different parts of the canal will be very different, but I imagine the density will be just the same at *a* as at *d*. For suppose *ab* and *de* to be joined, as in the figure, by a canal of an uniform diameter and regular shape, nowhere approaching near enough to *gcmnfh* to be affected by the repulsion of the particles within it. If the matter was not of the same density [at *a* and *d*] the matter therein could not be at rest, but there would be a continual current through the canal, which seems highly improbable.

COR. Let *C* be a conductor of electricity of any shape, *em* and *fn* wires extending from thence to a great distance. Let *a* and *b* be two equal bodies placed on those wires at such a distance from C as not to be sensibly affected by the electricity thereof, and let the conductor or wires be electrified by any part: the quantity of electric fluid in the bodies a and b will not be sensibly different, or they will appear equally electrified.

Case 1. Let the parallel planes Aa, Bb, &c., be continued infinitely. Let all infinite space except the space contained between Aa and Cc, and between Ee and Hh, be filled uniformly with particles repelling inversely as the square of their distance; let the space between Ee and Hh be filled with fluid of the same density, the particles of which can move from one part to another; and let the space between Aa and Cc be filled with matter whose density is to [that in] the rest of space as AD to AC.

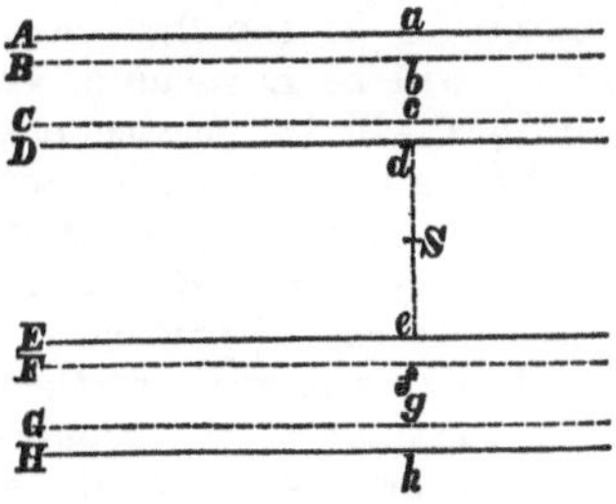

Take $EF = \frac{1}{2}CD$, and GH such that the matter between Ee and Ff when pressed close together, so that the particles touch each other, shall occupy the space between Gg and Hh.

The space between Ee and Ff will be a vacuum, that between Ff and Gg of the same density as the rest of space; and between Gg and Hh the particles will touch one another.

Case 2. Let everything be as in case the first, except that there is a canal opening into the plane Hh, by which the matter in the space EH is at liberty to escape; part of the matter will then run out, and the density therein will be everywhere the same as without, except in the space EF, which will be a vacuum, EF being equal to CD.

Case 3. Suppose now that a canal opens into the plane Aa by which the fluid in the space AC may escape. It will have no tendency to do so, for the repulsion of the redundant fluid in AC on a particle at a will be exactly equal to [the] want of repulsion of the space EH.

Case 4. Let now the space between Aa and Cc be filled with matter whose density is to the rest of space as AB to AC.

Then the space between Hh and Gg will be a vacuum, GH being equal to $\frac{1}{2}BC$. In the space EF the particles of matter will be pressed together so as to touch each other, the quantity of matter therein exceeding what is naturally contained in that space by as much as is driven out of the space GH; and in the space between Ff and Gg the matter will be of the same density as without.

Case 5. Suppose now that a canal opens into the plane Hh as in Case 2, then will matter run into the space EH, and the density will be everywhere the same as without, except in the space EF, where the particles will be pressed close together, the quantity of matter therein exceeding the natural quantity by as much as is naturally contained in the space *BC*.

Case 6. Suppose now that a canal opens into the plane *Aa*, the fluid will have no tendency to run out thereat.

Case 7. Let us now consider what will be the result if the repulsion of the particles is inversely as some other power of the distance between that of the square and the cube; and first let us suppose matters as in the first case. There will be a certain space, as *EF*, which will be a vacuum, and a certain space, as *FG*, in which the particles will be pressed close together, for if the matter is uniform in *EH*, all the particles will be repelled towards *H* if there is not a vacuum at *E*, nor the particles pressed close together at *G*, but only the density less at *E* than at *H*, then the repulsion of space *EH* at *E* will be less on [a] particle at *E* and greater on a particle at *H* than if the density was uniform therein, consequently on that account as well as on account of the repulsion of *AC* a particle at *E* or *H* will be repelled towards *H*, but if the space *EF* is a vacuum and the particles in *GH* pressed close together, then if the spaces *EF* and *GH* are of a proper size, a particle at *F* or *G* may be in equilibrio.

Case 8. If you now suppose a canal to open into the plane *Hh* as in the 3rd case, some of the matter will run out thereat, so that the whole quantity of matter in the space *EH* will be less than natural. For if not, it has already been shown that a particle at *H* will be repelled from *A*, but the quantity of matter which runs out will not be so much as the redundant matter in *AC*, for if there was, the want of repulsion of the space *EH* on a particle at *h* would be greater than the excess of repulsion of the space *AC*.

Case 9. Suppose now that a canal opens into the plane *Aa* as in Case 3; a particle at *a* will be repelled from *Dd*, but not with so much force as if there had been the natural quantity of fluid in the space *EH*, so that some of the fluid will run out at the canal, but not with so much force, nor will so much of the fluid run out as if there had been the natural quantity of fluid in *EH*.

Case 10. If you suppose matters to be as in the 4th case, then there must be a certain space adjacent to *Ee*, in which the particles will be pressed close together, and a certain space adjacent to *Hh* in which there must be a vacuum.

Case 11. If you suppose a canal to open into the plane *Hh*, some matter will run into the space *EH* thereby, so that the whole quantity of matter therein will be greater than natural.

The proof of these two cases is exactly similar to that of the two former.

Case 12. If you now suppose a canal to open into *Aa*, some fluid will run into it, but not with so much force nor in so great quantity as if the natural quantity of fluid had been contained in the space *Hh*.

I have supposed the planes *Aa*, &c. to be extended infinitely, because by that means I was enabled to solve the question accurately in the cases where the repulsion is supposed inversely as the square of the distance, which I could not have done otherwise, but it is evident that the phenomena will be nearly of the same kind if the planes are not infinitely extended.

For if the distance *ag* be small in respect of the length and breadth of the plane *Aa*, a particle placed at *a* will be repelled by the plane *Aa* with very nearly the same force as if the plane was infinitely extended.

It is plain that these 6 last cases agree very exactly with the laws of electricity laid down in the 3rd and 4th hypotheses [Thoughts... Art. 202].

If the lines *Bb* and *Dd* touch one another so that

[Here the MS. ends. Ed.]

Note 19, Art. 234.

Cavendish's Experiment on the Charge of a Globe between two Hemispheres.

This experiment has recently been repeated at the Cavendish Laboratory in a somewhat different manner.

The hemispheres were fixed on an insulating stand, so as to form a spherical shell concentric with the globe, which stood inside the shell upon a short piece of a wide ebonite tube.

By this arrangement, since during the whole experiment the potentials of the globe and sphere remained sensibly equal, the insulating support of the globe was never exposed to the action of any sensible electromotive force, and therefore had no tendency to become charged.

If the other end of the insulator supporting the globe had been connected to earth, then, when the potential of the globe was high, electricity would have crept from it along the insulator, and would have crept back again when, in the second part of the experiment, the potential of the globe was sensibly zero. In fact this was the chief source of disturbance in Cavendish's experiment. See Art. 512.

Instead of removing the hemispheres before testing the potential of the globe, they were left in their position, but discharged to earth. The effect on the electrometer of a given charge of the globe was less than if the hemispheres had been removed, but this disadvantage was more than compensated by the perfect security from all external electric disturbances afforded by the conducting shell.

The short wire which formed the communication between the shell and the globe was fastened to a small metal disk hinged to the shell, and acting as a lid to a small hole in it, so that when the lid and its wire were lifted up by means of a silk string, the electrode of the electrometer could be made to dip into the hole in the shell and rest on the globe within.

The electrometer was Thomson's Quadrant Electrometer.

The case of the electrometer, and one of the electrodes, were permanently connected to earth, and the testing electrode was also kept connected to earth, except when used to test the potential of the globe.

To estimate the original charge of the shell, a small brass ball was placed on an insulating stand at a distance of about 60 cm. from the centre of the shell.

The operations were conducted as follows :—

The lid was closed, so that the shell communicated with the globe by the short wire.

A Leyden jar was charged from a machine in another room, the shell was charged from the jar, and the jar was taken out of the room again.

The small brass ball was then connected to earth for an instant, so as to give it a negative charge by induction, and was then left insulated.

The lid was then lifted up by means of the silk string, so as to take away the communication between the shell and the globe.

The shell was then discharged and kept connected to earth.

The testing electrode of the electrometer was then disconnected from earth, and made to pass through the hole in the shell so as to touch the globe within without touching the shell.

Not the slightest deflexion of the electrometer could be observed.

To test the sensitiveness of the apparatus, the shell was disconnected from earth and connected to the electrometer. The small brass ball was then discharged to earth.

This produced a large positive deflexion of the electrometer.

Now in the first part of the experiment, when the brass ball was connected to earth, it became charged negatively, the charge being about $\frac{1}{54}$ of the original positive charge of the shell.

When the shell was afterwards connected to earth the small ball induced on it a positive charge equal to about one-ninth of its own negative charge. When at the end of the experiment the small ball was discharged to earth, this charge remained on the shell, being about $\frac{1}{486}$ of its original charge.

Let us suppose that this produces a deflexion D of the electrometer, and let d be the largest deflexion which could escape observation in the first part of the experiment.

Then we know that the potential of the globe at the end of the first part of the experiment cannot differ from zero by more than

$$\pm \frac{1}{486} \frac{d}{D} V,$$

where V is the potential of the shell when first charged.

But it appears from the mathematical theory that if the law of repulsion had been as $r^{-(2+q)}$, the potential of the globe when tested would have been by equation (25), p. 421,

$$0{\cdot}1478 \times qV.$$

Hence q cannot differ from zero by more than $\pm \frac{1}{72} \frac{d}{D}$.

Now, even in a rough experiment, D was certainly more than $300d$. In fact no sensible value of d was ever observed. We may therefore conclude that q, the excess of the true index above 2, must either be zero, or must differ from zero by less than

$$\pm \tfrac{1}{21600}.$$

Theory of the Experiment.

Let the repulsion between two charges e and e' at a distance r be

$$f = ee' \phi(r), \tag{1}$$

where $\phi(r)$ denotes any function of the distance which vanishes at an infinite distance.

The potential at a distance r from a charge e is

$$V = e \int_r^\infty \phi(r)\, dr. \tag{2}$$

Let us write this in the form

$$V = e \frac{1}{r} f'(r), \tag{3}$$

where

$$f'(r) = \frac{df(r)}{dr}, \tag{4}$$

and $f(r)$ is a function of r equal to $\int r \left[\int_r^\infty \phi(r)\, dr \right] dr$.

We have in the first place to find the potential at a given point B due to a uniform spherical shell.

Let A be the centre of the shell, a its radius, α its whole charge, and σ its surface-density, then

$$\alpha = 4\pi a^2 \sigma. \tag{5}$$

Take A for the centre of spherical co-ordinates and AB for axis, and let $AB=b$.

Let P be a point on the sphere whose spherical co-ordinates are θ and ϕ, and let $BP=r$, then

$$r^2=a^2-2ab\cos\theta+b^2. \tag{6}$$

The charge of an element of the shell at P is

$$\sigma a^2\sin\theta\,d\theta\,d\phi=\frac{1}{4\pi}a\sin\theta\,d\theta\,d\phi. \tag{7}$$

The potential at P due to this element is

$$\frac{1}{4\pi}a\frac{f'(r)}{r}\sin\theta\,d\theta\,d\phi, \tag{8}$$

and the potential due to the whole shell is therefore

$$V=\int_0^{2\pi}\int_0^{\pi}\frac{1}{4\pi}a\frac{f'(r)}{r}\sin\theta\,d\theta\,d\phi. \tag{9}$$

Integrating with respect to ϕ from 0 to 2π,

$$V=\int_0^{\pi}\tfrac{1}{2}a\frac{f'(r)}{r}\sin\theta\,d\theta. \tag{10}$$

Differentiating (6) with respect to θ,

$$rdr=ab\sin\theta\,d\theta. \tag{11}$$

Hence,

$$V=\int_{r_1}^{r_2}\frac{1}{2}\frac{a}{ab}f'(r)\,dr=\frac{1}{2}\frac{a}{ab}\{f(r_2)-f(r_1)\}, \tag{12}$$

the upper limit r_2 being always $a+b$, and the lower limit r_1 being $a-b$ when $a>b$, and $b-a$ when $a<b$.

Hence, for a point inside the shell

$$V=\frac{a}{2ab}\{f(a+b)-f(a-b)\}, \tag{13}$$

for a point on the shell itself

$$V=\frac{a}{2a^2}f(2a), \tag{14}$$

and for a point outside the shell

$$V=\frac{a}{2ab}\{f(b+a)-f(b-a)\}. \tag{15}$$

We have next to determine the potentials of two concentric spherical shells, the radius of the outer shell being a and its charge α, and that of the inner shell being b and its charge β.

Calling the potential of the outer shell A, and that of the inner B, we find by what precedes,

$$A = \frac{\alpha}{2a^2} f(2a) + \frac{\beta}{2ab} \{f(a+b) - f(a-b)\}, \tag{16}$$

$$B = \frac{\beta}{2b^2} f(2b) + \frac{\alpha}{2ab} \{f(a+b) - f(a-b)\}. \tag{17}$$

In the first part of the experiment the shells communicate by the short wire and are both raised to the same potential, say V.

Putting $A = B = V$ and solving equations (16), (17), we find for the charge of the inner shell

$$\beta = 2Vb \frac{bf(2a) - a\{f(a+b) - f(a-b)\}}{f(2a)f(2b) - \{f(a+b) - f(a-b)\}^2}. \tag{18}$$

In the original experiment of Cavendish the hemispheres forming the outer shell were removed altogether from the globe and discharged. The potential of the inner shell or globe would then be

$$B_1 = \frac{\beta}{2b^2} f(2b). \tag{19}$$

In the form of the experiment as repeated at the Cavendish Laboratory, the outer shell was left in its place, but was connected to earth, so that $A = 0$. In this case we find for the potential of the inner shell when tested by the electrometer

$$B_2 = V\left\{1 - \frac{a}{b}\frac{f(a+b) - f(a-b)}{f(2a)}\right\}. \tag{20}$$

Let us now assume with Cavendish, that the law of force is some inverse power of the distance, not differing much from the inverse square, that is to say, let

$$\phi(r) = r^{-(2+q)}, \tag{21}$$

then

$$f(r) = \frac{1}{1-q^2} r^{1-q}. \tag{22}$$

If we suppose q to be a small numerical quantity, we may expand $f(r)$ by the exponential theorem in the form

$$f(r) = \frac{1}{1-q^2} r\{1 - q\log r + \frac{1}{1.2}(q\log r)^2 - \&c.\}, \tag{23}$$

and if we neglect terms involving q^2, equations (19) and (20) become

$$B_1 = \tfrac{1}{2}\frac{a}{a-b} Vq\left[\frac{a}{b}\log\frac{a+b}{a-b} - \log\frac{4a^2}{a^2-b^2}\right], \tag{24}$$

$$B_2 = \tfrac{1}{2} Vq\left[\frac{a}{b}\log\frac{a+b}{a-b} - \log\frac{4a^2}{a^2-b^2}\right]. \tag{25}$$

Laplace [Mec. Cel. I. 2] gave the first direct demonstration that no function of the distance except the inverse square can satisfy the condition that a uniform spherical shell exerts no force on a particle within it.

If we suppose that β, the charge of the inner sphere, is always accurately zero, or, what comes to the same thing, if we suppose B_1 or B_2 to be zero, then

$$bf(2a) - af(a+b) - af(a-b) = 0.$$

Differentiating twice with respect to b, a being constant, and dividing by a, we find

$$f''(a+b) = f''(a-b),$$

or, if $a - b = c$,

$$f''(c+2b) = f''(c),$$

which can be true only if

$$f''(r) = C_0 \text{ a constant.}$$

Hence,
$$f'(r) = C_0 r + C_1,$$

and
$$\int_r^\infty \phi(r)\,dr = \frac{1}{r} f'(r) = C_0 + \frac{1}{r} C_1,$$

whence,
$$\phi(r) = C_1 \frac{1}{r^2}.$$

We may notice, however, that though the assumption of Cavendish, that the force varies as some inverse power of the distance, appears less general than that of Laplace, who supposes it to be any function of the distance, it is the most general assumption which makes the ratio of the force at two different distances a function of the ratio of those distances.

If the law of force is not a power of the distance, the ratio of the forces at two different distances is not a function of the ratio of the distances alone, but also of one or more linear parameters, the values of which if determined by experiment would be absolute physical constants, such as might be employed to give us an invariable standard of length.

Now although absolute physical constants occur in relation to all the properties of matter, it does not seem likely that we should be able to deduce a linear constant from the properties of anything so little like ordinary matter as electricity appears to be.

NOTE 20, ART. 272.

On the Electric Capacity of a Disk of sensible Thickness.

Consider two equal disks having the same axis, let the radius of either disk be a, and the distance between them b, and let b be small compared with a.

Let us begin by supposing that the distribution on each disk is the same as if the other were away, and let us calculate the potential energy of the system.

We shall use elliptical co-ordinates, such that the focal circle is the edge of the lower disk. In other words we define the position of a given point by its greatest and least distances from the edge of the lower disk, these distances being

$$a(\alpha+\beta) \text{ and } a(\alpha-\beta).$$

The distance of the given point from the axis is

$$r = a\alpha\beta, \tag{1}$$

and its distance from the plane of the lower disk is

$$z = a(\alpha^2-1)^{\frac{1}{2}}(1-\beta^2)^{\frac{1}{2}}. \tag{2}$$

If A_1 is the charge of the lower disk, the potential at the given point is

$$\psi = Aa^{-1}\operatorname{cosec}^{-1}\alpha, \tag{3}$$

or, if we write

$$\alpha^2 = \gamma^2+1, \tag{4}$$

$$\psi = Aa^{-1}\left(\frac{\pi}{2}-\tan^{-1}\gamma\right) \tag{5}$$

If A_2 is the charge of the upper disk, the density at any point is

$$\sigma = \frac{A_2}{2\pi a^2 p}, \tag{6}$$

where

$$p^2 = a^{-2}(a^2-r^2) = 1-\alpha^2\beta^2. \tag{7}$$

Putting $z=b$ in equation (2),

$$b^2 = a^2\gamma^2(1-\beta^2) \quad \text{or} \quad \beta^2 = 1-\frac{b^2}{a^2\gamma^2}. \tag{8}$$

Hence

$$p^2 = \frac{b^2}{a^2\gamma^2}-\gamma^2+\frac{b^2}{a^2}. \tag{9}$$

We have now to multiply the charge of an element of the upper disk into the potential due to the lower disk, and integrate for the whole surface of the upper disk,

$$\int 2\pi r\,dr\,\sigma\psi = A_1 A_2 a^{-1}\int_0^1\left(\frac{\pi}{2} - \tan^{-1}\gamma\right)dp$$

$$= A_1 A_2 a^{-1}\left(\frac{\pi}{2} - \int_0^1 \tan^{-1}\gamma dp\right). \tag{10}$$

Between the limits of integration we may write with a sufficient degree of approximation,

$$\tan^{-1}\gamma = \gamma = \frac{b}{a}\left\{1 + \left(\frac{b}{a}\right)^{\frac{1}{2}}\right\}\left\{p + \left(\frac{b}{a}\right)^{\frac{1}{2}}\right\}^{-1}. \tag{11}$$

At the centre of the disk $p = 1$ and

$$\gamma = \frac{b}{a}, \text{ which agrees with (9).}$$

At the circumference,

$$p = 0 \text{ and } \gamma = \left(\frac{b}{a}\right)^{\frac{1}{2}} + \tfrac{1}{8}\left(\frac{b}{a}\right)^{\frac{3}{2}} \text{ by (9),}$$

whereas the equation (11) gives

$$\gamma = \left(\frac{b}{a}\right)^{\frac{1}{2}} + \frac{b}{a},$$

so that when b is very small compared with a, the value of γ cannot differ greatly from that given by equation (11). Hence we may write the expression (10)

$$A_1 A_2 a^{-1}\left[\frac{\pi}{2} - \frac{b}{a}\left\{1 + \left(\frac{b}{a}\right)^{\frac{1}{2}}\right\}\log\left\{\left(\frac{a}{b}\right)^{\frac{1}{2}} + 1\right\}\right]. \tag{12}$$

The corresponding quantity for the action of the upper disk on itself is got by putting $A_1 = A_2$ and $b = 0$, and is

$$A_2^{\,2} a^{-1}\frac{\pi}{2}. \tag{13}$$

In the actual case $A_1 = A_2 = \frac{1}{2}E$, where E is the whole charge, and the capacity is

$$K > \frac{2a}{\pi - \frac{b}{a}\left\{1 + \left(\frac{b}{a}\right)^{\frac{1}{2}}\right\}\log\left\{\left(\frac{a}{b}\right)^{\frac{1}{2}} + 1\right\}}, \tag{14}$$

or, since in our approximation we have neglected $\left(\frac{b}{a}\right)^{\frac{1}{2}}$, our result may be expressed with sufficient accuracy in the form

$$K > \frac{2}{\pi}\left(a + \frac{1}{2\pi}b\log\frac{a}{b}\right), \tag{15}$$

showing that the capacity of two disks very near together is equal to that of an infinitely thin disk of somewhat larger radius.

If the space between the two disks is filled up, so as to form a disk of sensible thickness, there will be a certain charge on the curved surface, but at the same time the charge on the inner sides of the disks will disappear, and that on the outer sides near the edges will be diminished, so that the capacity of a disk of sensible thickness is very little greater than that given by (15).

We may apply this result to estimate the correction for the thickness of the square plates used by Cavendish. The factor by which we must multiply the thickness in order to obtain the correction for the diameter of an infinitely thin plate of equal capacity is $\frac{1}{2\pi}\log\frac{a}{b}$.

	$\frac{a}{b}$	$\frac{1}{2\pi}\log_e\frac{a}{b}$
Tin plate..................	600	1·017
Hollow plate	11	0·381
Portland stone, &c.......	30	0·540
Slate	75	0·686

The correction is in every case much smaller than Cavendish supposed.

Note 21, Arts. 277, 452, 473, 681.

Calculation of the Capacity of the Two Circles in Experiment VI.

The diameter of one of the circles was 9·3 inches, so that its capacity when no other conductor is in the field is $\frac{9\cdot 3}{\pi} = 2\cdot 960$. The distance between their centres was 36, 24, and 18 inches, which we may call c_1, c_2, and c_3.

The height of the centres of the circles above the floor was about 45 inches, so that the distance of the image of the circle would be about 90 inches and that of the image of the other circle would be about

$$r = (90^2 + c^2)^{\frac{1}{2}}.$$

Hence, if P is the potential of the circles when the charge of each is 1,

$$P = \frac{\pi}{2a} + \frac{1}{c} - \frac{2}{3}\frac{a^2}{c^3} + \&c. - 90^{-1} - r^{-1},$$

where the first term is due to the circle itself, the second and third to the other circle, as in Note 11, and the two last to the images of the two circles. We thus find for the three distances

$$P_1 = 0\cdot 3438, \qquad P_2 = 0\cdot 3567, \qquad P_3 = 0\cdot 3689.$$

The capacity is $2P^{-1}$, and the number of "inches of electricity," according to the definition of Cavendish, is $4P^{-1}$,

or 11·636, 11·212, 10·844,

for the three cases.

The large circle was 18·5 inches in diameter and its centre was 41 inches from the floor, so that its charge would be 12·69 inches of electricity.

Hence the relative charges are as follows:

	Calculated.	Measured by Cavendish, Art. 276.
The large circle	1·000	1·000
The two small ones at 36 inches	·917	·899
24	·884	·859
18	·855	·811

Note 22, Art. 283.

Electric Capacity of a Square.

I am not aware of any method by which the capacity of a square can be found exactly. I have therefore endeavoured to find an approximate value by dividing the square into 36 equal squares and calculating the charge of each so as to make the potential at the middle of each square equal to unity.

The potential at the middle of a square whose side is 1 and whose charge is 1, distributed with uniform density, is

$$4 \log (1 + \sqrt{2}) = 3{\cdot}52549.$$

In calculating the potential at the middle of any of the small squares which do not touch the sides of the great square I have used this formula, but for those which touch a side I have supposed the value to be 3·1583, and for a corner square 2·9247.

If the 36 squares are arranged as in the margin, and if the charges of the corner squares be taken for unity, the charges will be as follows:

A	B	C	C	B	A
B	D	E	E	D	B
C	E	F	F	E	C
C	E	F	F	E	C
B	D	E	E	D	B
A	B	C	C	B	A

A	B	C	D	E	F
1·000	·599	562	·265	·210	·201

and the capacity of a square whose side is 1 will be 0·3607.

The ratio of the capacity of a square to that of a globe whose diameter is equal to a side of the square is therefore 0·7214.

In Art. 654 Cavendish deduces this ratio from the measures in Art. 478 and finds it 0·73, which is very near to our result. If, however, we take the numbers given in Art. 478, we find the ratio 0·79. From Art. 281 we obtain the ratio 0·747.

The ratio of the charge of a square to that of a circle whose diameter is equal to a side of the square is by our calculation 1·133.

In Art. 648 Cavendish says that the ratio is that of 9 to 8 or 1·128, which is very close to our result, but in Arts. 283* and 682 he makes it 1·153.

The numbers in Art. 281 from which Cavendish deduces this would make it 1·1514.

The numbers given in Art. 478 would make it 1·176.

Cavendish supposes that the capacity of a rectangle is the same as that of a square of equal area, and he deduces this from a comparison of the square 15·5 with the rectangle 17·9 × 13·4.

It is not easy to calculate the capacity of a rectangle in terms of its sides, but we may be certain that it is greater than that of a square of equal area.

For if we suppose the electricity on the square rendered immoveable, and if we cut off portions from two sides of the square and place them on the other two sides so as to form a rectangle, we are carrying electricity from a place of higher to a place of lower potential, and are therefore diminishing the energy of the system.

If we now make the electricity moveable, it will re-arrange itself on the rectangle and thereby still further diminish the energy. Hence the energy of a given charge on the rectangle is less than that of the same charge on the square, and therefore the capacity of the rectangle is greater than that of the square.

Note 23, Arts. 288 and 542.

On the Charge of the Middle Plate of Three Parallel Plates.

The plates used by Cavendish were square, but for the purpose of a rough estimate of the distribution of electricity between the three plates we may suppose them to be three circular disks.

First consider two equal disks on the same axis, at a distance small compared with the radius of either.

If the disks were in contact, the distribution on each would be the same as on each of the two surfaces of a single disk, and it would be entirely on the outer surface.

* In Art. 283 of this book the number is printed 1·53. It should be 1·153.

If the distance between the disks is very small compared with their radii, the force exerted by one of the disks at any point of the other will be nearly but not quite normal to its surface. The component in the plane of the disk will be directed outwards from the centre, so that the density will be greater near the edge than in a single disk having the same charge, but as a first approximation we may assume that the sum of the surface-densities on both sides of any element of the disk is the same as if the other disk were away.

But the density on the outer surface of the disk will be increased, and the density on the inner surface diminished, by a quantity numerically equal to the normal component of the repulsion of the other disk divided by 4π, and the whole charge of the outer surface will be increased, and the whole charge of the inner surface diminished, by a quantity equal to the charge of that part of the other disk, the lines of force from which cut the disk under consideration.

Hence the charges of the inner and outer surfaces of the disk are

$$\frac{A}{a}\,\omega \text{ and } \frac{A}{a}\,(a-\omega)$$

respectively, where the value of the elliptic co-ordinate ω is that corresponding to the edge of the other disk.

If a is the radius of either disk, and c the distance between them,

$$\omega = \frac{1}{\sqrt{2}}\,(c\sqrt{4a^2+c^2}-c^2)^{\frac{1}{2}}.$$

If we now place another equal disk on the same axis at a distance c from one of them, the potential being the same for all three, the new disk will greatly diminish the charge of the surface of the disk which is next to it, but it will not have much effect on the charges of the other surfaces.

The result will therefore be that the charges of the two outer disks will together be greater, but not much greater, than that of a single disk at the same potential, but the charge of each of the surfaces of the middle disk will be the same as that of one of the inner surfaces of a pair of disks at distance c. Hence the charge of the middle disk will be to that of the two outer disks together as ω to a.

If we substitute for the square plates of twelve inches in the side disks of 13·8 inches diameter which would have nearly the same capacity, then if the distance between the outer disks is 1·15 inches, $c = {\cdot}575$ and $\omega = 1{\cdot}936$ and $a = 3{\cdot}5\,\omega$, or the charge of the middle disk would be 3·5 times greater if the other disks had been removed.

If the distance between the outer disks is 1·65 inches, $c = {\cdot}875$ and $\omega = 2{\cdot}293$, whence $a = 2{\cdot}2\,\omega$, or the charge of the middle disk would have been 2·2 times greater if the outer disks had been removed.

It is evident, however, that in the assumed distribution the potential is less at the edges of the outer disks than at their centres. The electricity will therefore flow more towards the edges of the outer disks, and, as this will raise the potential near the edge of the middle disk, the charge of the middle disk will be less than on our assumption. I have not attempted to estimate the distribution more approximately.

Cavendish found the charge of the middle disk $\frac{1}{7}$ and $\frac{1}{8}$ of what it would have been without the outer disks. This is much less than the first approximation here given, but much greater than Cavendish's own estimate, founded on the assumption that the distribution of electricity follows the same law in the three plates.

NOTE 24, ARTS. 338, 652.

On the Capacity of a Conductor placed at a finite distance from other Conductors.

Cavendish has not given any demonstration of the very remarkable formula given in Art. 338 for the capacity of a conductor at a finite distance from other conductors. We may obtain it, however, in the following manner.

If the distance of all other conductors is considerable compared with the dimensions of the positively charged conductor, C, whose capacity is to be tried, the negative charge induced on any one of the other conductors will depend only on the charge of the conductor C and not on its shape. This induced charge will produce a negative potential in all parts of the field; let us suppose that the potential thus produced at the centre of the conductor C is $-\frac{E}{x}$, where E is the charge of C and x is a quantity of the dimensions of a line.

If L is the capacity of C when no other conductor is in the field, then the potential due to the charge E will be $\frac{E}{L}$, and the potential, which arises from the negative charge induced on other conductors, will be $-\frac{E}{x}$, so that the actual potential will be $E\left(\frac{1}{L}-\frac{1}{x}\right)$.

Dividing the charge by the potential we obtain for the actual capacity

$$L\frac{x}{x-L},$$

or the capacity is increased in the ratio of x to $x-L$.

The idea of applying this result to determining the value of x by comparing the charges of bodies, the ratio of whose capacities is known, is entirely peculiar to Cavendish, and no one up to the present time seems to have attempted anything of the kind.

The height of the centre of the circles above the floor seems to have been about 45 inches. If we neglect the undercharge of other conductors and consider only the floor, x would be about 90 inches in modern measure, but as a capacity x is reckoned by Cavendish as $2x$ "inches of electricity," the value of x in "inches of electricity" would be 180.

If we could take into account the undercharged surfaces of the other conductors, such as the walls and ceiling, the "machine," &c., the value of x would be diminished, and it is probable that the value obtained from his experiments by Cavendish, $166\frac{1}{2}$, is not far from the truth.

Note 25, Arts. 360, 539, 666.

Capacities of the large tin Cylinder and Wires.

The dimensions of the cylinder are given more accurately in Art. 539. It was 14 feet 8·7 inches long, and 17·1 inches circumference. Its capacity when not near any conductor would be, by the formula in Note 12, 22·85 inches, and when its axis was 47 inches from the floor it would be 31·3 inches, or in Cavendish's language 62·6 inches of electricity. Cavendish makes its computed charge 48·4, and its real charge 73·6. See Art. 666. Now the charge of either of the plates D and E was by Art. 671, 26·3 inches of electricity, so that

$$\text{tin cylinder} = 1{\cdot}19\,(D + E).$$

The capacities of the different wires mentioned in Arts. 360 and 539 are, by calculation,

length.	diameter.	capacity.
29	$\frac{1}{6}$	2·67
22	$\frac{1}{6}$	2·09
37	·15	3·13
27·6	·15	2·46
20·8	·15	1·88
31	·15	2·71
24	·15	2·28

The ratio of the charge of the first of these wires to that of the second is 1·37.

Note 26, Art. 369.

Action of Heat on Dielectrics.

The effect of heat in rendering glass a conductor of electricity is described in a letter from Kinnersley to Franklin* dated 12th March, 1761. He found that when he put boiling water into a Florence

* Franklin's Works, edited by Sparks (1856), Vol. v., p. 367.

flask he could not charge the flask, and that the charge of a three pint bottle went freely through without injuring the flask in the least.

Franklin in his reply describes some experiments of Canton's on thin glass bulbs, charged and hermetically sealed and kept under water, showing "that when the glass is cold, though extremely thin, the electric fluid is well retained by it."

He then describes an experiment by Lord Charles Cavendish, showing that a thick tube of glass required to be heated to 400° F. to render it permeable to the common current.

A portion of a glass tube near the middle of its length was made solid, and wires were thrust into the tube from each end reaching to the solid part. The middle portion of the tube was bent, so that a portion, including the solid part, could be placed in an iron pot filled with iron-filings. A thermometer was put into the filings; a lamp was placed under the pot; and the whole was supported upon glass.

The wire which entered one end of the tube was electrified by a machine, a cork ball electrometer was hung on the other, and a small wire, reaching to the floor, was tied round the tube between the pot and the electrometer, in order to carry off any electricity that might run along upon the tube.

"Before the heat was applied, when the machine was worked, the cork balls separated at first upon the principle of the Leyden phial. But after the middle part of the tube was heated to 600, the corks continued to separate, though you discharged the electricity by touching the wire, the electrical machine continuing in motion. Upon letting the whole cool, the effect remained till the thermometer was sunk to 400."

Experiments on the conductivity of glass at different temperatures have been made by Buff*, Perry†, and Hopkinson‡.

Hopkinson finds that if B is the specific conductivity divided by the specific inductive capacity and multiplied by 4π, then for

$$\text{glass N°. 2,} \quad \log B = \bar{1}\cdot 35 + 0\cdot 0415\theta,$$

$$\text{glass N°. 7,} \quad \log B = \bar{4}\cdot 17 + 0\cdot 0283\theta,$$

where θ is the temperature centigrade.

Glass N°. 2 is of a deep blue colour; it is composed of silica, soda, and lime.

Glass N°. 7 is "optical light flint," density 3·2, composed of silica, potash, and lead; almost colourless, the surface neither "sweats" nor

* *Annalen der Chemie und Pharmacie*, xc. (1854), p. 257.

† *Proc. R. S.* 1875, p. 468.

‡ *Phil. Trans.* 167 (1877), p. 599.

tarnishes in the slightest degree. This glass at ordinary temperatures is sensibly a perfect insulator.

The conductivity of glass when heated makes it very difficult to determine its capacity as a dielectric. It appears from the experiments of Hopkinson on glasses of known composition, that the glasses made with soda and lime conduct more, and are also more subject to "electric polarization" and "residual charge" than those made with potash and lead.

Both the conductivity and the susceptibility to residual charge increase as the temperature rises, and this makes it very doubtful whether the apparent increase of dielectric capacity, which was observed by Cavendish and also by recent experimenters, is a real increase of the specific inductive capacity, or merely an effect of increased conductivity.

The experiments of Messrs Ayrton and Perry* on wax at different temperatures would seem to indicate a real increase of dielectric capacity, as well as of conductivity, as the temperature rises up to the melting point. During the process of melting the capacity decreases and at higher temperatures begins to increase again, but the conductivity continues to increase as the temperature rises.

Note 27, Art. 376.

Electrostatic capacity of different substances.

	Cavendish.	Boltzmann.	Wüllner.	Gordon.
Shellac	4·47		2·95 to 3·73	2·746
Rosin		2·55		
Rosin and bees'-wax ...	3·38			
Dephlegmated bees'-wax	3·7			
Plain bees'-wax	4			
	Schiller.			
Sulphur		3·84	2·88 to 3·21	2·579
Ebonite	2·21 to 2·76	3·15	2·56	2·284
Paraffin	1·81 to 2·47	2·32	1·96	1·994
Black caoutchouc	2·12			2·22
vulcanized	2·69			2·497

Note 28, Art. 383.

Capacity of a Cylindrical Condenser.

The rule by which Cavendish computed the charge of a condenser consisting of two cylindrical surfaces having the same axis is given at Art. 313.

* *Phil. Mag.* August, 1878.

If R is the external and r the internal radius, and l the length of the cylinders, then Cavendish's expression for the "computed charge" is $\frac{1}{2}\frac{R+r}{R-r}l$.

The true expression for the capacity is

$$\frac{1}{2}\,\frac{l}{\log R - \log r}$$

when the logarithms are Naperian.

We may express $\log R - \log r$ in the form of the series

$$2\frac{R-r}{R+r} + \frac{2}{3}\left(\frac{R-r}{R+r}\right)^3 + \frac{2}{5}\left(\frac{R-r}{R+r}\right)^5 + \&\text{c.},$$

and we thus find as an approximate value of the capacity

$$\frac{1}{4}l\frac{R+r}{R-r}\left\{1 - \frac{1}{3}\left(\frac{R-r}{R+r}\right)^2 - \frac{4}{45}\left(\frac{R-r}{R+r}\right)^4 - \&\text{c.}\right\}.$$

The first term agrees with Cavendish's rule, for the "capacity" is half the "inches of electricity," but the other terms show that Cavendish's rule gives too large a value for the computed charge.

The following table gives the charge as computed by Cavendish compared with that given by the correct formula.

	Cavendish.	True.	Observed charge by computed.
Flint jar	85·9	72·55	9·88
...... cylinder	87·1	73·59	8·83
Therm. I.	11·0	8·37	9·58
...... II.	11·1	7·84	10·29
Green cyl. 1	77·2	65·92	11·15
.... 2	76·6	61·54	11·22
.... 3	40·8	34·29	10·29

Note 29, Art. 437.

Electrical Fishes.

The fishes which are known to possess the power of giving electric shocks belong to two genera of Teleostean Fishes and one of Elasmobranch Fishes, and the position and relations of the electric organs are different in each.

In every instance, however, the electric organ may be roughly described as being divided in the first place into parallel prisms or columns by septa, which we may call (with reference to the organ, not the fish) longitudinal septa, and in the second place each column is divided transversely by diaphragms, the structure of which is different in the different families, but in every case the terminations of the nerves lie on that surface of each diaphragm which during the discharge becomes its negative surface.

In the large family of the Torpedos the electric organs are formed of a large number of short columns, the columns running from the belly to the back of the fish. The nerves terminate on the ventral surface of each diaphragm, and the electric discharge is from belly to back through the organ, or in other words, the back of the fish becomes positive with respect to the belly.

There seems to be but one species of Gymnotus. It is a long eel-like fish. Its electric organs consist of a smaller number of very long columns running from the tail to the head of the fish. The nerves terminate on the posterior surface of the diaphragms, and the electric discharge is from tail to head through the organ, or the head of the fish becomes positive with respect to the tail.

There are three species of Malapterurus which are known to be electrical. In these the electric organs run longitudinally. Bilharz, observing that the nerves appear to terminate in an expansion like the head of a nail on the posterior surface of the diaphragms, concluded that the electric discharge must be from tail to head through the organ, as in the Gymnotus. Ranzi* however, and afterwards, independently of him, Du Bois Reymond† found that the discharge is really from head to tail through the organ, so that the tail becomes positive with respect to the head, and Schultze, who had been led to believe, from a comparison of his own observations on the organs of pseudo-electric fishes with the drawings of Bilharz, that the nerves might pass through the diaphragms and terminate on their anterior surfaces, found, on examining the preparations sent him by Du Bois Reymond, that this was really the case in Malapterurus, so that we may now assert that in every known case the terminations of the nerves are on that side of each diaphragm which during discharge becomes negative.

The origin of the nerves which supply the electric organs is different in the three families.

In the Torpedos the electric nerves are derived from the posterior division of the brain. Irritation of this lobe produces an electric discharge of the organ, but no muscular contraction. Irritation of other parts of the brain produces muscular contractions, but not electric discharges, unless the disturbance produced affects the electric nerves.

In the Gymnotus the electric nerves arise from the whole length of the spinal cord, and in Malapterurus the electric organs are supplied by the 2nd and 3rd pair of spinal nerves.

The electric nerves are so called because they govern the discharges of the electric organ. No essential difference has been observed between

* *Nuovo Cimento*, Tomo II, Dicembre 1856, p. 447, quoted by Du Bois Reymond "Zur Geschichte der Entdeckungen am Zitterwelse," *Archiv fur Anatomie Physiologie*, &c. Leipzig, 1859, p. 210.

† *Monatsbericht d. k. Akad.* Berlin, 1858.

the electric phenomena in these nerves and those in other nerves. They must be classed, with respect to origin as well as function, among the motor nerves. The only difference is that their function is to govern the electric discharge of a peculiar organ, instead of the contraction of a muscle.

The experiments of Dr Davy* and those of Matteucci† shewed that the discharge of the Torpedo produces all the known phenomena of an electric discharge. Faraday‡ did the same for the Gymnotus, and Du Bois Reymond§ for the Malapterurus.

M. Marey|| has recently investigated some of the electrical phenomena of the discharge of the Torpedo. He employed three methods of indicating the discharge, the prepared leg of a frog, which is extremely sensitive to the feeblest current, but has the disadvantage that the time required for the contraction of the muscles, and still more the time required for their relaxation, is many times the period of the recurrence of the electric discharges of the Torpedo, so that the rapidly changing phases of the discharge cannot be distinguished by this method.

The second indicator used by Marey was the electromagnetic signal of M. Deprez, which can register 500 electric currents in a second by the motion of a tracing point over the smoked surface of a revolving cylinder. The action of this instrument was sufficiently prompt to register the number of the separate currents of which the "continued discharge" of the Torpedo consists. It was not, however, sufficiently sensitive to trace the curve of the intensity of the current when the strength of the current was less than that required to work the tracing point, and the trace therefore represents only the phases of greatest strength of current in each separate discharge.

M. Marey calls each separate discharge of the Torpedo an electric *flux*.

The whole discharge consists of a rapid succession of these fluxes, at the rate of from 60 to 140 per second, gradually decreasing in intensity, but remaining sensible sometimes for a second or a second and a half. In one of the tracings 120 fluxes may be counted quite distinctly, with a somewhat irregular continuation of feebler fluxes.

The electromagnetic signal, however, depending on the attraction of a soft iron armature, is acted on by a force varying nearly as the square of the strength of the current. It is therefore unable to respond to feeble currents, and it does not indicate the direction of the currents, even when improved in certain particulars by M. Marey.

The third indicator used by M. Marey was the capillary electrometer of M. Lippmann. In this instrument a capillary glass tube is filled in one part with mercury and in the other with dilute sulphuric acid. The pressure of the mercury is so adjusted that the division between the

* *Phil. Trans.*, 1834. † *Comptes Rendus*, 1836.
‡ *London Medical Gazette*, 1838. § *Berlin Monatsb.*, 1858.
|| *Travaux du Laboratoire de M. Marey*, III. (1877).

two liquids appears in the middle of the field of a microscope. The electrodes of the instrument are connected with the two liquids respectively, and when a small electromotive force acts from one electrode to the other, the surface of separation of the two liquids is seen to move in the same direction as the electromotive force, that is to say, the mercury advances if the electromotive force is from the mercury to the acid, and retreats if it is in the opposite direction.

This instrument, therefore, is admirably suited for the investigation of small electromotive forces, and the mass of the moving parts is so small that it responds most promptly to every variation of the electromotive force. Its only defect is that its range is limited to the electromotive force required to decompose the acid, and the electromotive force of the Torpedo, as we know, is of far greater intensity than this. M. Marey therefore used a shunt, so as to diminish the force acting on the electrometer to such a degree as to be within the working limits of the instrument.

He thus ascertained that the back of the fish is positive with respect to the belly, not only on the whole, but during every phase of each flux, and that it does not sink to zero between the fluxes.

The modern researches on the electric fishes would seem to point to the conclusion that the electric organ is not like a battery of Leyden jars in which electricity is stored up ready to be discharged at the will of the animal, but rather like a Voltaic battery, the metals of which are lifted out of the cells containing the electrolyte, but are ready to be dipped into them.

There seems to be no electric displacement in the organ till the electric nerve acts on it. The energy of the electric discharge which then takes place is not supplied to the organ by the nerve; the nerve only sets up an action which is carried on by the expenditure of energy previously supplied to the organ by the materials which nourish it.

During the discharge certain chemical changes take place in the organ. These changes involve a loss of intrinsic energy, and the chemical products found in the organ after repeated electric discharges are similar to the products found in muscles after they have performed mechanical work.

The organ, by repeated discharges, becomes incapable of responding to stimulation, and can only recover its power by the gradual process by which it is nourished.

Faraday proposed to try whether sending an artificial current through the Gymnotus would exhaust the organ, if sent in the direction of the natural discharge, or would restore it more rapidly to vigour if sent in the opposite direction. The only experiments on the effect of electricity on electric fishes seem to be those of Dr Davy, who found that an artificial current did not excite the electric organs of the Torpedo, though it had an effect on the muscles, but less than on those of other fishes, and of Du Bois Reymond, who found that Malapterurus was very slightly affected by induction currents passing through the water of his tub, though they were strong enough to stun and even to kill other fishes. When the induction currents were made very strong, the fish swam about till he had placed his body transverse to the lines of discharge, but did not appear to be much annoyed by them *.

The most valuable experiments hitherto made are probably those of Dr Carl Sachs, who went out to Venezuela in 1876 for the express purpose of studying the Gymnotus in its native rivers, with all the resources of Du Bois Reymond's methods. Dr Sachs lost his life in an Alpine accident in 1878, and as he did not himself publish his researches, it is to be feared that their results are lost to science.

Note 30, Art. 560.

Excess of redundant fluid on positive side above deficient fluid on negative side of a coated plate.

When two equal disks have the same axis, the first being at potential V and the other connected to the earth, the algebraic sum of the charges of the two disks is just half the charge of the two disks together if they were both raised to potential V.

If the two disks are very near each other, the charge of the two together is very little greater than that of one by itself at the same potential.

Hence the excess of the redundant fluid above the deficient, when one of the disks is raised to potential V and the other connected with the earth, is very little greater than $\frac{1}{\pi} aV$, where a is the radius. (See Note 4.)

Note 31, Art. 573.

Intensity of the Sensation produced by an Electric Discharge.

Cavendish tried this and several other experiments (Arts. 406, 573, 597, 610, 613) to determine in what way the intensity of the sensation of an electric shock is affected by the two quantities on which the physical properties of the discharge depend, namely the quantity of redundant fluid discharged, and the degree of electrification before it is discharged, the resistance of the discharging circuit being supposed constant.

He seems to have expected (Art. 597) that the strength of the shock would be "as the quantity of electricity into its velocity," or in modern language, as the product of the quantity into the mean strength of the current of discharge. Since the electromotive force acting on the body of the operator is measured by the product of the strength of the current into the resistance of the body, which we may

* A somewhat extensive account of the subject is given in a dissertation, *De' Pesci elettrici e pseudoelettrici*, per Stefano St. Sihleanu (di Bucuresti, Romania), Napoli, 1876.

suppose constant, Cavendish's hypothesis would make the intensity of the shock proportional to the work done by the discharge within the body.

According to this hypothesis, if a jar charged to a given degree produces a shock of a certain intensity, then a charge equal to n times the charge of this jar, communicated to n^2 similar jars, and discharged through the same resistance, would give a shock of equal intensity.

By the experiment recorded in Arts. 406 and 573, in which $n=2$, it appeared that the shock given by four jars charged with the electricity of two jars, was rather greater than that of a single jar.

In the experiment in Art. 610 Cavendish compared the shock of jar 1 electrified to $2\frac{1}{2}$, with that of $B+2A$ electrified to the same degree and communicated to the whole battery. Here the capacity of $B+2A$ was equal to 6 times jar 1, and that of the whole battery was 154 times jar 1, so that 6 times the quantity of electricity communicated to 154 jars gave a shock of about the same strength, though as Cavendish remarks, "as there is a good deal of difference between the sensations of the two, it is not easy comparing them."

Here 154 is the $2\frac{7}{8}$ power of 6, so that the shock seems to depend rather more on the quantity of electricity than on the degree of electrification. This is the only experiment which Cavendish has worked out to a numerical result.

By the other experiments recorded in Art. 610, $34\frac{1}{2}$ communicated to 7 rows, gives a shock equal to 22 communicated to one row. This would make the number of jars as the 4·3 power of the charges. By Art. 613 the number of jars would be as the 3·3 power of the charge.

Cavendish had not the means of producing a steady current of electricity, such as we now obtain by means of a Voltaic battery, so that he could not discover the most important of the facts now known about the physiological action of the current, namely, that the effects of the current, whether in producing sensations, or in causing the contraction of muscles, depend far more on the rapidity of the changes in the strength of the current than on its absolute strength. It is true that a steady current, if of sufficient strength, produces effects of both kinds, but a current so weak that its effect, when steady, is imperceptible, produces strong effects, both of sensation and contraction, at the moments when the circuit is closed and broken.

But although this may be considered as established, I am not aware of any researches having been made, from the results of which it would be possible to determine, from the knowledge of the physical character of two electric discharges, which would produce the greater physiological effect.

The kind of discharges most convenient for experiments of this kind is that in which the current is a simple exponential function of the time, and of the form

$$x = Ce^{-\frac{t}{\tau}},$$

where x is the strength of the current at the time t, C its strength at the beginning of the discharge, and τ a small time, which we may call the time-modulus.

In this case the whole physical nature of the discharge is determined by the values of the two constants C and τ. The intensity of the sensation produced by the discharge through our nerves is, therefore, some function of these two constants, and if we had any method of ascertaining the numerical ratio of the intensities of two sensations, we might determine the form of this function by experiments. We can hardly, however, expect much accuracy in the comparison of sensations, except in the case in which the two sensations are of the same kind, and we have to judge which is the more intense.

According to Johannes Müller, the sensation arising from a single nerve can vary only in one way, so that, of two sensations arising from the same nerve, if one remains constant, while the other is made to increase from a decidedly less to a decidedly greater value, it must, at some intermediate value, be equal in all respects to the first.

In the ordinary mode of taking shocks by passing them through the body from one hand to the other, the sensations arise from disturbances in different nerves, and these being affected in a different ratio by discharges of different kinds, it becomes difficult to determine whether, on the whole, the sensation of one discharge or the other is the more intense.

I find that when the hands are immersed in salt water the quality of the sensation depends on the value of τ.

When τ is very small, say 0·00001 second, and C is large enough to produce a shock of easily remembered intensity in the wrists and elbows, there is very little skin sensation, whereas when τ is comparatively large, say 0·01 second, but still far too small for the duration of discharge to be directly perceived, the skin sensation becomes much more intense, especially in any place where the skin may have been scratched, so that it becomes almost impossible so to concentrate attention on the sensation of the internal nerves as to determine whether this part of the sensation is more or less intense than in the discharge in which τ is small.

There are two convenient methods of producing discharges of this type.

(1) If a condenser of capacity K is charged to the potential V, and discharged through a circuit of total resistance R (including the body of the victim),

$$C = \frac{V}{R}, \qquad \tau = KR.$$

The whole quantity discharged is $Q = C\tau = VK$, and if r is the resistance of the body of the victim, the work done by the discharge in the body is

$$W = \tfrac{1}{2} QV \frac{r}{R} = \tfrac{1}{2} V^2 K \frac{r}{R}.$$

(2) If the current through the primary circuit of an induction coil is y, the coefficient of mutual induction of the primary and secondary coils M, that of the secondary circuit on itself L, and the resistance of the secondary circuit R, then for the discharge through the secondary circuit when the primary circuit is broken,

$$C = \frac{M}{L} y, \qquad \tau = \frac{L}{R},$$

$$Q = \frac{M}{R} y, \qquad W = \frac{1}{2} \frac{M^2 y^2}{L} \frac{\tau}{R}.$$

I first tried the comparison of shocks by means of an induction coil, in which M was about 0·78 and L about 52 earth quadrants, and in which the resistance of the secondary coil was 2710 Ohms. By adding some German silver wire to the primary coil, its resistance was made up to nearly 1 Ohm, and the primary thus lengthened, another wire of the same resistance, and a variable resistance Q were made into a circuit. One electrode of the battery was connected to the junction of the two equal resistances, and the other was connected alternately to the two ends of the resistance Q, so that the current through the primary was varied in the ratio of the primary P to $P + Q$, while the resistance of the battery-circuit remained always the same. When the smaller primary current, y, was interrupted, I took the secondary discharge through my body directly, but when the larger current, y', was interrupted, I made the secondary discharge pass through a capillary tube filled with salt solution as well as my body.

The resistance between my hands when both were immersed in salt-water was 1245 Ohms, making with the secondary coil a resistance of 3955 in the secondary circuit, so that the time-modulus of the discharge was $\tau = 1{\cdot}3 \times 10^{-3}$ seconds.

The resistance of the first capillary tube was 370000, so that when it was introduced $\tau = 1{\cdot}4 \times 10^{-5}$.

By a rough estimate of the comparative intensity of the shocks I supposed them to be of equal intensity when $y' = 8{\cdot}4y$, and therefore if we suppose that two shocks remain of equal intensity when C varies as τ^p, $p = 0{\cdot}468$.

By another experiment in which a tube was used whose resistance was 450000, $p = 0{\cdot}534$.

When the shocks at breaking contact were nearly equal, that at making contact was very much more intense with the small primary current and small secondary resistance than with the large primary current and large secondary resistance.

I then compared the discharges from two condensers of 1 and 0·1 microfarads capacity respectively, charging them with a battery of 25 Leclanché cells, the electromotive force of which was about 36 Ohms.

The resistance of the discharging circuit for the microfarad was 11200 Ohms, including my body, so that

$$\tau = 1{\cdot}12 \times 10^{-2} \text{ seconds.}$$

The resistance of the discharging circuit of the tenth of a microfarad was 3600, so that $\tau' = 3{\cdot}6 \times 10^{-4}$.

The values of C were inversely as the resistances, so that if the two shocks were, as I estimated them, nearly equal, the value of p would be 0·670.

This experiment was much more satisfactory and more easily managed than that with the induction-coil, and I thought it desirable to apply the same method to the comparison of the contractions of a muscle when its nerve was acted on by the discharge. I therefore availed myself of the kindness of Mr Dew-Smith, who prepared for me the sciatic nerve and gastrocnemius muscle of a frog, and attached the preparation to his myograph. The discharge was conducted through about 0·4 cm. of the nerve by means of Du Bois Reymond's unpolarizable electrodes, the resistance of the electrodes and nerve being 35000 Ohms. When the electrodes were in contact their resistance was 23000, leaving about 12000 as the resistance of the nerve itself.

I used two condensers, one 0·1 microfarad, and the other an air-condenser of 270 centimetres capacity in electrostatic measure, or about 3×10^{-4} microfarads.

The first was charged by one cell and the second by 25. The resistances were arranged so that the contractions produced in the muscle were much less than a third of a maximum contraction. The discharges were made alternately every 15 seconds, and when the resistances were 35000 and 140000 respectively, the alternate contractions as recorded on the myograph were as follows:

Small condenser.	Large condenser.
144	146
147	148
147	147
146	146
147	145

Here the time-modulus was $1{\cdot}05 \times 10^{-5}$ seconds for the small condenser and $1{\cdot}4 \times 10^{-2}$ for the large one, and the values of C were as 1 to 100, so that $p = {\cdot}640$.

If we suppose that Cavendish took the shocks through pieces of metal held in his hands, the resistance of the circuit would depend on the state of his skin. He occasionally used a piece of apparatus, which he nowhere describes, but which he names in three places* a shock-melter.

From Art. 585 it would appear that it was filled with salt water, even when fresh water was the subject of the experiment, and from Art. 637 Cavendish seems to have considered it his last resource as a method of receiving shocks. I therefore think that it must have

* Arts. 585, 622, 637. See facsimile at p. 326.

been an apparatus by which his hands were well wetted with salt water, so that the resistance of his body would be between 1000 and 2000 Ohms.

The capacity of his battery of 49 jars was 321000 glob. inc., which comes to rather less than half a microfarad.

The discharges of this through 2000 Ohms would have a time-modulus of about one-thousandth of a second.

The following table gives the different results obtained by Cavendish and by myself, with the time-modulus of the discharges compared. The quantity p is such that the ratio of the initial strength of the two discharges is inversely as the p power of the ratio of the time-moduli when the shocks are equal in intensity, or

$$\frac{C_1}{C_2} = \left(\frac{\tau_1}{\tau_2}\right)^{-p}, \quad \frac{Q_1}{Q_2} = \left(\frac{\tau_1}{\tau_2}\right)^{1-p}, \quad \frac{W_1}{W_2} = \left(\frac{\tau_1}{\tau_2}\right)^{1-2p}.$$

The number of jars among which a quantity of electricity must be divided in order to give a shock of a given intensity through a given resistance, varies as the $\frac{1}{1-p}$ power of the quantity of electricity.

Cavendish's experiments.

	τ_1	τ_2	p
Art. 573	0·0000065	0·000026	0·5 +
... 610	0·0000065	0·001	0·652
... do.	0·00014	0·001	0·767
... 613	0·00014	0·00042	0·697

Experiments by the Editor.

Induction-coil	0·000014	0·0013	0·468
do.	0·000011	0·0013	0·534
Condensers	0·00036	0·0112	0·670

Experiments on the prepared nerve and muscle of a frog.

	0·00001	0·014	0·640

This value of p does not differ much from 0·652, the only result which Cavendish has deduced in a numerical form from his experiments.

The most unaccountable of all the results arrived at by Cavendish is one which seems to have perplexed him so much that he has left the account of the experiments among which it occurs in a very imperfect state. He found (Arts. 639, 644) that the shock of a Leyden jar taken through a long thin copper wire produced a more intense sensation than when it was taken from the jar directly.

As in some of the experiments the wire was wound on a reel, and therefore the self-induction of the current might produce an oscillatory discharge, the physiological effects of which might be different from those of the simple discharge; I charged two Leyden jars to the same potential, using Thomson's Portable Electrometer as a gauge electrometer, and took the discharge of one through the secondary wire of an induction-coil, the resistance of which was about 1000 Ohms, and that of the other through an ordinary resistance-coil of 1000 Ohms.

In every trial I found that the sensation was more intense when taken through the ordinary resistance-coil than when taken through the induction-coil, and it is manifest that in the latter case the current begins and ends much less abruptly, so that the result is quite in accordance with the modern theory, that the sensation depends on the rapidity with which the strength of the current changes. I am, therefore, quite unable to account for the opposite result obtained by Cavendish. At the same time it is quite impossible that Cavendish could be mistaken in this comparison of the intensity of his sensations, for he had more practice than any other observer in comparing them, and he repeated this experiment many times.

The only apparent objection to the experiment is that the resistance of the copper wires was only 430 in one case and only 1000 in the other, whereas the resistance of a man's body, from one hand to the other, varies from about 1000 when the hands are thoroughly wet, to about 12000 when they are dry, so that the resistance of the copper was small compared with the possible variations of the resistance of Cavendish's body.

The resistances of the tubes filled with solutions of salt, &c., were very much greater, being from 20000 to 900000.

Note 32, Arts. 398, 576, 687.

Comparison of the Resistance of Iron Wire and Salt Water.

Cavendish never published the method by which he made this comparison, but the result given in Art. 398 seems to have been accepted by men of science on Cavendish's bare word, without any question as to how it was obtained.

It appears from Art. 576 that Cavendish made his body and the iron wire the branches of a divided circuit, and then tried how many inches of salt water must be put in the place of the iron wire, so that the shock might appear of the same strength.

By Matthiessen's experiments on the resistance of metals, the resistance of an iron wire of the dimensions given by Cavendish would be about 196 Ohms. As this is much less than that of a man's body from hand to hand, it would have made hardly any difference to the shock whether Cavendish took it through his body alone, or through his body and the iron wire in series.

By using the iron wire as a shunt and increasing the discharge so as to obtain a shock of easily remembered intensity, Cavendish was enabled to compare the wire with a column 5·1 inches long of saturated solution of salt.

By this experiment the resistance of saturated solution of salt is 355400 times that of iron.

By the statements in Art. 398, that the resistance of rain-water is 400,000,000 times that of iron wire, and 720 times that of a saturated solution of sea-salt, the resistance of saturated solution would be 555555 times that of iron wire.

It is true that this result given by Cavendish does not agree with the only experiment he has recorded, but we must remember that it is the only result which he published, and therefore he must have thought it the best he had.

By Kohlrausch's experiments on salt solutions combined with Matthiessen's on metals, the resistance of saturated solution of salt is 451390 times that of annealed iron, when both are at 18°C. The ratio of the resistances would agree with that given by Cavendish at a temperature of about 11°C.

The coincidence with the best modern measurements is remarkable.

NOTE 33, ART. 619.

Conductivity of Solutions of Salt.

According to the measurements of Kohlrausch* the electric conductivity k, of saturated solution of sodium chloride, the conductivity of mercury at 0° C. being taken as unity, is given by the equation

$$10^8 k = 1259\,(1 + 0{\cdot}0308t + 0{\cdot}000146t^2).$$

When the temperature is near 18° C., we may use the equation

$$10^8 k = 2015 + 45{\cdot}1\,(t - 18).$$

Saturated solution at 18° contains according to Kohlrausch 26·4 per cent. of salt. Cavendish's saturated solution contained $\frac{1}{3{\cdot}78}$ of salt, which is equivalent to 26·45 per cent.

Kohlrausch finds that saturated solution of salt is one of the best standard substances for the comparison of the resistance of other electrolytes. Its conductivity seems to be sensibly the same, whether it is made with chemically pure salt or with the ordinary salt of commerce. The temperature coefficient is also smaller than that of many other electrolytes.

* Wiedemann's *Annalen* Bd. VI. (1879) p. 51.

For other solutions of sodium chloride he finds that at 18°

$$10^8k = 13650p - 22700p^2,$$

where p is the proportion, by weight, of the salt to the whole solution.

For the particular solutions examined by Cavendish we have

p	10^8k	resistance in terms of sat. sol.	resistance found by Cavendish.	
$\frac{1}{3\cdot78}$	2015	1	1	sat. sol.
$\frac{1}{12}$	980	2·56	1·91	salt in 11
$\frac{1}{30}$	430	4·69	3·97	salt in 29
$\frac{1}{70}$	190	10·58	8·8	salt in 69
$\frac{1}{143}$	94	21·44	15·75	salt in 142
$\frac{1}{150}$	90	22·39	20·05	salt in 149
$\frac{1}{1000}$	13·65	147·6	93·02	salt in 999
$\frac{1}{3000}$	4·55	442·9	340·85	salt in 2999

Note 34, Art. 626.

Conductivity of other Solutions.

The substances mentioned by Cavendish are easily identified, with the exception of "calc. S. S. A." and "f. alk. D." The weights of the quantities furnish no indication, for they are so large as to show that a dilute solution was used. The letters A and D probably indicate the bottles in which the solutions were kept.

The expression f. alk. or fixed alkali occurs in several parts of Cavendish's writings, especially in the manuscripts lithographed by Mr Vernon Harcourt in the *Report of the British Association* for 1839. It certainly means pearl ashes or carbonate of potash. The full title seems to have been *alkali fixum vegetabile*, as distinguished from *alkali fixum fossile*, which is sodic carbonate, and other writers seem to have used the expression fixed alkali for either of these, but Cavendish always uses the expression as a synonym for pearl ashes, and distinguishes potassic hydrate by the name of "sope leys."

The conductivity as determined by Cavendish agrees much better with potassic carbonate than with potassic hydrate, the conductivity of which is much greater.

It seems likely that calc. S. S. was sodic carbonate, and the conductivity would agree very well with this explanation, only it is difficult to find among the names in use at the time any which could be written in this form. Mr Maine has suggested *Calcined Salsola Soda.* The burnt seaweed from the shores of the Mediterranean, from which soda was often extracted was, I believe, called salsola, but I doubt whether the word soda was then in use.

The weights of the other substances are, when reduced to pennyweights, not very far from the equivalent numbers now received, hydrogen being taken as the unit.

The most remarkable exception is common salt itself, the solution of which was one in 29, and therefore in 1116 there were 37·2 parts of salt. Now the equivalent of NaCl is 58·5, which is very much greater.

Besides this the conductivity of a solution of salt in 29 of water would be much less in comparison with that of the other solutions than would appear from Cavendish's results, whereas if we assume that the molecular strength of the salt solution was really the same as that of the other solutions, the numbers do not differ much from those given by Kohlrausch.

The following table shows the results obtained by Cavendish and by Kohlrausch.

Name given by Cavendish.	Modern Symbol.	Weight used by Cavendish.	Modern equivalent.	Conductivity found by Cavendish. (Sea Salt = 1.)	Conductivity found by Kohlrausch. (NaCl = 1.)
Sea Salt	$NaCl$	37·2 ?	58·5	1·00	1·00
Sal Sylvii	KCl	74	74·5	1·08	1·21
Sal Ammoniac	NH_4Cl	51	53·5	1·13	1·17
Calcined Glauber's Salt	$\frac{1}{2}Na_2SO_4$	69	71	0·696	0·95
Quadrangular Nitre	$NaNO_3$	89	85	0·887	0·91
Calc. S.S.	$\frac{1}{2}Na_2CO_3 ? + xH_2O$	346	$83 + 18x$	0·852	0·72
f. alk.	$\frac{1}{2}K_2CO_3 + xH_2O$	139	$99 + 18x$	0·819	0·96
Oil of Vitriol	$\frac{1}{2}H_2SO_4$	48	49	0·783	1·23
Spirit of Salt	$\frac{1}{2}HCl + xH_2O$	130	$36·6 + 18x$	1·72	1·97

The theory of the electric resistance of electrolytes has been put on an entirely new footing by M. F. Kohlrausch, who has not only measured the resistance of a large number of solutions of different strengths and at different temperatures, but has discovered that the conductivity of a dilute solution of any electrolyte in water is the sum of two quantities, which we may call the specific conductivities of the components of the electrolyte, multiplied by the number of electro-chemical equivalents of the electrolyte in unit of volume of the solution. (Since the components of an electrolyte are not themselves electrolytes, it is manifest that they can have no actual conductivity, but the number to which we may give that name is such that when any two ions are actually combined into an electrolyte, the conductivity of the electrolyte depends on the sum of their respective numbers.)

Kohlrausch has also calculated the actual average velocity in millimetres per second with which the components are carried through the solution under an electromotive force of one volt per millimetre; and on the hypothesis that the components are charged with the electricity which travels with them, he has calculated the force in kilogrammes weight which must act on a milligramme of the component in order to make its average velocity in the solution one millimetre per second.

It appears to me that the simplest measure of the specific conductivity of an ion is the *time* during which we must suppose the electric force to act upon it so as to generate twice its actual average velocity. If we suppose that all the molecules of the ion are acted on by the electromotive force, but that each of them is brought to rest by a collision with a molecule of the opposite kind n times in a second, then the average velocity will be half that which the force can communicate to the molecule in the nth part of a second.

According to the theory of Clausius, it is only a small proportion, say $1/p$, of the molecules, which, at any given instant, are dissociated from molecules of the other kind, so as to be free to move under the action of the electromotive force, so that we must suppose each of the free molecules to continue free for a time pT; but since the proportion of free molecules to combined ones is quite unknown, the only definite result we can obtain from Kohlrausch's data is a certain very small time T, such that if the electromotive force acted on the molecules of the component during the time T, it would impress on them a velocity twice their actual average velocity.

Since the time T is very small, it is more convenient to speak of the molecule being brought to rest n times in a second, and to calculate n.

Salts with univalent acids.	$n \times 10^{-10}$	$T \times 10^{18}$
H	15941	6273
K	2354	42480
NH_4	5297	18880
Na	6131	16310
Li	30214	3310
Ag	1030	97087
Cl	2551	39200
Br	1030	97087
I	637	156986
F	7848	12740
CN	3433	29129
NO_3	1569	63735
ClO_3	1324	75529
$C_2H_3O_2$	3286	30432
$\frac{1}{2}$Ba	2207	45310
$\frac{1}{2}$Sr	3581	27925
$\frac{1}{2}$Ca	8681	11520
$\frac{1}{2}$Mg	16180	6180
$\frac{1}{2}$Zn	6817	14670
$\frac{1}{2}$Cu	4806	20807

Univalent Metals with bivalent acids.	$n \times 10^{-10}$	$T \times 10^{18}$
H_2	26732	3741
K_2	2844	35160
$(NH_4)_2$	6719	14883
Na_2	8730	11455
Li_2	55430	1804
Ag_2	1275	78431
SO_4	2305	43384
CO_3	4071	24564
Bivalent Metals with SO_4.		
Mg	26480	3776
Zn	11281	8865
Cu	11772	8496
SO_4	4218	23708

NOTE 35, ART. 654.

On the Ratio of the Charge of a Globe to that of a Circle of the same Diameter.

The true value of this ratio is $\frac{1}{2}\pi = 1{\cdot}570796....$

Cavendish has given several different values as the results of his experiments.

In the account of his experiments, which represents his most matured conclusions, he states this ratio as 1·57 (Art. 237).

All the other values, however, either as stated by Cavendish or as deducible from his experiments, are lower than this.

In Art. 281 the charge of the globe of 12·1 inches diameter being 1, that of a circle 18·5 inches diameter is given as ·992. The ratio of the charge of a globe to that of a circle of equal diameter as deduced from this is 1·542.

In Art. 445 the charge of the globe is compared with that of a pasteboard circle of 19·4 inches diameter. Cavendish gives the actual observations but does not deduce any numerical result from them, which shows that he did not attach much weight to them. As they seem to be the earliest measurements of the kind, I have endeavoured to interpret the observations by assuming that the positive and negative separations were equal when the observations are qualified in the same words by Cavendish.

I thus find 14·2 or 14·3 for the charge of the globe, and 15·2 for that of the circle, and from these we deduce for the ratio of the charge of a globe to that of a circle of equal diameter 1·5054.

In Art. 456 the ratio as deduced by Cavendish from the observations on the globe and the tin circle of 18·5 inches diameter is 1·56.

From the numerical data given in the same article, the ratio would be 1·554.

Cavendish evidently thought the result given here of some value, for he quotes it in the foot-note to Art. 473.

Another set of observations is recorded in Art. 478, from which we deduce the ratio 1·561.

It appears by a comparison of Arts. 506 and 581 that Cavendish, at the date of the latter article (which is doubtful), supposed the ratio to be 1·5. (See foot-note to Art. 581.)

At Art. 648 the ratio is stated as 1·54.

At Art. 654 measures are given from which we deduce 1·542 and 1·37.

The numbers in Art. 682 are the same as those in Art. 281.

ALPHABETICAL INDEX.

The references are to the Articles.